APPLIED STATICS AND STRENGTH OF MATERIALS

Thomas Burns

Delmar Publishers

I T P   An International Thompson Publishing Company

Albany • Bonn • Boston • Cincinnati • Detroit • London • Madrid
Melbourne • Mexico City • New York • Pacific Grove • Paris
San Francisco • Singapore • Tokyo • Toronto • Washington

NOTICE TO THE READER

Cover Design: Barry Littman

Delmar Staff
Publisher: Robert Lynch
Acquisitions Editor: John Anderson
Developmental Editor: Jeanne Mesick
Project Editor: Thomas Smith
Production Coordinator: Karen Smith
Art and Design Coordinator: Cheri Plasse

Copyright © 1997
By Delmar Publishers
a division of International Thomson Publishing Inc.

The ITP logo is a trademark under license

Printed in the United States of America

For more information, contact:

Delmar Publishers
3 Columbia Circle, Box 15015
Albany, New York 12212-5015

International Thomson Publishing Europe
Berkshire House 168-173
High Holborn
London WC1V7AA   England

Thomas Nelson Australia
102 Dodds Street
South Melbourne, 3205   Victoria, Australia
Nelson Canada
120 Birchmount Road
Scarborough, Ontario   Canada MlK 5G4

Delmar Publishers' Online Services
To access Delmar on the World Wide Web, point your browser to:
http://www.delmar.com/delmar.html
To access through Gopher:
gopher://gopher.delmar.com
(Delmar Online is part of "thomson.com", an Internet site with information on more than 30 publishers of the International Thomson Publishing organization.) For more information on our products and services:
email: info @delmar.com or call 800-347-7707

International Thomson Editores
Campos Eliseos 385, Piso 7
Cot Polanco   11560 Mexico D F Mexico

International Thomson Publishing GmbH
Königswinterer Strasse 418
53227 Bonn   Germany

International Thomson Publishing Asia
221 Henderson Road
#05-10 Henderson Building   Singapore 0315

International Thomson Publishing-Japan
Hirakawacho Kyowa Building, 3F
2-2-1 Hirakawacho
Chiyoda-ku, Tokyo 102   Japan

1 2 3 4 5 6 7 8 9 10   XXX   02 01 00 99 98 97 96

Library of Congress Cataloging-in-Publication Data
Burns, Thomas (Thomas M.), 1961-
    Applied statics and strength of materials / Thomas Burns.
       p.   cm.
    Includes index.
    ISBN 0-8273-6959-X
    1. Structural design.   2. Statics.   3. Strength of materials.
    4. Strains and stresses.   I. Title.
TA658.B82   1997
624.1'71--dc20
                                     96-3965
                                      CIP

CONTENTS

PREFACE

This text has been written for students of engineering technology, building construction, architecture, and related technical disciplines who are taking their initial courses in the behavior of structures. The design of any structure begins with an understanding of the structure from two distinct perspectives—externally and internally. From an external perspective, the designer must first understand how applied forces cause a structure to react. This knowledge of external forces is the foundation of structural behavior, which forms the first step in structural design. The designer must ultimately know how a structure will bend, deflect, or displace under the affect of the applied loads and subsequent reactions. This study of the internal behavior of a structure will allow a designer to focus on the behavior of specific materials under load. The concept of statics relating to the external behavior of a structure while strength of materials focuses on the internal behavior, is an important tool in setting the framework for these subjects.

I have taught statics and strength of materials to the Civil Engineering Technology students at Cincinnati State for the past nine years and have written this text to fill a need for a readable, progressive, and integrated text on statics and strength of materials. Teaching these courses repeatedly over the years, I have been reminded of the recollections of Washington Roebling, the builder of the Brooklyn Bridge, on his academic experience in higher education:

"I am still busy trying to forget the heterogeneous mass of unusable knowledge that I could only memorize, not really digest."

I believe the experience of Washington Roebling in many ways mirrors the experience of today's students in statics and strength of materials. This may in part be due to the student's inability to see the framework in which these topics are to be addressed. They have a tendency to view a truss problem as different from a simple beam problem because the framework of equilibrium has not been successfully reinforced. It is all too easy for the student to become frustrated with the texts that are difficult to read or explain the topics in the same old, standard manner. A text must be a tool that provides the necessary reinforcement for guidance and learning when the student leaves the classroom. Hopefully, you will find this text is such a tool.

I have attempted to provide easy-to-read, yet extensive, coverage of the most essential topics that are normally covered in these two topics. The text provides the necessary information required for the students of the aforementioned programs to successfully grasp the concepts of statics in Chapters I through 7. Equilibrium is repeatedly stressed throughout this section and a standard format for problem-solving is introduced and used throughout the numerous example problems. Chapters 8 through 17 focus on topics generally covered in strength of materials courses, with Chapters 15 through 17 focusing on design concepts. Beam, column, and connection design is explored with coverage on steel, concrete, and timber as structural materials. The steel design portions address both the Allowable Stress Design philosophy as well as the more recent Load and Resistance Factor Design method.

The text assumes that the student has a thorough background in algebra and trigonometry and is comfortable with using the fundamentals of each. The use of integral calculus can be incorporated throughout several topics in statics and strength of materials—such methods are included in Appendix E. The text provides many of the standard steel section tables (in both U.S. and SI units) in Appendix A, as well as some of the common AISC beam loading tables in Appendix C.

The examples and exercises are presented in both U.S. customary system of units as well as SI units. These examples and exercises strive to integrate and convey the fundamental ideas found in the learning process for each subject. Therefore, a detailed, step-by-step procedure is given in every example, to facilitate the students' understanding of basic concepts. Again, the author's emphasis on providing a tool of learning is paramount in the structure of the text.

In closing, I have always believed that true learning takes place only when a student can make the connection from the fundamental concept to the application of this concept in the problem-solving process. This book will hopefully provide a pathway for that connection to take place. There are many people whom I wish to thank for their help in making this book possible. I will always be thankful to my Mom and Dad, for giving me the values of hard work and personal responsibility while instilling in me the importance of life-long learning. I would also like to thank my colleagues of the Civil Engineering Technology program at Cincinnati State for reminding me that there can still be a bastion of quality in an environment that is increasingly driven by numbers. Also, my gratitude is extended to my editor, Jeanne Mesick, for her work in making this text a reality.

Acknowledgments

Doug Broad
Nash Community College
Rocky Mount, NC

John Crigler
Bishop State Community College
Mobile, Al

Ellis Dale Hart
NE Louisiana University
Monroe, LA

Dr. Maurice Ades
Aiken Technical College
Martinez, GA

Dave Hanna
Ferris State University
Big Rapids, MI

Tom Rogers
Umpqua Community College
Oakland, OR

To my wife and children;
Kim, Allie, Mike, Tommy

1

INTRODUCTION TO STATICS

1.1   Introduction

Engineering mechanics is the science that studies the effects of forces acting on rigid bodies and structures. Engineering mechanics has many distinct fields of study and this text covers two of these; statics and strength of materials. **Statics** is the study of how forces act and react on rigid bodies which are at rest or not in motion. This study is the basis for the engineering principles which guide the design of all structures, since before we can begin to design any structure we must first know the forces applied to it. **Strength of materials** is the study of the physical changes which a body undergoes as it is acted upon by forces. The study of strength of materials will allow measurement of the deflection or expansion of the members or structures which we may design.

Structures stand up due to the interaction of forces which take place within their particular environment. In 1686, Issac Newton presented three laws of motion describing the effects of forces on rigid bodies. Of particular interest to the study of statics was his third law of motion which states: *For every action there is always an equal and opposite reaction.* This principle serves as the backbone for statics, since the structures are in equilibrium and usually motionless. A floor joist does not crash downward when loaded because there are equal and opposite forces (at its bearing points) supporting it. Structures should react with equal and opposite intensity to the forces that act upon them. The major focus in the study of statics is to develop the ability to determine the **magnitude** and **direction** of these reactions.

The study of statics has many applications in daily life. Anyone who has used a ladder should have a deep appreciation for the laws that govern the stability of the ladder. Hopefully the reactions generated by the ground surface and the wall surface are sufficient to maintain equilibrium. If these forces cannot be developed, an accident will likely occur (Figure 1-1). Another example of the laws of statics would be the reaction of the floor on a chair. The floor's reactions on the chair legs are keeping a person from falling through the floor as they sit. Even something as insignificant as opening a door is governed by rotational aspects, which will be discussed in the next chapter. Indeed, understanding the laws of statics is fundamental to acquiring the knowledge to design and build structures of every type.

1.2   Use of Units in Statics

The basic quantities which are used in statics are those of **force** and **length**. A force is an action on a body which tries to move, stop, or change its direction. Forces are **vector** quantities, meaning that forces have both magnitude and direction. We cannot refer to a force simply by magnitude any longer. We must also describe the direction of the force (up or down) and as well as its point of application on the body.

Length is a **scalar** quantity, meaning it has magnitude only and references the position of a body or a force in a particular system. The position may be relative to other bodies or with respect to other forces acting on the body.

Two common unit systems are the United States customary system (U.S.) and the International System of Units (SI). The U.S. system measures force in units of

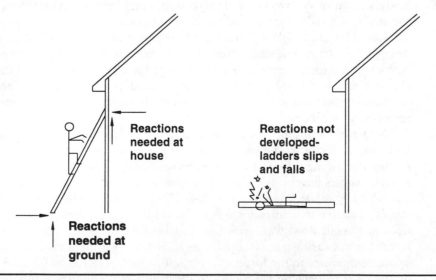

Reactions needed at house

Reactions not developed- ladders slips and falls

Reactions needed at ground

Figure 1-1   Importance of Reactions in Everyday Use

pounds or kips (1 kip = 1000 pounds) while length is measured in feet or inches. Engineers and technicians in the United States are comfortable with this system of measurements. Much of the rest of the world, however, uses the SI system. In the SI system, the fundamental unit of length is the meter and the fundamental unit of mass is the kilogram. The unit of force is referred to as a newton (N). A newton is defined as follows:

$$\text{Unit Force} = \text{Unit Mass} \times \text{Unit Acceleration}$$

$$1 \text{ Newton} = 1 \text{ Kilogram} \times 1 \text{ meter/second}^2$$

An important difference between the U.S. customary system and the SI system is that the U.S. system is based on gravity; where force (pounds, kips, etc.) is the basic unit. The SI System is an *absolute system;* where mass (kilogram) is the basic unit. In order to calculate weight, the SI System would multiply the mass of an object by the Earth's gravitational acceleration constant, which is 9.81 meters per second squared (m/sec^2).

Federal and many state agencies are beginning to mandate metric usage in the development of all their projects and construction documents. Use of metric units in all construction projects for the U.S. government's General Services Administration (GSA) started in 1994 and the department of transportation for many states have likewise incorporated the SI system. Because of this trend, the student should be familiar with the SI system.

This book contains example problems in both the U.S. customary units and the SI system. The exercise problems at the end of each chapter contain many SI unit problems as well. As the student becomes more familiar with the SI system, its usage will seem much less cumbersome. The SI system will employ prefixes that represent factors of ten with their basic units. Such common prefixes are shown in Table 1-1.

Typical quantities that designers use are length, force, and stress. U.S. customary units and their SI counterparts are shown in Table 1-2, along with the conversion factor from U.S. to SI units. This will help the today's designer in the crossover between both systems. The metric equivalents of the many standard

Table 1-1   Common SI System Prefixes

Factor	Preface	Symbol
$.000001 = 10^{-6}$	micro	μ
$.001 = 10^{-3}$	milli	m
$1000 = 10^3$	kilo	K
$1000000 = 10^6$	mega	M
$1000000000 = 10^9$	giga	G

Table 1-2 SI System Conversions

Quantity	U.S.	SI	Conversion Factor from US to SI	Conversion Factor from SI to US
Length	Foot	meter	.3048	3.281
	Inch	meter	.0254	39.37
Force	Pound	newton	4.448	.224
	KIP	kilonewton	4.448	.224
Stress	psi	Pascal (N/m^2)	6895.	.00145
	ksi	Megapascal	6.895	.145
	psf	Kilopascal	.478	20.92
Area	sq. in.	sq. mm.	645.2	.00155
Moment of Inertia	in.^{4}	mm^4	416,231	.0000024

steel sections are also included in the tables found in Appendix A, for use in the examples and exercises found throughout the text.

1.3   Numerical Accuracy and Dimensional Analysis

In this day and age of electronic calculators and computers, the accuracy of our answers *seems* to be without limit. In reality, the accuracy of our answer can be no greater than the least accurate piece of data used in obtaining that answer. This point brings up the discussion of **significant figures**. Significant figures can be counted left-to-right across a number, not including any zeros which end the number, unless the zeros are to the right of the decimal. When multiplying or dividing, the number of significant figures in the answer should be the least amount of significant figures in any piece of data. To demonstrate significant figures let's do the following. Suppose a distance is given as 62.30 ft. and we need to divide it by 4.1. The first number contains three significant figures and the second contains two. As we punch these numbers into our calculators, the "answer" which may appear (depending on the decimal setting on the calculator) is 15.19512195, which "appears" to be accurate to 10 significant figures. Of course this is ridiculous, because an answer can only be as accurate as the least accurate piece of data used in acquiring it. In this case, two significant figures should be used and the answer should be 15.

In statics and strength of materials, most answers do not need more than three significant figures, since loads on structures vary widely and can only be estimated.

The topic of significant figures deals only with the accuracy of the information presented. Perhaps of greater importance than significant figures, is the need for dimensional analysis. **Dimensional analysis** is the process of carrying the units of the data through the mathematical operations. This process is an important tool that can be used as a check to see if our answer is in the correct units and possibly catch unwanted mistakes. The following shows the use of dimensional analysis:

$$60 \text{ miles/hour} \times 5280 \text{ ft./mile} \times 1 \text{ hour/3600 seconds} = 88 \text{ ft./second}$$

It is highly recommended that dimensional analysis be used so that unit mistakes can be minimized. Many professional engineering organizations require checking of calculations by dimensioanl analysis as a means of quality control.

1.4   Basic Trigonometric and Mathematical Function Review

Many people's fear of engineering mechanics is largely related to their fear of math or "math anxiety." To understand and utilize the fundamental principles of statics and strength of materials, a solid background of algebra and trigonometry is required. To solve a problem in statics, one must be able to "use" the necessary mathematical principles. Should these mathematical principles in themselves be a problem, the student will be working a math problem inside a statics problem. This shifts the focus away from the statics problem and reduces the effort towards understanding statics. Should the basic laws of trigonometry and algebraic principles need to be refreshed, the following brief review is presented.

Trigonometry

Trigonometry is the branch of mathematics that deals with the relationships between the sides of a triangle. A triangle is a polygon that contains three sides and whose interior angles always sum up to 180°. Trigonometry focuses on the relationships that exist in right triangles and in oblique triangles.

Right triangles are those that contain one 90°, or right angle (Figure 1-2). Right triangle trigonometry involves learning the relationships that exist between the sides of the right triangle and the interior angles. In general, for right triangles, there are three trigonometric functions that must be learned. Recognize for the triangle shown; side y is OPPOSITE to the angle θ, side x is ADJACENT to the angle θ, and side r is the HYPOTENUSE.

Basic Trigonometric Identities

$\sin \theta = y/r = $ opposite side/hypotenuse

$\cos \theta = x/r = $ adjacent side/hypotenuse

$\tan \theta = x/r = $ adjacent side/hypotenuse

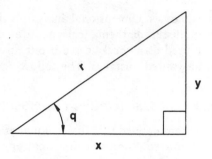

Figure 1-2   Right Triangle Nomenclature

It should be realized that the sine, cosine, and tangent function are merely ratios of the various sides of this triangle. To illustrate this idea, get out a calculator and take the sine of 36°. The calculator should display .5877852. What does this number mean? It simply means that the ratio of the opposite side to the hypotenuse is .5877852 or, if the hypotenuse was 1 unit long , the opposite side would be .5877852 of a unit long.

The Pythagorean theorem is another widely used relationship in right triangle trigonometry. This theorem simply states that the hypotenuse squared is the sum of the two other sides squared.

Pythagorean Theorem

$$r^2 = x^2 + y^2$$

Inverse trigonometric functions are used when the sides of a right triangle are known and the interior angles need to be found. This operation is the opposite of what was previously shown and is designated by using the word "inverse" or "arc" in front of sin, cos, tan, etc. Referring to Figure 1-2, the following could be stated:

Inverse Trigonometric Functions

$$\theta = \arccos{(x/r)} = \cos^{-1}{(x/r)}$$

$$\theta = \arcsin{(y/r)} = \sin^{-1}{(y/r)}$$

$$\theta = \arctan{(y/x)} = \tan^{-1}{(y/x)}$$

The following example will illustrate these trigonometric concepts.

EXAMPLE 1.1

For the right triangles shown below, calculate the unknown sides and/or interior angles.

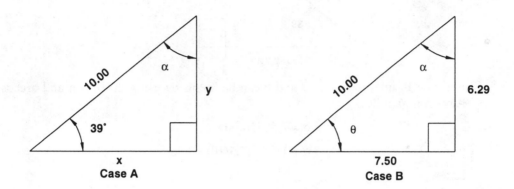

For case A, sides x and y can be found by using the sine and cosine function as follows:

$$\sin 39° = y/10.00 \qquad \text{Therefore, } y = 10.00 \sin 39° = 6.29$$

$$\cos 39° = x/10.00 \qquad \text{Therefore, } x = 10.00 \cos 39° = 7.77$$

Using the Pythagorean theorem as a check:

$$R = \sqrt{(6.29)^2 + (7.77)^2} = 10.0 \text{ (as it should)}$$

The interior angle α can be found by knowing that the three interior angles in every triangle must add up to 180°. Therefore:

$$\text{angle } \alpha = 180° - 90° - 39° = 51°$$

Alternately, the interior angle α can be found by using the arcsin function, since the sides are now known. This can be done as follows:

$$\sin \alpha = x/10.00 = \text{opposite/hypotenuse}$$

$$\sin \alpha = 7.77/10.00 = .777$$

$$\alpha = \arcsin (.777) = 51°$$

(It should be noted that inverse trigonometric relationships, such as arcsine, may be listed as inverse sin or $\sin^{-1}$ on some calculators.)

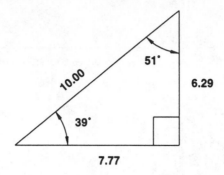

For case B, interior angles α and θ can be found by using the arcsin and arccos functions as follows:

$$\sin \alpha = 7.50/10.00$$

$$\alpha = \arcsin (7.5/10) = 48.6°$$

Likewise,

$$\cos \theta = 7.50/10.00$$

$$\theta = \arccos (7.5/10) = 41.4°$$

Notice that all the interior angles sum up to 180°.

Oblique triangles are those that do not contain a 90° interior angle (**Figure 1-3**). Such triangles cannot use the previously mentioned trigonometric functions or the Pythagorean theorem. The two laws that govern oblique triangles are the cosine law and the sine law, which are actually derived from the previously discussed right triangle trigonometry. The **cosine law**, illustrated in **Figure 1-3**, is as follows:

Cosine Law

$$a^2 = b^2 + c^2 - 2bc \cos A$$

$$b^2 = a^2 + c^2 - 2ac \cos B$$

$$c^2 = a^2 + b^2 - 2ab \cos C$$

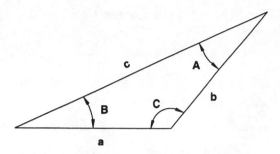

Figure 1-3   Oblique Triangle Nomenclature

Recognize that the angle C is opposite of side c and angle A is opposite of side a. Notice that all angles are opposite from the side to which they correspond. Another law frequently used on oblique triangles is the **sine law**, which is given as follows:

Sine Law

$$a/\sin A = b/\sin B = c/\sin C$$

The following example will illustrate the use of the sine law and cosine law on oblique triangles.

EXAMPLE 1.2

Calculate the unknown sides and interior angles for the oblique triangle shown below.

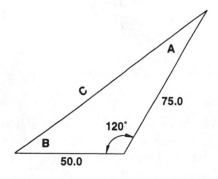

The side needed to be found is side c, across from the 120° angle. Use the cosine law as follows:

$$c^2 = a^2 + b^2 - 2ab \cos C$$

$$c^2 = (50.0)^2 + (75.0)^2 - (2)(50.0)(75.0)\cos 120°$$

$$c^2 = 11875, \text{ therefore } c = 109$$

The unknown interior angles can be found by utilizing the sine law as follows:

$$a/\sin A = c/\sin C$$

$$50.0/\sin A = 109/\sin 120°$$

$$\sin A = 50/(109/\sin 120°) = .397$$

$$A = \sin^{-1}(.397) = 23.4°$$

The other unknown interior angle can be found by utilizing the sine law in the same manner.

$$b/\sin B = c/\sin C$$

$$75.0/\sin B = 109/\sin 120°$$

$$\sin B = 75.0/(109/\sin 120°) = .596$$

$$B = \sin^{-1}(.596) = 36.6°$$

Again notice in the figure below that all the interior angles add up to 180°

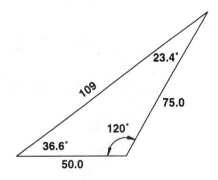

Algebra

The solution of statics problems involves writing algebraic equations and being able to solve them. The development of such equations is what a majority of this course is typically directed towards, therefore it is assumed that the ability to manipulate algebraic equations is already an acquired skill. The following is a review of the fundamentals of algebraic equations.

An algebraic term typically consists of two parts; constants and variables. Constants are quantities within the term that do not change, usually numbers like 5 or π. Variables represent quantities that can change and are usually represented by a letter such as x or y. The degree of a variable is the power to which it is raised; such as squared, cubed, etc. The following are algebraic terms:

$$3x$$

$$\pi\, r^2$$

$$3.67\, y^3$$

Addition and subtraction of algebraic terms are performed by adding and subtracting similar terms. The terms that contain the same variables are the only ones that can be added to or subtracted from. This is done as follows:

$$3x + 4y + 2y + 1.5x$$

is the same as

$$4.5x + 6y$$

When equations containing algebraic expressions are written, it is convenient to group the variables on one side of the equation. This is done by adding or subtracting similar values from each side. This is done as follows:

$$15 - 3x + 9y^2 = 30 - 1.5x + 2y^2$$
$$\underline{+3x - 9y^2} + 15 - 3x + 9y^2 = \underline{+3x - 9y^2} + 30 - 1.5x + 2y^2$$
$$15 - \underline{30} = 1.5x - 7y^2 + 30\ \underline{-30}$$
$$-15 = 1.5x - 7y^2$$

Multiplication and division of algebraic expressions are performed by using techniques assumed to be acquired by the student at this point. However, one of the most important points to review is the multiplication and division of algebraic expressions by a constant. Remember that each of the terms has to be multiplied or divided by the constant. Let's look at the following algebraic expression:

$$-15 = 1.5x - 7y^2$$

Assume that it was desired for this to be expressed in terms of x. First we would add $7y^2$ to each side as follows:

$$-15 + 7y^2 = 1.5x - 7y^2 + 7y^2$$
$$-15 + 7y^2 = 1.5x$$

Next we would divide both sides of the equation by 1.5 to get the following:

$$(-15 + 7y^2)/1.5 = 1.5x/1.5$$

$$-10 + 4.66y^2 = x$$

Such manipulation of equations is very common in statics and should be more thoroughly reviewed, if needed.

1.5   Vectors in Statics

As was mentioned previously, statics is the study of how forces act and react on rigid bodies which are in equilibrium. In order to understand statics, one must first have an understanding of force as it is applied to a body. Force is a **vector** quantity, which means it is a quantity that has both magnitude and direction. A force has no meaning to us if it is described in magnitude only. For instance, if a problem stated that a 1000-pound-force acted on a beam, the probable question that should come up would be "Where does it act?" or "What direction does it act in?" Forces must be defined by answering these questions of *what* and *where*.

Since forces are vectors, it is important to understand how to use vectors in a mathematical fashion. This will be the subject of Chapter 2 and it will be paramount to the student's success.

1.6   Summary

Statics is the study of how forces act and react on bodies which are in equilibrium. This study forms the basis for all types of design since a designer must first know the forces which act on a structure that is presently under design. Forces are vector quantities which are typically given in units of pounds, kips, or newtons. Vector quantities have both magnitude and direction. To study forces, the student will have to possess a thorough knowledge of algebraic manipulations and be comfortable using trigonometric formulas. Solution of statics problems will involve the use of such principles.

EXERCISES

1. For the right triangle shown below, determine the hypotenuse (side r) if:
    a. $x = 6$ ft. and $y = 9$ ft.
    b. $x = 100$ mm and $\theta = 41°$
    c. $y = 4$ in. and $\beta = 63°$
    d. $x = 250$ mm and $\beta = 33°$

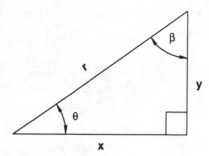

2. For the right triangle shown in Exercise 1, determine the interior angles θ and β if:
   a. $x = 5$ ft. and $y = 6.5$ ft.
   b. $x = 16$ m and $y = 4$ m
   c. $x = 2$ ft. and $y = 22$ ft.
   d. $x = 10.5$ and $y = 6$ ft.
3. For the oblique triangle shown below, determine side c if:
   a. $a = 30$ mm, $b = 59$ mm, and $C = 110°$
   b. $a = 6$ ft., $b = 7$ ft., and $C = 95°$
   c. $b = 6$ m, $B = 32°$, $C = 115°$
   d. $a = 5$ in., $b = 7$ in., $C = 93°$

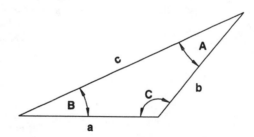

4. Determine the interior angles for the oblique triangle used in exercise 3 if:
   a. $a = 40$ mm, $b = 89$ mm, and $c = 120$ mm
   b. $a = 60$ ft., $b = 55$ ft., and $c = 105$ ft.
   c. $a = 50$ in., $b = 65$ in., and $c = 90$ in.
5. A surveyor is stationed 100 ft. from the side of a multi-story building and reads a zero angle along a horizontal line, which is 5 ft. up the side of the building. The surveyor then turns a vertical angle of 77° and sights the top of the building. How tall is the building?
6. A builder is laying out a rectangular room that is being built for an addition. If the dimensions of the two sides are 9'3" and 16'8", what should

be the corner to corner dimension measure such that the addition will be "square"?

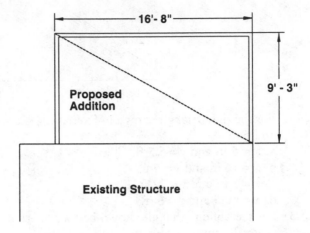

7. A piece of land is being purchased for a development with a restriction that any developed lot can be no closer than 600 ft. from a power line. From the monument shown, locate the corner of the first lot point that could meet this restriction along the eastern boundary line.

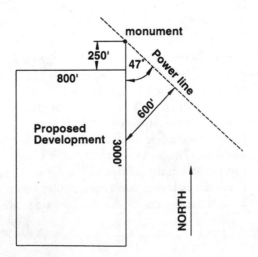

8. The frame below carries a load as shown. Determine the distance between the supports.

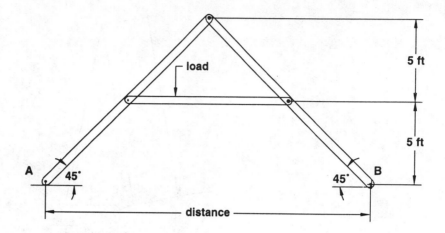

9. A boom is supported as shown below. Calculate the distance between supports and the length of the cable.

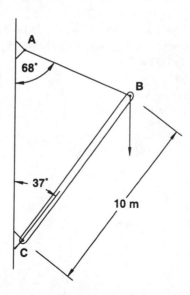

10. The building is 400 ft. tall and has suffered damage from a recent earthquake. If a surveyor is stationed 150 ft. away, sets a zero angle on the bottom corner of the building, and turns the angle shown to the top of the

building; what is the angle at which the building is leaning with respect to vertical?

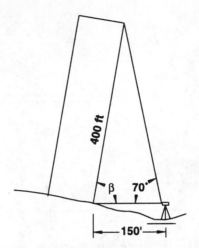

11. Add and subtract the following algebraic expressions:
     a. $3x + 16 + 7y^3 = 14y^3 - 2 - 12x$
     b. $3x + 3x^2 - 7x^3 = 2.4x^2 + 2.4x^3$
     c. $144 + 2a - 3ab^2 = 2a^2b + 200 + 2a$
12. Describe the following algebraic expressions in terms of x:
     a. $3x - 33 = 7y$
     b. $6x^2 - 6y - 144 = y^2$
     c. $100 + 3.5x^2 = 255$
     d. $120y + 2x = 100$

CHAPTER

2

FORCES AND FORCE SYSTEMS

2.1 Introduction

In the most general of terms, a force is the action of one body on another. A force affects a body by tending to cause movement of that body in both an external and internal manner. Movement of a body in external sense can be viewed as movement of the complete body, while movement in an internal sense would only involve the stretching and deforming of the body. The subject of statics concerns itself with the external view of forces on a body. Statics addresses the concept of how forces make the complete body react. (Does it move to the right? The left? Up? Down?) The subject of strength of materials concerns itself with the action of a force on a body and how the forces initiates changes to the body itself. Therefore, a force can be defined as the push or pull on an object which typically causes some type of reaction.

Forces are vectors, and because of this they have both magnitude and direction. This is extremely important to realize because an analysis of a structure cannot be completed by simply knowing the magnitude of forces without the direction of application. For instance, it would be impossible to solve for the reactions of a loaded beam if the only fact given is that the beam had 500 lbs acting on it. Does the 500 lbs act vertically downward, at a 45° angle, or horizontally? This has to be clarified before a solution process can begin. Another characteristic which defines a force is the point of application or line of action. These will need to be clarified before a solution can be found. It is not enough to know that the force is 500 lbs and it acts downward on the beam. We must know if it acts at the center point or if it acts 3 feet from the right support. Therefore, to properly define a force acting upon an object, it is necessary to define the following:

- Magnitude
- Direction
- Point or Line of Application

These characteristics are illustrated in Figure 2-1.

2.2   Techniques of Finding Resultants

In solving statics problems, often it is advantageous to replace all the forces which act on a body with a single force. This single force, called a **resultant**, must produce the *same movement and effects that all the original forces would produce on the body*. The resultant of several forces can most easily be thought of as the sum or addition of these forces. However, since forces are not scalar quantities, the addition of several forces into a single resultant must be accomplished using trigonometric principles. This is the focus of the following paragraphs in this section. The ability to find resultant forces is extremely important because it allows the solution process to be greatly simplified.

Vector Addition Using Trigonometric Functions

Some force systems have resultants which are easily found by the application of trigonometric principles. One of these methods involves the addition of two vectors which act through the same point. The technique to find the resultant is commonly referred to as the **triangle law**. This law simply stated as follows:

Triangle Law

The resultant of two vectors can be found by placing the two vectors (which act at the same point) in a tip-to-tail fashion and completing the triangle with a vector.

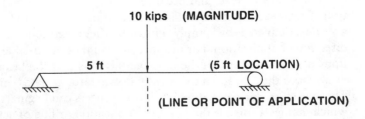

Figure 2-1   Necessary Characteristics of a Force

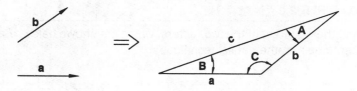

Figure 2-2   Application of the Triangle Law

The resultant vector is the sum of the original two vectors and therefore replaces the original two vectors. This is illustrated in Figure 2-2.

This resultant could be found by applying the sine and cosine laws, which were reviewed in Chapter 1. By treating the resultant side of the previously shown triangle as the unknown side, and knowing the magnitudes of the others sides and the included angle, the magnitude of the resultant can be solved using the cosine law. The direction of the resultant as referenced off vector *a* can be found using the sine law. The cosine law and sine law are repeated as follows:

Cosine Law

$$a^2 = b^2 + c^2 - 2bc \cos A$$

$$b^2 = a^2 + c^2 - 2ac \cos B$$

$$c^2 = a^2 + b^2 - 2ab \cos C$$

Sine Law

$$a/\sin A = b/\sin B = c/\sin C$$

The following example illustrates finding the resultant of two forces.

EXAMPLE 2.1

Find the resultant of the two vectors, which are shown below, using the previously mentioned mathematical techniques.

To begin, the vectors should be arranged in a tip-to-tail fashion, which may be done as shown below.

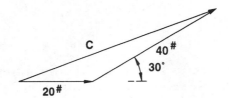

The magnitude of the resultant can be found through the use of the cosine law as follows:

$$c^2 = a^2 + b^2 - (2)(a)(b) \cos C$$

$$c^2 = 20^2 + 40^2 - 2(20)(40)\cos 150°$$

$$c = \sqrt{3386} = 58.2 \text{ lbs}$$

Next, the direction of this resultant can be found off horizontal by simply referencing it off the horizontal vector. This means solving for angle B of the triangle, using the sine law.

$$b/\sin B = c/\sin C$$

$$40/\sin B = 58.2/\sin 150°$$

$$\sin B = 40 \sin 150°/58.2 = .806$$

$$B = \sin -1(.806) = 20.1°$$

Therefore, the resultant of the two vectors is a vector force of 58.2 lbs, making a counterclockwise angle of 20.1° with the horizontal.

The addition of vectors that act through the same point (concurrent vectors) can be expanded to include more than two vectors. The technique to find the resultant is commonly referred to as the **polygon method**. This method simply expands on the previously mentioned triangle law and is as follows:

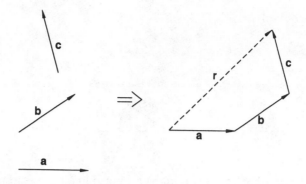

Figure 2-3   Application of the Polygon Law

| **Polygon Method** |
| The resultant of three or more vectors can be found by placing the vectors (which act at the same point) in a tip-to-tail fashion and completing the polygon |

The vector which completes the polygon is the resultant of the vectors and it extends from the tail of the first vector to the tip of the last vector. The resultant vector is the sum of the original vectors and therefore replaces the original vectors. This is illustrated in Figure 2-3 below.

The polygon method can also be accomplished in the trigonometric fashion that was shown in Example 2.1, although it requires finding an intermediate resultant, r'. This intermediate resultant is to be the sum of the two preceding vectors and will subsequently be added to the next vector to find another intermediate resultant. This process will repeat itself until the last vector has to be added to the last intermediate resultant. The following example will illustrate this method.

EXAMPLE 2.2

Find the resultant for the three vectors shown below, which are concurrent. Express the resultant in an angle off the horizontal.

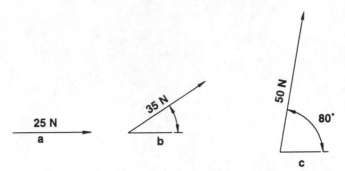

To begin, the vectors should be arranged in a tip-to-tail fashion and the polygon completed with the resultant. This is shown below.

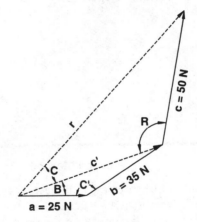

The magnitude of the final resultant will have to be found by first finding an intermediate resultant, c'. This intermediate resultant can be found by adding the first two vectors through the use of the cosine law as follows:

$$c'^2 = a^2 + b^2 - (2)(a)(b)\cos C'$$

$$c'^2 = (25)^2 + (35)^2 - 2(25)(35)\cos 147°$$

$$c' = \sqrt{3317.7} = 57.6 \text{ N}$$

Next, the direction of this resultant can be found off horizontal by simply referencing it off the horizontal vector. This means solving for angle B of the triangle, using the sine law.

$$b/\sin B = c/\sin C'$$

$$35/\sin B = 57.6/\sin 147°$$

$$\sin B = 35 \sin 147°/57.6 = .33$$

$$B = \sin^{-1}(.33) = 19.3°$$

The intermediate resultant of the first two vectors is a force (c') of 57.6 newtons making a counterclockwise angle (B) with the horizontal of 19.3°. This intermediate resultant is then added to the last vector (50 newtons, 80° off horizontal) to find the resultant of all three vectors, as shown below:

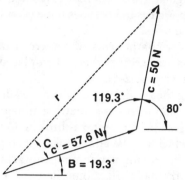

The final resultant can be found again through the use of the cosine law as follows:

$$r^2 = c^2 + c'^2 - (2)(c)(c')\cos R$$

$$r^2 = (50)^2 + (57.6)^2 - 2(50)(57.6)\cos 119.3°$$

$$r = \sqrt{8640} = 92.9 \text{ N}$$

Next, the direction of this resultant can be found off horizontal by simply referencing it off the 57.6 N vector. This means solving for angle C of the triangle, using the sine law, and adding that angle to angle B (the angle which the 57.6 N vector lies off horizontal). This is done as follows:

$$c/\sin C = r/\sin R$$

$$50/\sin C = 92.9/\sin 119.3°$$

$$\sin C = 50 \sin 119.3°/92.9 = .469$$

$$C = \sin^{-1}(.469) = 28°$$

Adding this angle to angle B (19.3°) gives a final resultant of 92.9 N at 47.3° counterclockwise off horizontal.

As demonstrated by the previous example, adding more than two vectors together using the polygon method can be rather time-consuming. The next section will explore an alternate method of adding multiple vectors together, which many find to be more appealing.

2.3   Resolution into Rectangular Components

Most people are predisposed to think of things in a rectangular or right-angle fashion. Most of our streets, most of our buildings, and most of our mapping is laid out in rectangular grids. Therefore, it should come as no surprise that when we deal with force vectors many times it is easiest to think of them in terms of their rectangular components.

The rectangular components of a vector can most simply be viewed as the horizontal and vertical vectors that, when placed in tip-to-tail fashion, form the vector in question (Figure 2-4).

Remembering the triangle law, which was discussed in section 2.2, resolving a vector into its components is somewhat similar. Instead of adding two vectors together to get a resultant, a vector is broken down into its two rectangular component vectors, which can be added together to give the resultant. This is simple to perform since the rectangular components actually form a right triangle which, of course, follow the basic trigonometric rules. For Figure 2-5, these rules are as follows:

Rectangular Components of a Vector

X or horizontal component $= R \cos \theta$ or $R \sin \phi$

Y or vertical component $= R \sin \theta$ or $R \cos \phi$

The following example will demonstrate how any vector can be broken down into its rectangular components.

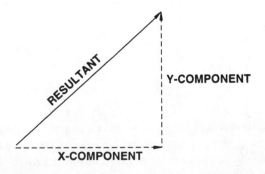

Figure 2-4   Vector Viewed as a Sum of Rectangular Components

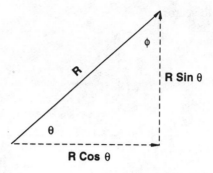

Figure 2-5   Calculation of Rectangular Components

EXAMPLE 2.3

Find the rectangular components for each of the following vectors.

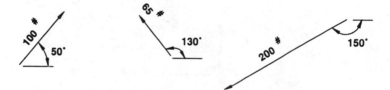

For the 100 lb vector:

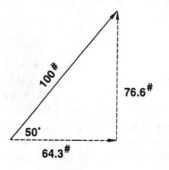

$$X = 100 \cos 50° = 64.3 \text{ lb}$$

$$Y = 100 \sin 50° = 76.6 \text{ lb}$$

For the 65 lb vector:

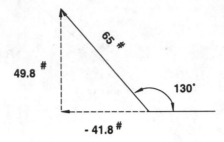

$X = 65 \cos 130° = -41.8$ lb (the negative indicates to the left)

$Y = 65 \sin 130° = 49.8$ lb

For the 200 lb vector:

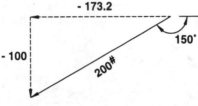

$X = 200 \cos 210° = -173.2$ lb

$Y = 200 \sin 210° = -100$ lb

Notice on the last vector, 210° was substituted since electronic calculators typically use angles based on the quadrant system. Alternately, one could have used the –150° angle, recognizing that most electronic calculators consider clockwise angles as negative.

The use of rectangular components is particularly helpful when trying to find a resultant of several vectors. By breaking down each vector into their own horizontal and vertical components, one can quickly find the resultant vector by using addition and simple trigonometric functions. When several vectors are broken down into their rectangular components, all the horizontal components can be added in a "scalar" fashion, since the vectors lie in the same line. Similarly, the vertical components can also be added in a scalar fashion, for they as well lie in the a same line. By considering the sum of all the x components, ΣX, and the sum of all the y components, ΣY, we can see that these are the legs of a right triangle (Figure 2-6).

Since the sum of all the x components form one leg of the right triangle and the sum of all the y components form the other leg of the right triangle, the result-

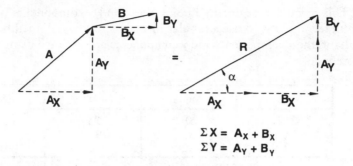

$$\Sigma X = A_X + B_X$$
$$\Sigma Y = A_Y + B_Y$$

Figure 2-6 Application of the Method of Components

ant is simply the hypotenuse. This can easily be found using the Pythagorean theorem, as previously discussed in Chapter 1.

The direction can also easily be found by utilizing the inverse tangent function and knowing the sides of the triangle as follows:

$$\alpha = \tan^{-1}(\Sigma Y/\Sigma X)$$

The resultant of several vectors that pass through the same point can very quickly be found by using the above principles. This technique is usually referred to as the **method of components**. The following example will illustrate the ease with which this technique can be applied.

EXAMPLE 2.4

Using the same three vectors that were used in Example 2.2, find the resultant using the method of components. Express the vector in an angle off the horizontal.

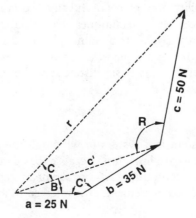

This technique begins by finding the x and y components of each vector. A helpful technique is to use a tabular format as follows. Remember that to get the resultant, the x components and y components are summed using the Pythagorean theorem.

Vector	Angle	X	Y
25 N	0	25	0
35 N	33	29.35	19.06
50 N	80	8.68	49.24
TOTAL		63.03	68.30

The resultant is found as follows:

$$R = \sqrt{(63.03)^2 + (68.30)^2} = 92.9 \text{ N}$$

The direction of this resultant can be found off horizontal as follows:

$$\alpha = \tan^{-1}(\Sigma Y / \Sigma X)$$

$$\alpha = \tan^{-1}(68.30/63.03)$$

$$\alpha = 47.3°$$

Notice that these are exactly the same answers as found in Example 2.2, although the process is much less tedious.

2.4   Types of Force Systems

As the student begins to apply the laws of statics to the solution of real problems, a helpful tool is the ability to discern the type of force system which is currently at hand. This ability may prove useful since certain force systems are more easily solved by specific solution techniques. The initial system determination should be whether a system is coplanar or noncoplanar. This distinction would mean that the forces lie in the same plane (i.e. two-dimensional) or does not (i.e. three-dimensional). Besides this basic distinction, the following system characteristics may also be assigned:

Concurrent

A concurrent system is a system in which all the forces intersect at one specific point. Such systems are intriguing because they can be solved easily by mathematical or graphical techniques. A concurrent system cannot have parallel forces within it. A good rule to remember is referred to as the **three-force principle**.

> ### Three-Force Principle
>
> If three nonparallel forces act on a body, the lines of action of the three forces must be concurrent, for equilibrium to occur.

Nonconcurrent

A force system in which all of the forces acting in it do not act or have their lines of action acting through a single, specific point. This system may necessitate the use of rotational equilibrium to solve for unknowns.

Parallel

A special case of the nonconcurrent force system involves a force system in which all of the forces involved in it are parallel to each other. None of the forces or their lines of action intersect. These systems will necessitate the use of rotational equilibrium to solve for unknowns.

The determination of system type can shed some light on what steps to pursue in the solution process.

2.5   Finding the Resultant of Concurrent Force Systems

As discussed in the previous section, a concurrent system is one in which all the forces or their lines of action intersect at one specific point. The resultant of these force systems can be found easily by trigonometric or graphical techniques; trigonometric techniques include the law of sines, the law of cosines, as well as the Pythagorean theorem. The resultant of such a system can also be found through a graphical technique such as the triangle law or the polygon law. These were covered in Sections 2.2 and 2.3.

The following examples illustrate techniques for finding resultants of concurrent force systems.

EXAMPLE 2.5

A car was driving over a shallow trench and broke down. Because of the excavation, the car cannot be pulled from the front, but instead, has to be pulled by the two chains as shown below. If 100 lbs is being applied to chain A, what force must be exerted of chain B to move the car in a straight path? Consider the car weight as negligible and the surface to be friction-free.

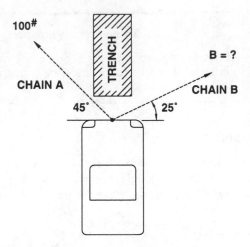

Problem Identification

The only two forces in the system act through the same point and therefore are concurrent. To keep the car in a straight path, the resultant of these forces would have to be perpendicular to the front of the car.

Solution

Considering the two forces in the chains as vectors, one possible solution would be to look at the resultant as the sum of the two vectors, possibly by using the triangle law. If the location of the resultant has been set (which it has) the triangle law could be illustrated as follows:

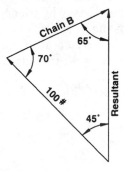

Using this triangle, one can easily solve for the unknown force in chain B, by using the law of sines as follows:

$$100 \text{ lbs}/\sin 65° = B/\sin 45°$$

$$B = 100(\sin 45°)/\sin 65° = 78 \text{ lb}$$

Therefore to keep the car in a straight path, 78 lbs would need to be applied to chain B.

Alternatively, one may also view this problem in a slightly different way. To keep the car in a straight path would mean that the x components of chain A and B would have to cancel out, thereby letting their y components combine to pull the car along the straight line.

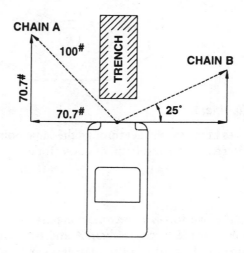

Breaking the 100 lb force in chain A into its rectangular components reveals the x component to be 70.7 lbs. Therefore, the x component in chain B would have to be equal but opposite to keep the car straight. Knowing the line of action in chain B, we could use the simple trigonometric relationship as illustrated below.

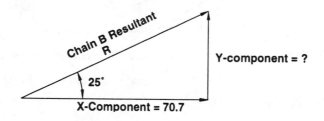

$$\cos 25° = 70.7 \text{ lb}/R$$

$$R = 70.7 \text{ lb}/\cos 25° = 78 \text{ lb}$$

As we can see, the answer is exactly the same using either technique.

EXAMPLE 2.6

A boat is being pushed with a force at the rear and pulled with a force at the front, as shown below. What will be the resultant magnitude and direction that the boat moves? Consider the boat weight as negligible and the surface to be friction-free.

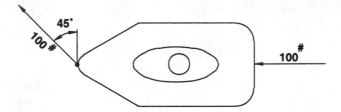

Problem Identification

The two forces in the system act through the same point and therefore are concurrent. To determine the resultant of these forces we need to add the vectors together.

Solution

Considering the two forces as vectors, one possible solution would be to look at the components of the forces in chain A and B. By reconciling these forces into their components, the method of components can be used to easily find the resultant forces magnitude and direction.

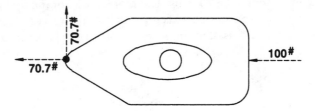

The x component and y component of the rear force can be found as follows:

$X = 100$ lbs cos $0° = -100$ lb (negative sign indicates force acts to the left)

$Y = 100$ lbs sin $0° = 0$ lbs

The x component and y component of front force can be found as follows:

$X = 100$ lbs cos $45° = -70.7$ lb (negative sign indicates to the left)

$Y = 100$ lbs sin $45° = 70.7$ lb

Therefore, the sum of the x components, ΣX, is -170.7 lb and the sum of the y components, ΣY, is 70.7 lb. The magnitude of the resultant can be found using the Pythagorean theorem as follows:

$$\text{Resultant} = \sqrt{\Sigma X^2 + \Sigma Y^2}$$

$$R = \sqrt{(-170.7)^2 + (70.7)^2} = 184.8 \text{ lb}$$

The direction can also easily be found by utilizing the inverse tangent function and knowing the sides of the triangle, as follows:

$$\alpha = \tan^{-1}(\Sigma Y/\Sigma X)$$

$$\alpha = \tan^{-1}(70.7/-170.7) = -22.5°$$

Therefore, the resultant force is 184.8 lb at an angle of −22.5°, as shown below. The negative angle indicates a clockwise angle off the horizontal reference.

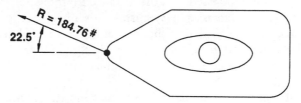

2.6   Defining the Types of Movement: The Concept of Moment

The type of potential movements a body may undergo is an essential concept that must be grasped before the student can move on to other topics involved in the field of statics. To understand how a body may move under the influence of applied forces is outlined in the following paragraphs.

Forces have the tendency to move bodies in one of two ways. The first is referred to as **translational movement**. This movement is in a straight-line path under the influence of forces. Such movement occurs in many everyday applications, and can easily be viewed right now by pushing your book across your desk. Such movement is translational because the book tends to follow the line of the push.

If your push was not "centered" along one side of the book, however, you may have noticed another type of movement while pushing the book across the desk. You may have noticed that the book also started to spin or twist as it moved across.(This can be more apparent if you try to push with one finger instead of using a whole hand). This spinning or twisting is evidence of the other movement which may take place. This type of movement is **rotational movement** (Figure 2-7).

Rotational movement turns a body about some point or axis and is caused by a "rotational force" that is referred to as a **moment**. A moment is caused by a force tending to produce a rotation about a point or an axis. The only possible way that this can occur is to have the force applied at some distance from the point or axis under consideration. A good example of this would be a revolving door which spins or rotates around its center shaft. As a force is applied at some distance away from the center shaft, the door rotates allowing a person to open the door. If someone pushed on the center shaft, no rotation would occur and the door would not open because the force was not applied at some distance away from the axis of rotation (Figure 2-8). Practically everyone has an inclination to use the principles of engineering mechanics, whether they are aware of them or not!

The magnitude of a "rotational force" or moment is dependent on two items—the magnitude of the force and the *perpendicular* distance between the point or axis of rotation and the force involved (Figure 2-9). This can be formalized in the equation shown as follows:

$$\text{Moment} = \text{Force} \times \text{Distance} \qquad \text{(Eq. 2–1)}$$

It must be stressed that the distance from the point (or axis) must be perpendicular to the force. An easy mistake for a beginner to make is to use the wrong distance when calculating the moment of a force. Units of moment are typically inch-pound ("#), foot-pound ('#), inch-kip ("K), or foot-kip ('K) in U.S. customary units; while in the SI system, units of newton-meter (N-m) or kilonewton-meter (KN-m) are common.

It must be recognized from the following discussion that a force acting through a point or axis cannot cause a moment about that point or axis (Figure 2-10). Such force would only have the tendency to cause translational movement; much in the same manner as pushing on the center shaft of the revolving door. The following example will demonstrate further the concept of moment.

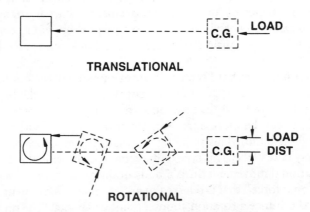

Figure 2-7 Translational versus Rotational Movement

EXAMPLE 2.7

The frame shown below is loaded under the two separate scenarios. Calculate the moment caused about point *A* at the base of the frame.

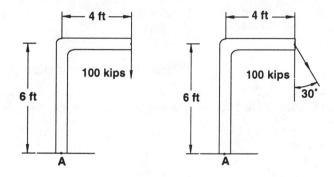

Problem Identification

In each case calculate the moment caused about point *A* by calculating the force multiplied by the perpendicular distance.

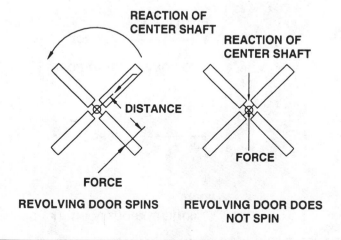

Figure 2-8   Example of a Moment on a Revolving Door

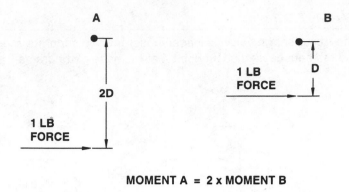

MOMENT A = 2 x MOMENT B

Figure 2-9   Calculation of a Moment about a point

Solution

In the first case, the line of action of the 100 lb force is 4 ft. away from point *A*. Therefore, the moment is as follows:

$$M_A = 100 \text{ kips} \times 4 \text{ ft.} = 400 \text{ ft.-kips in a clockwise direction}$$

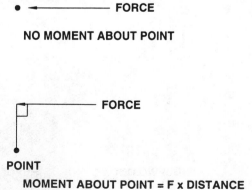

Figure 2-10   Examples of a Moment about a Point

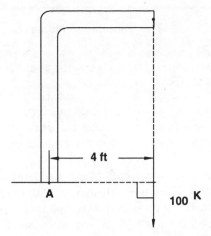

In the next case, the moment is found in a similar manner; except that the perpendicular distance to the 100 lb force is more difficult to find. Using trigonometric principles, we can find that the perpendicular distance is as follows:

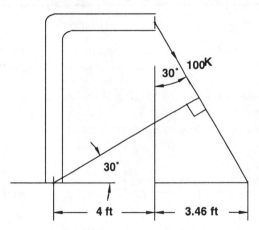

distance = 7.46 cos 30 = 6.46 ft.

Calculating the moment is then:

$$M_A = 100 \text{ kips} \times 6.46 \text{ ft.} = 646. \text{ ft.-kip clockwise.}$$

2.7   The Principle of Moments

The principle of moments, sometimes referred to as Varignon's theorem, will help in calculating the moment created by a force for which the perpendicular distance may be difficult to determine. This principle is as follows:

The Principle of Moments

The moment created by a force about a point is exactly the same moment which is created by the components of that force about the point.

This should come as no surprise, since much of the discussion in the chapter to this point has centered on how a vector can be replaced by its rectangular components. This logic can be extended to the calculation of moments, and thus we have the principle of moments.

This principle can be very useful. Many problems which have inclined forces are easier to work with when using their rectangular components because dimensioning is usually carried out in a rectangular system (horizontal and vertical). Therefore, by utilizing the principle of moments, we may sometimes circumvent the tedious process of finding the distance to an inclined force.

Another principle used frequently in determining moments about a point is transmissibilty. This principle is as follows.

Transmissibility

A force can be assumed to act anywhere along its line of action with no external change to the body itself.

This principle simplifies the finding of the perpendicular distance which will be needed to calculate a moment. The following example will repeat the second part of the previous example to illustrate the simplicity of this principle.

EXAMPLE 2.8

Recalculate the moment about point *A* for the frame used in the second part of Example 2.7.

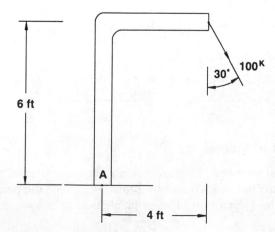

Problem Identification

Calculate the moment about point A. Since the force is inclined, it would be advantageous to used the principle of moments.

Solution

The 100 kip force acting at a 30° angle shown can be replaced by vertical and horizontal components, as follows:

$$\text{Horizontal} = 100 \sin 30° = 50 \text{ kips}$$
$$\text{Vertical} = 100 \cos 30° = 86.6 \text{ kips}$$

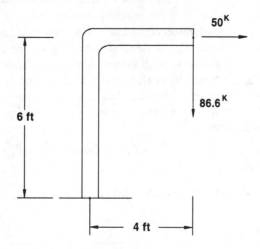

Using the principle of moments, the moment about point A could be found as follows:

$$M_A = (50 \text{ kips})(6 \text{ ft.}) + (86.6 \text{ kips})(4 \text{ ft.}) = 646. \text{ ft.-kips in the clockwise direction}$$

Notice that this is the same answer given in the previous example, although the solution did not involve calculation of a perpendicular distance. Also be aware that the moments caused by each component were added, since they both acted in the same direction (clockwise).

2.8 Finding the Resultant of Parallel Force Systems

The ability to find the resultant of parallel force systems can be very useful when dealing with beams. This is because many loads on beams are often parallel and by resolving forces into one resultant a problem can be simplified. A resultant of a parallel force system must have the same net effect on the body as the parallel system which it is replacing. That means the translational characteristics, as well as the rotational characteristics, must be equal.

Because parallel forces all act in the same direction, the translational characteristic is rather easily defined as follows:

$$R = F_1 + F_2 + F_3 + F_i \ldots$$

where

$$R = \text{the resultant force}$$

$$F_1 \text{ through } F_i = \text{all the forces}$$

The resultant must be the sum of all the forces acting on the body. For instance, if a 10 lb force and a 60 lb force act downward on a body, the resultant would have to be a 70 lb force acting downward.

The rotational characteristic is defined by finding the location on the body where the resultant would need to act, to give the body equivalent rotational behavior to the actual parallel system. This means that a point on the body is chosen and the moment about that point is calculated from the forces in the actual system. The resultant must act in the location where the moment caused by the resultant is equivalent to the moment produced by the actual system. This is defined as follows:

$$Rx = F_1 x_1 + F_2 x_2 + F_3 x_3 + F_i x_i \ldots$$

$$x = (F_1 x_1 + F_2 x_2 + F_3 x_3 + F_i x_i \ldots)/R$$

This is illustrated in Figure 2-11 and also in the following example.

EXAMPLE 2.9

The body shown below is acted upon by a group of parallel forces. Find the resultant (magnitude and location) of the resultant.

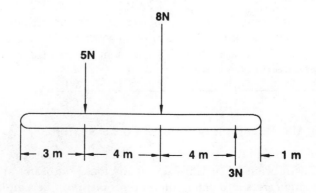

Problem Identification

Find the magnitude and location of the force which can replace the parallel system shown.

Solution

Considering translational effects, the magnitude of the resultant can be found by simply summing the forces involved. This can be done as follows:

$$R = F_1 + F_2 + F_3 + F_i...$$
$$R = 5N + 8N - 3N = 10N$$

This 10N force has to act at a particular point that gives the resultant the same rotational characteristics as the real system. If a point on the left edge of the body is chosen as the reference point, the moments from the real system would be as follows:

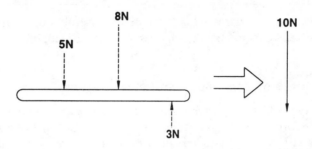

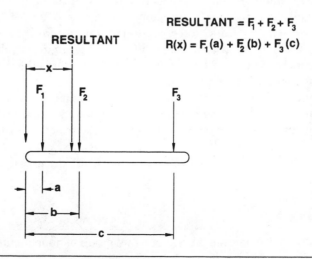

RESULTANT = $F_1 + F_2 + F_3$

$R(x) = F_1(a) + F_2(b) + F_3(c)$

Figure 2-11 Resultant of Parrallel Forces

$$M_L = +5N(3\ m) + 8N(7\ m) - 3N(11\ m) = +38\ N\text{-}m\ \text{(clockwise)}$$

The resultant would have to therfore act as follows:

$$x = +38\ N\text{-}m/10N = 3.8\ m\ \text{from the left edge.}$$

This is shown in the figure below.

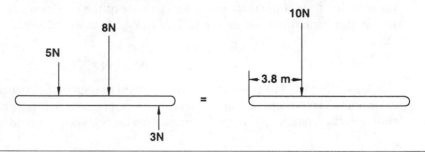

Resultants of parallel loads can be very useful in other applications such as in finding centroids of shapes, which is the focus of Chapter 6.

2.9   Resultants of Nonconcurrent Force Systems

Many typical force systems dealt with in statics are nonconcurrent, which means that the forces do not all intersect at a common point. The resultant of such a system must have the characteristics of a vector; namely magnitude, direction, and point of application.

The resultant of such a system is found by breaking the given forces into their rectangular components, thus ending up with two sets of parallel forces. Once these two sets of parallel forces are found (the sum of the vertical forces and the sum of the horizontal), the magnitude of the resultant is calculated by using the Pythagorean theorem. The direction of the resultant is then found by utilizing the tangent function. The final part of this procedure is to find the location of the resultant. This is performed in the same manner as with the parallel system resultant—by calculating a moment caused by the actual force system relative to a point and finding where the resultant has to act in order to give the same moment.

The following example will further illustrate the use of this technique.

EXAMPLE 2.10

The body shown below is acted upon by a group of nonconcurrent forces. Find the resultant (magnitude, direction, and location).

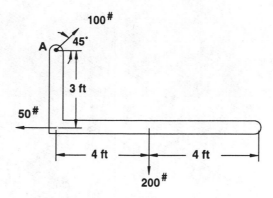

Problem Identification

Find the magnitude, direction, and location of the force which can replace the nonconcurrent system shown.

Solution

Begin by breaking down the inclined force into its rectangular components. This allows the x and y components to be added in an algebraic fashion. From a translational perspective, the magnitude of the resultant can be found by simple summing the forces involved and then applying the Pythagorean theorem. This can be done as follows:

$$R_x = F_{1x} + F_{2x} + F_{3x} + F_{ix} \ldots$$
$$R_x = -50 \text{ lb} + 70.7 \text{ lb} = 20.7 \text{ lb}$$
$$R_y = F_{1y} + F_{2y} + F_{3y} + F_{iy} \ldots$$
$$Ry = -200 \text{ lb} + 70.7 \text{ lb} = -129.3 \text{ lb}$$

Therefore:

$$R = \sqrt{(20.7)^2 + (-129.3)^2} = 130.9 \text{ lb}$$

The direction of the resultant can be found using the inverse tangent function as follows:

$$\alpha = \tan^{-1}(R_y/R_x)$$

$\alpha = -80.9°$ (the negative angle indicates clockwise off the horizontal reference)

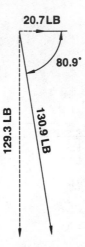

The point of application for this resultant can be located in a similar fashion to that used on parallel forces. To begin, chose a point and calculate the moments about that point caused by the forces in the actual system. In this example let's use point A.

$$M_A = +50 \text{ lbs}(3 \text{ ft.}) + 200 \text{ lbs } (4 \text{ ft.}) = +950 \text{ ft.-lbs (clockwise)}$$

Therefore, the resultant force (130.9 lbs at an angle of $-80.9°$) has to cause this same moment. Using the theory of transmissibility, assume that the resultant acts at a point which would cancel the x component out because its line of action acts through point A.

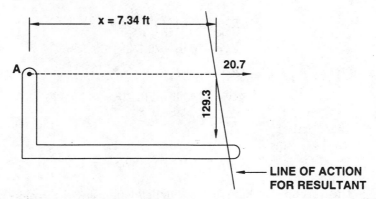

Therefore, the distance away from point A which the y component acts can be calculated as follows:

$$129.3 \text{ lbs}(x) = 950 \text{ ft.-lbs}$$

$$x = 7.34 \text{ ft.}$$

Therefore, the resultant is 130.9 lbs at a 80.9° angle acting on a line which intersects a point 7.34 ft. right of point *A*.

2.10   Summary

Forces acting on bodies tend to cause movement in a translational or straight-line path. The overall direction of movement is due to the sum of all the force effects on the body. This sum is termed a resultant. Besides moving bodies along a straight line, forces may also tend to cause rotation. Such rotation is caused by forces acting a distance away from a point or axis. This rotational force is referred to as a moment. Different force systems may be encountered when solving problems and therefore, it is important to have knowledge on the techniques used to add forces together. The addition of forces together must consider both translational and rotational effects and is referred to as finding the resultant.

EXERCISES

1. The vectors shown below are concurrent about a point. Find their result- ant using the polygon method. Find the resultant both graphically and mathematically.

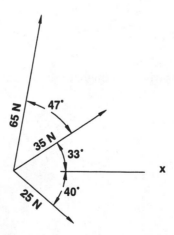

2. The vectors shown below are concurrent about a point. Find their result- ant using the polygon method. Find the resultant both graphically and mathematically.

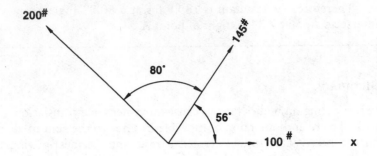

3. Find the rectangular components of the vectors shown below.

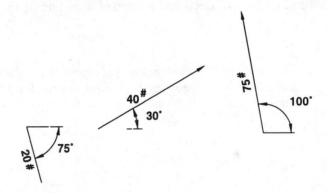

4. Add the following rectangular components to find the resultant.

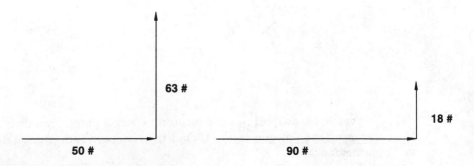

5. Using the method of components, find the resultant of the following concurrent forces.

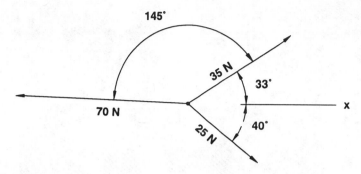

6. Using the method of components, find the resultant of the following concurrent forces.

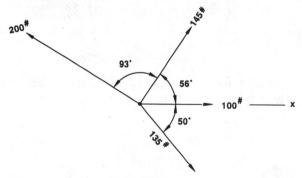

7. A hook has the forces applied to it as shown below. Calculate the resultant force.

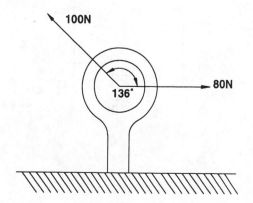

8. A nail needs to be removed from a piece of wood by a force applied at a 40° angle as shown. If a vertical force of 80 lbs is needed to overcome friction, what force needs to be used?

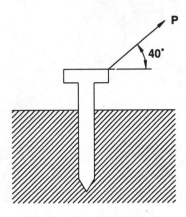

9. A bracket has forces applied to it as shown below. Calculate the resultant force applied.

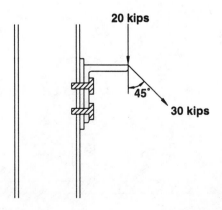

10. A tower is subjected to the forces as shown. Calculate the magnitude of F_2 which will result in the tower being placed only in compression.

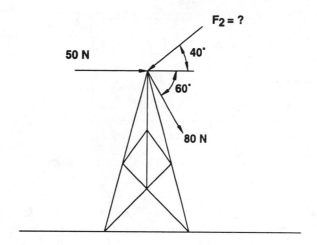

11. For the tower in the previous problem, what force in F_2 would place the tower under no net loads at all?

12. Three plates are connected using welds with concurrent forces applied as shown. Calculate the resultant of the forces.

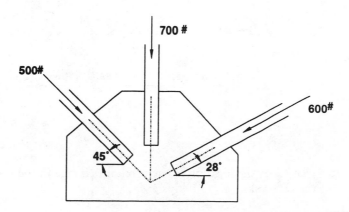

13. Find the resultant of the parallel force system shown below. Reference the resultant from point B.

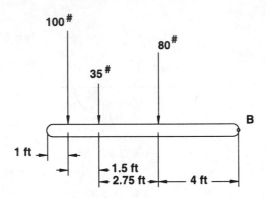

14. A train engine has the axle loads as shown. Calculate the resultant of the loads and reference the location from the front axle of the train. All loads shown are in kips.

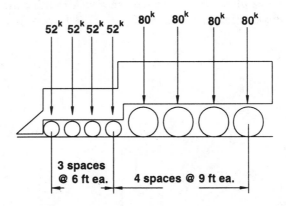

15. How does the location of the resultant change in the previous problem if the rear axles carry 60 kips each.

16. The beam shown below is loaded with a group of parallel forces. Calculate the resultant and locate with respect to the left support and also the center of the beam.

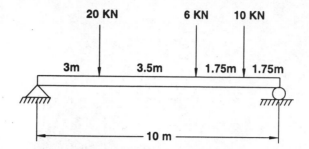

17. The beam below is loaded with the three forces shown. Where would an additional 10 kip force (downward) need to be located such that the resultant force lies directly in the center of the beam?

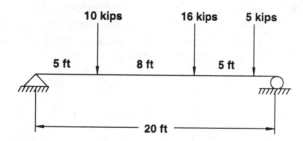

18. Where would an additional 2 kip (downward) force need to lie in the previous problem to have the resultant force act in the center of the beam?

19. A frame is subjected to the forces as shown. Calculate the resultant and locate with respect to point A.

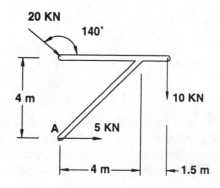

20. The retaining wall shown is subjected to the vertical and lateral forces given. Calculate the resultant force and reference with respect to point *A* in the middle of the footing.

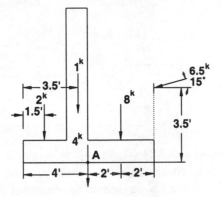

21. If the retaining wall in the previous exercise did not have the 6.5 kip force on it, where would the resultant act and what would its magnitude be?
22. Calculate the resultant of the forces on the bracket shown. Reference the resultant with respect to point *A*.

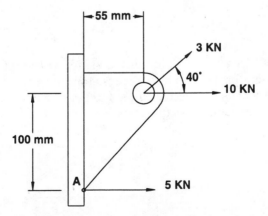

23. If the inclined force acting on the bracket in the previous exercise acted on a 20° angle, where would the resultant act?
24. Find the resultant of the forces and applied moment on the truss shown below. Reference the resultant with respect to point *B*.

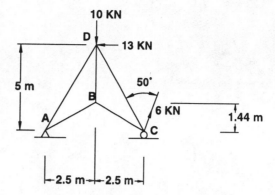

25. For the truss in the preceding exercise, what would the resultant be if the 13 KN force did not exist? Where would it be located with respect to point *B*?

3

EQUILIBRIUM OF FORCES SYSTEMS

3.1   Introduction

All of the great structures of the world must resist applied forces through the development of counteracting forces, typically referred to as **reactions**. As such, the structure is said to be in **equilibrium**. Equilibrium on a body will exist when the resultant of all the forces on the body equals zero. Equilibrium is necessary for the body to remain at rest, which is typical of many structures that are normally encountered (Figure 3-1). This text will consider only bodies at rest, since the behavior of a moving body is usually covered in a branch of engineering mechanics referred to as dynamics.

Equilibrium is the fundamental principle that underlies all of the solution techniques currently used. Equilibrium can simply be stated as follows:

General Concept of Equilibrium

1. The sum of all forces on a body must equal zero, or $\Sigma F = 0$.
2. The sum of all moments about any point relative to the body must equal zero, or $\Sigma M = 0$.

Figure 3-1 All structures Based on Principles of Equilibrium, Piscataqua River Bridge. (Courtesy of Bethlehem Steel Corperation)

This concept requires that a body in equilibrium must have the sum of all forces and moments equal zero, such that no translational or rotational movement occurs. If the forces or moments which act on a body do not equal zero, then the body would begin to move and equilibrium would no longer exist (Figure 3-2).

Typically the equation of force equilibrium is broken down into rectangular components for use in planar systems. It is simpler to think about forces in terms of their x and y components since these can be readily added. Therefore, the concept of two-dimensional equilibrium is most often expressed by the three basic equations as follows:

$$\Sigma F_x = 0$$

$$\Sigma F_y = 0$$

$$\Sigma M = 0$$

When dealing with three dimensional bodies, the force and moment equations are modified slightly to include the out-of-plane effects in the z direction and are expressed as follows:

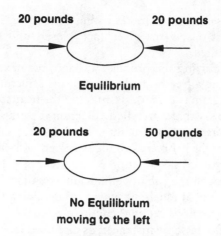

Figure 3-2 Collinear Forces in Equilibrium and not in Equilibrium

$$\Sigma F_x = 0$$

$$\Sigma F_y = 0$$

$$\Sigma F_z = 0$$

$$\Sigma M_x = 0$$

$$\Sigma M_y = 0$$

$$\Sigma M_z = 0$$

The topic of three-dimensional equilibrium will be discussed later in this chapter and the focus at the present will be on the two-dimensional cases. Since we now have some understanding of the concepts behind equilibrium, the next section will discuss a format for the techniques involved in problem-solving.

3.2   Problem-Solving and the Solution Process

Solving any problem is a process that contains both individual and universal characteristics. The characteristic of first identifying a problem is a universal characteristic, although what follows may in large part be a function of an individual's judgment. Statics is the study of how forces act and react on or within a structure and this requires the solution of unknown forces. This solution of the unknown forces is the specific problem presented to the student. The following paragraphs will create a framework that will outline the process to be used in the problem-solving process.

Every problem must begin with an **identification** of the desired problem that is to be solved. In statics, the solution will typically involve finding unknown support reactions or member forces within a structure. To correctly identify the problem, one must review the given facts closely and be clear as to what force or reaction is desired.

After identifying the problem at hand, the next part of the solution process will be drawing a **free-body diagram** of the structure involved. The concept of a free-body diagram is discussed further in the next section, although at this time it will suffice to say that a free-body diagram is a graphical representation of *all* the forces which act on that structure. The free-body diagram *must be correctly represented,* since the problem cannot be solved without it.

The last part of the solution process will be solving for the unknown forces, primarily through the use of **equilibrium equations.** Although other methods are available, the use of algebraic equations involving the equilibrium concepts (discussed in the previous section) can be used in all cases. (Graphical solution processes are useful in some instances, as discussed later in this chapter, but for other situations are extremely tedious.)

Therefore, the basic framework for solving statics can be summarized as follows:

1. Identify the problem. Review all facts closely and focus on the desired unknown.
2. Draw a free-body diagram of the structure involved.
3. Solve for the unknowns, using (typically) algebraic equations of equilibrium.

3.3   Construction of Free-Body Diagrams and Support Conditions

To solve a statics problem, one of the most fundamental skills needed is the ability to graphically represent all the forces which act on the body under consideration. This representation of all the forces acting on a body is what is known as a **free-body diagram.**

To draw a free-body diagram, it is necessary to isolate the body from its surroundings or support conditions. This "pulling away" of the body is necessary to isolate the body so the unknown forces can be visualized. When isolating a body, it is necessary to replace the actual supports with the reactions that they may act on the body. Typical support conditions are shown in Figure 3-3, and are supposed to represent the possible forces which could be generated for these support conditions. Although the support forces (often termed reactions) could potentially exist in a particular case, the student should realize that not every case contains all support reactions shown.

Many free-body diagrams could be drawn in just one structure. As shown in Figure 3-4, several free-body diagrams can be produced from just one structure. Many structures are more easily solved by breaking the problem into separate free-body diagrams.

Type of Support	Reactions and Free-Body	Reaction Description
PIN (Same as Smooth Surface)	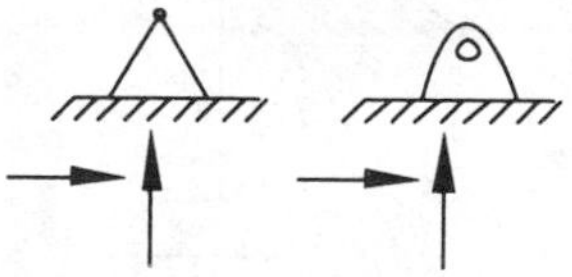	Potential reaction normal to contact surface and parallel to the contact surface
ROLLER (Same as Rough Surface)	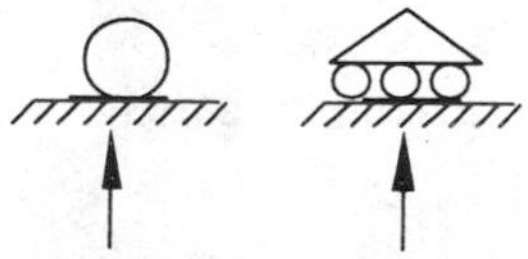	Potential reaction normal to contact surface
FIXED	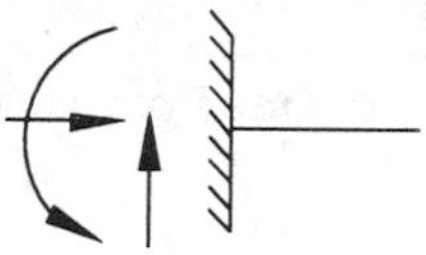	Potential reactions normal and parallel to contact surface with an additional moment reaction
CABLE (Same as Link)		Potential reaction in the direction of cable of link

Figure 3-3   Modeling of Typical Support Conditions for use in Free-Body Diagrams

After constructing the free-body diagram, the solution process will usually proceed by applying algebraic equations. When using algebraic equations, a sign convention for the forces and moments must be consistently used during the solution procedure. Any sign convention can be used, and be correct, as long as it remains constant during that problem. This book will utilize the following:

Sign Convention for Text

Forces:

- To the right (+), to the left (–)
- Upwards (+), Downwards (–)

Moments:

- Clockwise Rotation (+), Counterclockwise (–)

This convention is purely arbitrary and can be changed as long as it remains consistent throughout a solution.

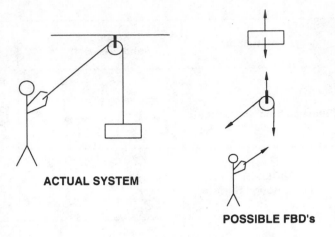

ACTUAL SYSTEM

POSSIBLE FBD's

Figure 3-4 Concepts of Free-Body Diagrams

A common misconception is that the sign conventions between forces and moments have to be somehow related—that is, a postive force has to cause a positive moment. This is *not* the case. *Please realize that positive forces can cause either positive or negative moments and that a negative force can create either a positive or negative reaction.*

3.4 Equilibrium of Colinear Force Systems

The most basic force system to study is the colinear force system. **Colinear force systems** have all the forces within that system acting in the same line (Figure 3-5). Equilibrium occurs when the sum of all forces acting along that line equal zero. The solution process will follow the framework which was outlined in section 3.2, with the algebraic equation written for the forces along the line of action. This equation is as follows:

$$\Sigma F_{\text{line}} = 0$$

Colinear forces exist in special members which are referred to as two-force members. **Two-force members** are members under the action of two equal colinear forces acting along the length of the member. These two-force members are under either tension or compression forces and are used frequently in buildings and other structures. These members will be discussed further in Chapter 4.

The following example will demonstrate the solution of a colinear force system.

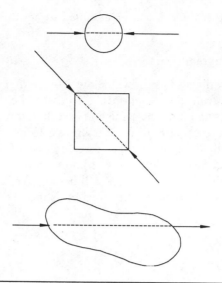

Figure 3-5   Colinear Force Systems

EXAMPLE 3.1

A 100 lb weight is hanging from a rope which is attached to the ceiling as shown. Calculate the tension force created in the rope.

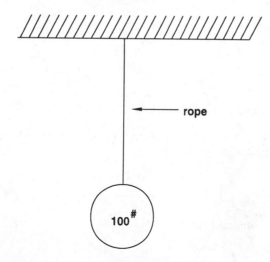

Problem Identification

The task in this problem is to solve for the unknown tension in the rope. We will refer to the unknown tension force in the rope as T. The other force given is the weight of 100 lbs which can be assumed to act downwards under gravitational effects.

Free-Body Diagram

The free-body diagram is shown below. Notice that the weight and the rope have been isolated from their support. All forces— the weight and the tension in the rope— are acting along the same line and therefore, the force system is colinear. The direction of the 100 lb force is known to be in correct direction, while the direction of the unknown tension force can be arbitrarily chosen. When solving the equations, the direction of the unknown forces will be verified. The algebraic sign of the solved unknown force will be positive when the direction is correctly assumed and negative when the direction is incorrectly assumed.

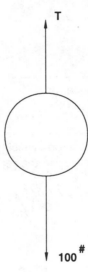

Algebraic Equations of Equilibrium

Solution of the unknown reactions now involves the writing of the algebraic equations of equilibrium as follows:

$$\Sigma F_{line} = 0;$$

$$+T - 100 \text{ lb} = 0$$

Simplifying and solving:

$$T = 100 \text{ lb}$$

Therefore the tension in the rope is 100 lbs. Notice the unknown tension force, T, had a positive algebraic sign in the solved equation. This indicates that the direction that was initially chosen when drawing the free-body diagram was correct.

3.5   Equilibrium of Concurrent Force Systems

In Chapter 2, it was explained that concurrent forces are those that act through a common point. Such forces can be added together to form a single resultant. Many problems that are encountered happen to be concurrent force systems. A useful reminder of concurrence is that a force system that only contains three non-parallel forces is concurrent in order to be in equilibrium. Equilibrium can only occur when the sum of all forces in the x and y direction equal zero and the moment about any point relative to the body is likewise zero.

The solution of unknown forces in a concurrent force system can be found using the principles of equilibrium in conjunction with algebraic equations of equilibrium, trigonometric solutions, and graphical techniques. Each of these will be discussed in the following paragraphs.

Algebraic Equations

The solution of equilibrium conditions through the use of algebraic equations can be used in all types of problems, including concurrent force problems. Since all forces in a concurrent system intersect through a single point, it is typically advantageous to consider a free-body diagram of the system at this point. The primary equilibrium equations to be utilized for coplanar, concurrent cases are as follows:

$$\Sigma F_x = 0$$

$$\Sigma F_y = 0$$

All forces acting in a concurrent system are usually broken down into their x and y components and then summed to solve for the unknown forces.

Trigonometric Techniques

Concurrent force systems are unique because the resultant of the system has to also pass through the point of concurrence. Remembering that the resultant of a concurrent system in equilibrium is zero, it would then follow that the sum of all forces in a concurrent system would have to likewise be zero. Therefore, the forces in a concurrent system can be represented vectorially by a closed shape and, by using trigonometric laws, the solution of unknowns can be performed (Figure 3-6).

A concurrent system composed of three forces can be represented by a force triangle, with the unknowns easily being solved by the application of the law of sines, law of cosines, or trigonometric functions. A concurrent force system composed of four or more forces can be represented by a force polygon. The solution of the force polygon by the use of mathematical equations is much more tedious and is usually not recommended over the solution by using algebraic equations.

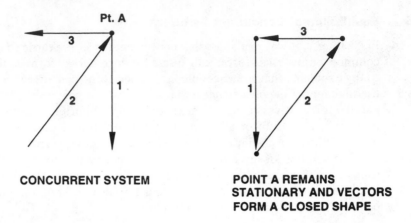

| CONCURRENT SYSTEM | POINT A REMAINS STATIONARY AND VECTORS FORM A CLOSED SHAPE |

Figure 3-6 Equilibrium of Concurrent Force Systems

Graphical Techniques

The use of graphical techniques to solve concurrent force system is a valid and sometimes helpful method. As mentioned previously, the concurrent forces can be arranged as vectors placed tip-to-tail, forming a closed figure (either a triangle or a polygon). If the figure was drawn to scale, the unknown forces can be solved by measuring the length of the unknown force vector.

The following example will illustrate the use of all of these techniques.

EXAMPLE 3.2

For the given problem shown below, calculate the tension force in the cables AB and BC.

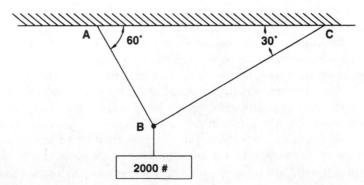

Problem Identification

The task in this problem is to solve for the unknown tension in the two cables, which can be viewed as support conditions. Since all forces pass through point B, this is a concurrent force system. The 2000 lb downward weight is known, as well as the angle of the supporting cables.

Free-Body Diagram

It is generally advantageous to separate the body from its supports, as shown below. All forces, including the tension in the cables, are included in the free-body diagram. A solution by each of the techniques will be shown below.

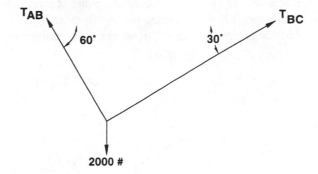

Solution by Algebraic Equations of Equilibrium

Since concurrent problems utilize only the equilibrium equations with regard to the summation of forces in the x and y directions, the solution of the unknown tension reactions involves the writing of the algebraic equations of equilibrium as follows:

$$\Sigma F_x = 0$$

$$-T_{AB} \cos 60° + T_{BC} \cos 30° = 0$$

Rearranging this equation:

$$T_{BC} \cos 30° = T_{AB} \cos 60°$$

or

$$T_{BC} = .577\, T_{AB}$$

$$\Sigma F_y = 0$$

$$+T_{AB} \sin 60° + T_{BC} \sin 30° - 2000 = 0$$

Substituting the relationship found in the $\Sigma F_x = 0$ equation ($T_{BC} = .577\, T_{AB}$), the equation can be solved as follows:

$$+T_{AB} \sin 60° + (.577\, T_{AB}) \sin 30° - 2000 = 0$$

Solving for T_{AB}, we find the following:

$$.866 \, T_{AB} + (.577 \, T_{AB})(.50) = 2000$$

$$T_{AB} = 1733 \text{ lbs.}$$

Substituting this answer into the $\Sigma F_x = 0$ equation we find:

$$T_{BC} = .577 \, T_{AB}$$

$$T_{BC} = .577 \, (1733 \text{ lbs}) = 1000 \text{ lbs.}$$

Solution by Trigonometric Techniques

Since the concurrent force system contains three forces, these three forces can be represented by vectors which, when placed tip-to-tail, form a closed triangle. The resultant of such a system is zero. Drawing such a triangle may be done as follows:

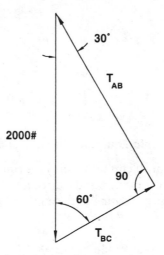

Using the law of sines, the solution for the unknowns T_{AB} and T_{BC} is readily handled.

$$2000/\sin 90° = T_{AB}/\sin 60°$$

$$T_{AB} = 2000 \text{ lb } (\sin 60°/\sin 90°) = 1732 \text{ lbs}$$

and

$$2000/\sin 90° = T_{BC}/\sin 30°$$

$$T_{BC} = 2000 \text{ lb } (\sin 30°/\sin 90°) = 1000 \text{ lbs}$$

Solution by Graphical Techniques

This solution involves the construction of a force triangle, similar to what was done in the previous method. The solution is measured from the diagram using a predetermined scale. Using a scale of 1 inch = 500 lbs and drawing a force triangle, the unknown vector can be scaled and the solution found as follows:

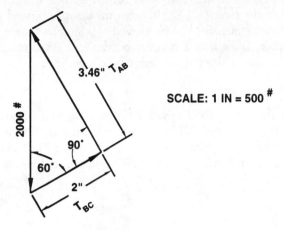

SCALE: 1 IN = 500 #

The next example further illustrates the solution process of another concurrent force system.

EXAMPLE 3.3

For the structure shown below, find the tension in the cable and the compression force in the strut. Neglect the member's weight.

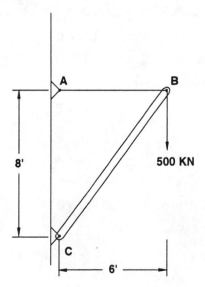

Problem Identification

The task in this problem is to solve for the tension in the cable (T_{AB}) and the compression in the strut (C_{BC}). Although the strut is pinned at its support, the compression in the member can be viewed as a singular force acting along the axis of the member, since it is a two force member. (Two-force members always have their force acting along the length of the member).

Free-Body Diagram

To draw the free-body diagram it is necessary to separate the body from its supports as shown below. All forces, including the tension in the cable and the compressive force in the strut, are included in the free-body diagram. A solution by the technique of algebraic equations will be shown below.

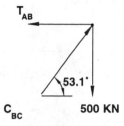

Solution by Algebraic Equations of Equilibrium

Concurrent problems utilize only the force equations of equilibrium in the x and y directions. The solution of the unknown reactions involves writing the algebraic equations of equilibrium as follows:

$$\Sigma F_x = 0$$

$$-T_{AB} \cos 0° + C_{BC} \cos 53.1° = 0$$

Rearranging this equation, we find:

$$C_{BC} \cos 53.1° = T_{AB} \cos 0°$$

or

$$T_{AB} = .60 \, C_{BC}$$

$$\Sigma F_y = 0$$

$$+C_{BC} \sin 53.1° - 500 \, KN = 0$$

Solving for C_{BC} we find:

$$C_{BC} = 625 \, KN$$

Substituting the answer for C_{BC} back into the $\Sigma Fx = 0$ equation:

$$T_{AB} = .60 \; C_{BC}$$

$$T_{AB} = .60(625 \text{ KN}) = 375 \text{ KN}$$

As an alternative to this solution, the forces in these members could be found by treating this system as if it was not concurrent and looking only at the support conditions as stated. This topic is discussed in the upcoming section.

3.6   Equilibrium of Parallel Force Systems

A special type of force system in which none of the applied forces on the body intersects is referred to as a parallel force system. The resultant of a parallel force system which is in equilibrium will be found to be equal to zero (Figure 3-7). Earlier in this chapter, these ideas were expressed by the following equations when dealing with coplanar problems:

$$\Sigma F_x = 0$$

$$\Sigma F_y = 0$$

$$\Sigma M = 0$$

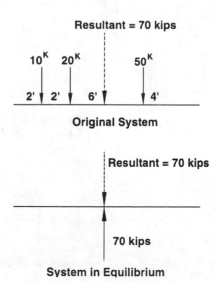

Figure 3-7   Resultant of a Parallel Force System

In parallel forces systems, equilibrium cannot be proven without the use of a moment equation. Therefore, the use of a moment equation is always used in a parallel type of system.

A common example of parallel force system problem is that of a beam loaded under typical gravity loads. Loads may be **concentrated (point) loads** or **distributed loads.** Concentrated loads may be due to a piece of equipment, furniture, or column which is applied on a beam; while distributed loads may be caused by floor loads, roof loads, or even the beam's own weight. Many times distributed loads are assumed to be of constant intensity and are referred to as **uniformly distributed loads** (Figure 3-8). Uniformly distributed loads may be due to items such as the selfweight of an individual beam or the flooring it supports; as well as the applied loading from wind, snow or other live loads. Another type of uniform load is a **uniformly varying load** . This type of load is "spread out" over the length of a beam and varies in intensity. A good example of this type of load is the lateral earth pressure which is applied to a retaining wall. Such a load varies from an intensity of zero at the ground surface, to a maximum at the bottom of the retaining wall.

In parallel force systems, as in the other systems, the problem is to solve for the unknown reactions. The solution process will consist of identifying the problem, drawing a free-body diagram, and writing algebraic equations to solve for the unknowns. As the solution process involving algebraic equations unfolds, the student must be especially aware of the sign convention. Again, this text refers to forces upward and rightward as positive; while those forces acting down or to the

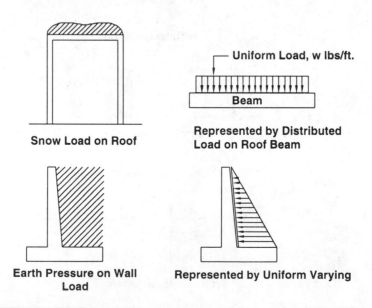

Figure 3-8   Examples of Uniform Loads on Structures

left as negative. Moments which cause clockwise rotation will be positive; while those causing counterclockwise rotation will be assigned a negative value. The following examples demonstrate the solution of various parallel force type problems.

EXAMPLE 3.4

Calculate the reactions at the supports for the beam shown below. The supports are a pin support at end *A* and a roller support at end *B*.

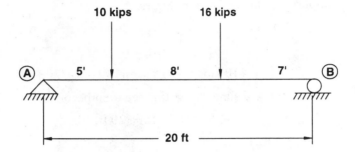

Problem Identification

The task in this problem is to solve for the support reactions at *A* and *B*. A pin and a roller have been specified as the support conditions.

Free-Body Diagram

The free-body diagram is shown below. All forces, including all possible support reactions, are included in the free-body diagram. The directions of the applied forces are known to be in correct direction, while the direction of the forces for the unknown reactions can be arbitrarily chosen. When solving the equations, the direction of the unknown forces will be verified. This occurs by the algebraic sign of the solved force being positive when the direction is correctly assumed, and negative when the direction is incorrectly assumed.

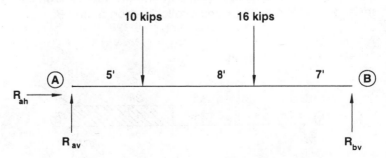

Algebraic Equations of Equilibrium

Solution of the unknown reactions involves the writing of the algebraic equations of equilibrium as follows:

$$\Sigma F_x = 0$$

$$+R_{ah} = 0$$

$$\Sigma F_y = 0$$

$$+R_{av} + R_{bv} - 10 \text{ kips} - 16 \text{ kips} = 0$$

Simplifying, this equation becomes:

$$+R_{av} + R_{bv} = 26 \text{ kips}$$

$$\Sigma M_a = 0$$

$$+10 \text{ kips}(5') + 16 \text{ kips}(13') - R_{bv}(20') = 0$$

Rearranging and simplifying, this equation becomes:

$$258 \text{ kip-ft.} = (20 \text{ ft.}) \, R_{bv}$$

Solving for R_{bv}:

$$R_{bv} = 12.9 \text{ kips}$$

Substituting $R_{bv} = 12.9$ kips into the $\Sigma F_y = 0$ equation:

$$+R_{av} + (12.9 \text{ kips}) = 26 \text{ kips}$$

$$R_{av} = 13.1 \text{ kips.}$$

Comments

Notice the signs for the solution of the unknown reactions were positive, meaning that the directions which were assumed in the free-body diagram were correct.

EXAMPLE 3.5

Calculate the reactions at the supports for the beam shown below. The supports are at pin support at end *A* and a roller support at end *B*.

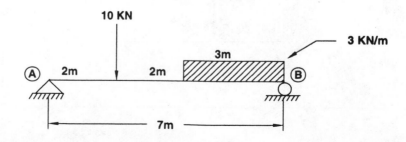

Problem Identification

The task in this problem is to solve for the support reactions at A and B. A pin and a roller have been specified as the support conditions. In this problem, a uniformly distributed load of 3 KN/m is also used.

Free-Body Diagram

The free-body diagram is shown below. All forces, including all possible support reactions, are included in the free-body diagram. A resultant load of 9 KN replaces the 3 KN/m uniform load which acts over the 3 m distance. This resultant of 9 KN acts in the center of the uniform load and will provide the same behaviors as the actual uniform load. This resultant is used to simplify the calculation of the moment produced by the uniform load. The directions of the given forces are known to be in correct direction, while the direction of the forces for the unknown reactions can be arbitrarily chosen.

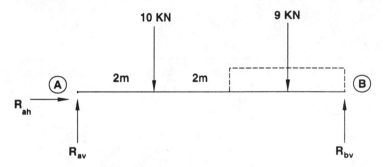

Algebraic Equations of Equilibrium

Solution of the unknown reactions now involves the writing of the algebraic equations of equilibrium as follows:

$$\Sigma F_x = 0$$

$$+R_{ah} = 0$$

$$\Sigma F_y = 0$$

$$+R_{av} + R_{bv} - 10 \text{ KN} - 9 \text{ KN} = 0$$

Simplifying, the equation becomes:

$$+R_{av} + R_{bv} = 19 \text{ KN}$$

$$\Sigma M_a = 0$$

$$+10 \text{ KN}(2 \text{ m}) + 9 \text{ KN}(5.5 \text{ m}) - R_{bv}(7 \text{ m}) = 0$$

Simplifying and rearranging, this equation becomes:

$$69.5 \text{ KN-m} = (7 \text{ m}) \, R_{bv}$$

Solving for R_{bv}:

$$R_{bv} = 9.9 \text{ KN}$$

Substituting $R_{bv} = 9.9$ KN into the $\Sigma F_y = 0$ equation:

$$+R_{av} + (9.9 \text{ KN}) = 9.1 \text{ KN}$$

$$R_{av} = 9.1 \text{ KN}.$$

Comments

Realize that the algebraic signs for the unknowns again came out positive in the solution, meaning that the directions of the unknown reactions were correctly assumed.

EXAMPLE 3.6

Calculate the reactions at the supports for the beam shown below. The supports are a pin support at end A and a roller support at end B.

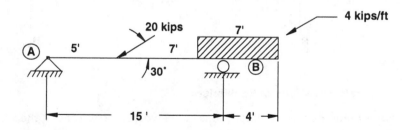

Problem Identification

In this problem the support reactions at A and B are needed. A pin and a roller have been specified as the support conditions. A diagonal load at a 30° angle is used as well as a uniformly distributed load of 4 kips/ft.

Free-Body Diagram

The free-body diagram is shown below. All forces, including all possible support reactions, are included in the free-body diagram. A resultant load of 28 kips replaces the 4 kips/ft. load which acts over the 7 ft. distance. This resultant of 28 kips acts in the center of the uniform load and will provide the same behavior as the original load. The diagonal load of 20 kips at a 30° angle is broken down into its rectangular components (10 kips vertical, 17.32 kips horizontal) acting at the same location. The directions of the applied forces are known to be in correct direction, while the direction of the forces for the unknown reactions can be arbitrarily chosen.

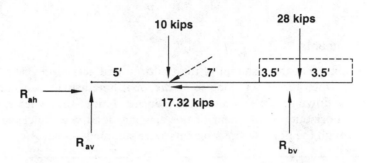

Algebraic Equations of Equilibrium

Solution of the unknown reactions now involves the writing of the algebraic equations of equilibrium as follows:

$$\Sigma F_x = 0$$

$$+R_{ah} - 17.32 \text{ kips} = 0$$

Solving the equation yields:

$$R_{ah} = 17.32 \text{ kips}$$

$$\Sigma F_y = 0$$

$$+R_{av} + R_{bv} - 10 \text{ kips} - 28 \text{ kips} = 0$$

Simplifying this equation becomes:

$$+R_{av} + R_{bv} = 38 \text{ kips}$$

$$\Sigma M_a = 0$$

$$10 \text{ kips}(5 \text{ ft.}) + 28 \text{ kips}(15.5 \text{ ft.}) - R_{bv}(15 \text{ ft.}) = 0$$

Simplifying and rearranging, this equation becomes:

$$484 \text{ kip-ft.} = (15 \text{ ft.})R_{bv}$$

Solving for R_{bv}:

$$R_{bv} = 32.3 \text{ kips}$$

Substituting $R_{bv} = 32.3$ kips into the $\Sigma F_y = 0$ equation:

$$+R_{av} + (32.3 \text{ kips}) = 38 \text{ kips}$$

$$R_{av} = 5.7 \text{ kips}.$$

Comments

Realize that the resultant of the uniform load acts in the center of the load and that when forces act along some diagonal line, usually it is convenient to break these down into their x and y components. Notice that when summing moments, the horizontal component of the diagonal load was cancelled out since it acts through the point where moments were summed—point A.

3.7   Equilibrium of Nonconcurrent Force Systems

A nonconcurrent force system is one in which all the forces do not intersect at a given point of concurrence. Many force systems are nonconcurrent and, in fact, we have already discussed at length one type of nonconcurrent force system—the parallel force system in section 3.6.

Solving for reactions in nonconcurrent systems, the solution follows the previously discussed format of identifying the problem, drawing a free-body diagram and writing algebraic equations. Because not all forces in every nonconcurrent system will be parallel, other concepts may be helpful in the solution of these problems.

One such concept is the principle of **transmissibility**. This principle was discussed in section 2.6 and states that a force acting on a body can act anywhere along its line of action without changing the external effect of the force on that body. Transmissibility is very useful in the application of moment equations which will be used in the solution of unknown forces for some nonconcurrent force systems.

Another concept which is frequently used in nonconcurrent systems is the principle of moments or Varignon's theorem which was discussed in section 2.7. This stated that the sum of moments from the components of a force will create the same moment about a point (or axis) as the force itself would produce. It is useful to use components of a force since they have already been used in the ΣF_x and ΣF_y equations. An illustration of this concept is shown in Figure 3-9.

The following examples illustrate the solution of unknown forces in nonconcurrent systems.

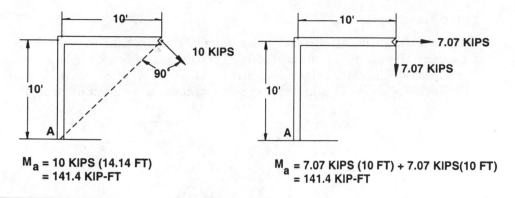

M_a = 10 KIPS (14.14 FT)
  = 141.4 KIP-FT

M_a = 7.07 KIPS (10 FT) + 7.07 KIPS(10 FT)
  = 141.4 KIP-FT

Figure 3-9   Varignon's Theorem: Moment About a Point Due to a Force is Equal to the Moment Caused by the Components of That Force

EXAMPLE 3.7

The ladder shown below carries the weight of two people at the locations as shown. If the base of the ladder (point *A*) is considered a rough surface and the top of the ladder (point *B*) is considered a smooth surface, determine the reactions at each.

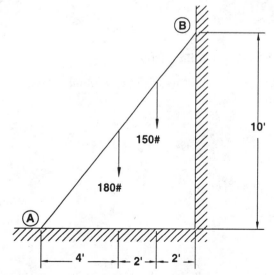

Problem Identification

The task in this problem is to solve for the support reactions at *A* and *B*. A rough surface is one which is able to develop a frictional component, therefore its reactions are represented in a similar fashion to a pin. A smooth surface cannot develop this frictional component and therefore its only reaction is perpendicular to the contact surface, much like that of a roller.

Free-Body Diagram

All forces, including all possible support reactions, are included in the free-body diagram. The downwards loads of 150 and 180 lbs are applied in the locations as specified. The directions of these applied are known to be in correct direction, while the direction of the forces for the unknown reactions can be arbitrarily chosen.

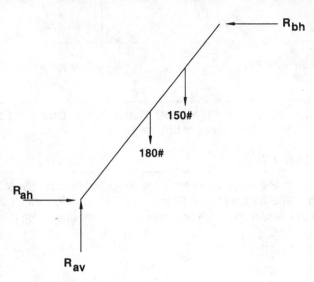

Algebraic Equations of Equilibrium

The solution of the unknown reactions involves writing algebraic equations of equilibrium as follows:

$$\Sigma F_x = 0$$

$$+R_{ah} - R_{bh} = 0$$

$$\Sigma F_y = 0$$

$$+R_{av} - 180 \text{ lbs} - 150 \text{ lbs} = 0$$

Simplified:

$$+R_{av} = 330 \text{ lbs}$$

$$\Sigma M_a = 0$$

$$+180 \text{ lbs}(4 \text{ ft.}) + 150 \text{ lbs}(6 \text{ ft.}) - R_{bh}(10 \text{ ft.}) = 0$$

Simplifying and rearranging this equation becomes:

$$1620 \text{ lb-ft.} = (10 \text{ ft.}) R_{bh}$$

Solving for R_{bh}:

$$R_{bh} = 162 \text{ lbs}$$

(Notice that the positive answer means that the direction as assumed in the free-body diagram was correct).

Substituting $R_{bh} = 162$ lbs back into the $\Sigma F_x = 0$ equation:

$$+R_{ah} - (162\ \text{lb}) = 0$$

$$R_{ah} = 162\ \text{lbs}$$

Comments

Notice that when moments were taken about point A, that the distances to the 180 and 150 lb forces are the perpendicular distances to their individual lines of action. This example uses the previously discussed principle of transmissibility. Many students make the mistake of using the distance *along the ladder* when summing moment, which is wrong because these distances are not perpendicular to the force.

EXAMPLE 3.8

Determine the reactions on the truss structure as shown below. The support at A is a pin and the support at B is a roller.

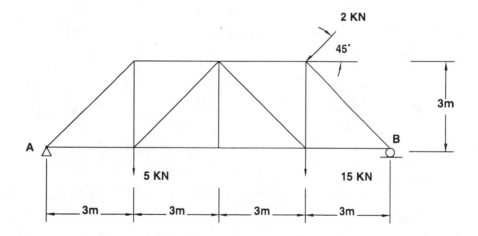

Problem Identification

The problem is to solve for the unknown support reactions. The system is a non-concurrent force system.

Free-Body Diagram

The following free-body diagram separates the truss from its supports and replaces the supports with the potential forces which act at the pin and the roller. Also notice that the inclined force was broken into its rectangular components which will be beneficial in the solution of our algebraic equations.

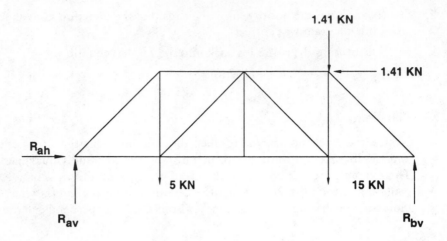

Algebraic Equations of Equilibrium

Solving of the unknowns involves the writing and solution of algebraic equations as follows:

$$\Sigma F_x = 0$$

$$+R_{ah} - 1.4 \text{ KN} = 0$$

Solving:

$$R_{ah} = 1.4 \text{ KN}$$

$$\Sigma Fy = 0$$

$$+ R_{av} + R_{bv} - 5 \text{ KN} - 15 \text{ KN} - 1.4 \text{ KN} = 0$$

$$+R_{av} + R_{bv} = 21.4 \text{ KN}$$

$$\Sigma M_A = 0$$

$$+5 \text{ KN}(3 \text{ m}) + 15 \text{ KN}(9 \text{ m}) + 1.4 \text{ KN}(9 \text{ m}) - 1.4 \text{ KN}(3 \text{ m}) - R_{bv} (12 \text{ m}) = 0$$

Solving for R_{bv}:

$$R_{bv} = 13.2 \text{ KN}$$

Substituting $R_{bv} = 13.2$ KN back into the ΣF_y equation:

$$+R_{av} + (13.2) = 21.4 \text{ KN}$$

$$R_{av} = 8.2 \text{ KN}.$$

Comments

Notice that all the of the assumed directions on the free-body diagram were found to be correct, since all of the unknowns solved to be positive.

EXAMPLE 3.9

The body shown below is supported by a pin at A and a frictionless roller at B. Find the reactions at these two points with the loads applied as given.

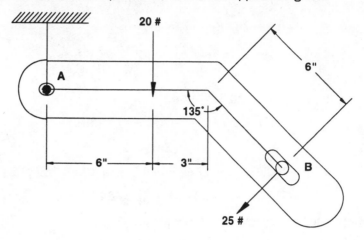

Problem Identification

Solve for the pin reaction at A and the roller reaction at B. Notice that the roller reaction is at an angle which is not in the typical x, y coordinate system.

Free-Body Diagram

The free-body diagram represents the pin forces at A and the 20 lb force as is typically done, however, the roller reaction and the 25 lb force must be broken down into its rectangular components. The roller reaction will be expressed in terms of R_b and since the angle of this support is 45°, then x and y components of R_b will both be .707 R_b.

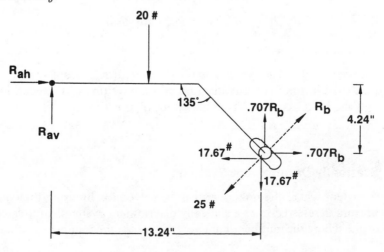

Algebraic Equations of Equilibrium

Solving for the unknowns is performed as follows:

$$\Sigma F_x = 0$$

$$+R_{ah} + .707\ R_b - 17.67\ \text{lbs} = 0$$

$$R_{ah} + .707\ R_b = 17.67\ \text{lbs}$$

$$\Sigma F_y = 0$$

$$+\ R_{av} + .707\ R_b - 20\ \text{lbs} - 17.67\ \text{lbs} = 0$$

$$+R_{av} + .707\ R_b = 37.67\ \text{lbs}$$

$$\Sigma M_A = 0$$

$$+20\ \text{lb}(6\ \text{in.}) + 17.67\ \text{lb}(4.24\ \text{in.}) + 17.67\ \text{lbs.}(13.24\ \text{in.})$$

$$-\ .707 R_b(4.24\ \text{in.}) - .707\ R_b(13.24\ \text{in.}) = 0$$

Simplifying and rearranging this equation:

$$428.87\ \text{lb-in.} = 12.36\ R_b$$

Solving for R_b:

$$R_b = 34.7\ \text{lbs}$$

Substituting $R_b = 34.7$ lbs back into the ΣF_y equation:

$$+R_{av} + .707(34.7) = 37.67\ \text{lb}$$

$$R_{av} = 13.14\ \text{lb}$$

Substituting $R_b = 34.7$ lbs back into the ΣF_x equation:

$$R_{ah} + .707\ (34.7) = 17.67\ \text{lb}$$

$$R_{ah} = -\ 6.86\ \text{lb}$$

Comments

Notice that the negative sign on the solution of R_{ah} means that the assumed direction used in the free-body diagram and the writing of the equation are incorrect and the force actually acts in the opposite direction.

3.8   Statically Determinant Systems

A system where the unknowns can be solved for by using the equations of equilibrium is referred to as a statically determinant system. An indeterminate system is one where the number of unknowns exceeds the number of equations which can

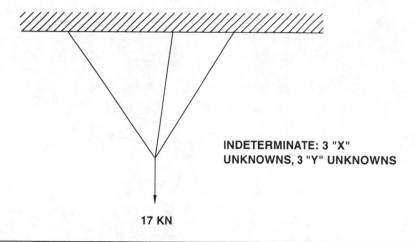

INDETERMINATE: 3 "X"
UNKNOWNS, 3 "Y" UNKNOWNS

17 KN

Figure 3-10   Case of Indeterminancy

be utilized. This can occur, for instance, where a system has four unknowns and we have only three equations of equilibrium. Such a problem would be referred to as being indeterminate to the first degree. A problem can also be indeterminate when unknowns exceed the available equations on equilibrium if one of the three equation of equilibrium cannot be readily used. This may happen, for instance, in a concurrent problem, as shown in Figure 3-10, where the moment equation for equilibrium may not be readily available for use.

It is advantageous to recognize the difference between statically determinant and indeterminate systems to know if the standard procedures as outlined in this book can be applied to a given problem. Indeterminate problems will require further study as to the solution processes which must be applied. Such problems are beyond the scope of this book.

3.9   Equilibrium of Concurrent, Non-Coplanar Force Systems

Forces which do not lie in the same plane are non-coplanar. Concurrent , non-coplanar forces arise in a number of brackets, tripods, and space truss arrangements. This section explores the solution of such systems.

When a force is non-coplanar it is referred to as being in space or part of a space structure. A force in space may consist of components in the x, y, and z directions. This can best be seen by imagining a box with a force, F, stretching from its origin to the opposite corner as seen in Figure 3-11. The components of the force, F, are referred to as F_x, F_y and F_z and form the edges of this box. The angles between the force, F, and these components are called the direction angles of the force and are called out as θ_x, θ_y, θ_z. The relationship between the force, F, and each component is simply as follows:

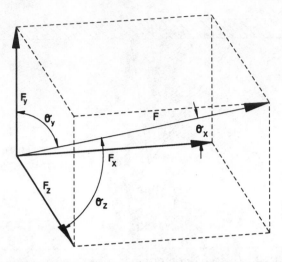

Figure 3-11 Vector F Expressed in Terms of it x, y, and z Components

$$F_x = F \cos \theta_x$$

$$F_y = F \cos \theta_y$$

$$F_z = F \cos \theta_z$$

Cosines of angles, θ_x, θ_y, and θ_z are typically referred to as **direction cosines**. However, when these direction cosines do not lie in one specific plane , it is hard to get their values unless we approach the components of vector F in a slightly different manner. In Figure 3-12, imagine force F being made up of an intermediate vector referred to as F' and the F_y vector. The intermediate vector F' is simply the vector sum of F_x and F_z or as stated:

$$F' = F_x + F_z$$

From trigonometry, realize that:

$$F_x = F' \cos \theta_2 = F \cos\theta_1 \cos \theta_2$$

$$F_z = F' \sin \theta_2 = F \cos\theta_1 \sin \theta_2$$

The vector F can then be found by adding the intermediate vector, F', with the component F_y. Realize that F_y can be expressed as:

$$F_y = F \sin \theta_1$$

The relationship between the direction cosines and the components as broken down into intermediate vectors is as follows:

$$\cos \theta_x = \cos \theta_1 \cos \theta_2$$

$$\cos \theta_y = \sin \theta_1$$

$$\cos \theta_z = \cos \theta_1 \sin \theta_2$$

Using the Pythagorean theorem, it can be found that the square of the force, F, is equal to the sum of the squares of the individual components. This can be expressed as follows:

$$F^2 = F_x^2 + F_y^2 + F_z^2$$

Since $F_x^2 = F^2(\cos^2 \theta_x)$, then the above relationship could be expressed as follows:

$$F^2 = F^2(\cos^2 \theta_x) + F^2(\cos^2 \theta_y) + F^2(\cos^2 \theta_z)$$

The solution of concurrent, non-coplanar systems is based on methods previously developed with concurrent systems earlier in this chapter—namely, that the sum of the forces about the point of concurrence has to be to zero. The solution is found by the following three equilibrium equations:

$$\Sigma F_x = 0$$

$$\Sigma F_y = 0$$

$$\Sigma F_z = 0$$

By breaking down forces into their x, y and z components using the direction angles, a solution based on three equation can take place. The resultant of a con-

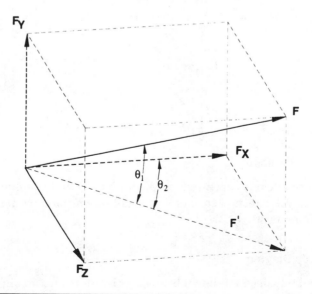

Figure 3-12   Concepts of Direction Angles θ_1, and θ_2

current system in equilibrium must equal zero. The sum of the components of a force system in equilibrium must also equal zero. The procedure for the solution of concurrent, non-coplanar systems is illustrated in the following examples.

EXAMPLE 3.10

For the derrick shown below, calculate the forces in the cables *OA* and *OB*, and in the strut *OC*.

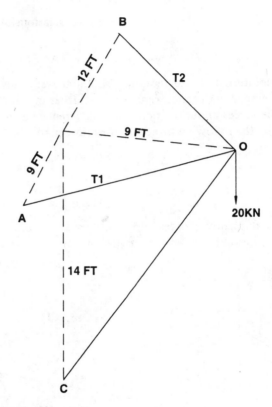

Problem Identification

The force system is concurrent through point O and is a non-coplanar or space structure. Use direction cosines and break down the three forces into their x, y, and z components.

Free-Body Diagram

The free-body diagram shown below consists of the tension in the two cables (*T1* and *T2*) , the compression in the strut (C) and the load of 20 KN.

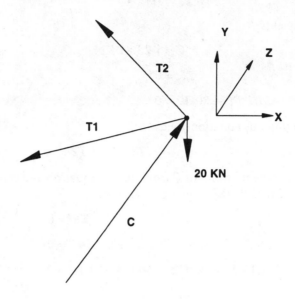

Solution by Algebraic Equations

Before algebraic equations are applied, it is necessary to calculate the components of each individual force. The 20 KN force only has a y component since it lies along the y axis while the cables ($T1$ and $T2$) have no y component since they are perpendicular to the y axis. Similarly, the compression force in the strut, C, does not have a z component since it is perpendicular to the z axis. The components for the forces are computed as follows:

For $T1$:

$$F_x = -T1 \cos 45° \quad = -.707\ T1$$

$$F_z = -T1 \cos 45° \quad = -.707\ T1$$

For $T2$:

$$F_x = -T2 \cos 53.1° = -.60\ T2$$

$$F_z = T2 \cos 36.9° \ = .80\ T2$$

For C:

$$F_x = C \cos 57.3° \quad = .54\ C$$

$$F_y = C \cos 32.7° \quad = .84\ C$$

Placing these in the three equations of force equilibrium, we find the following:

$$\Sigma F_x = 0$$

$$-.707\ T1 - .60\ T2 + .54C = 0$$

$$\Sigma F_y = 0$$

$$+.84\ C - 20\ KN = 0$$

Solving yields:

$$C = +23.81 \text{ KN}$$

$$\Sigma F_z = 0$$

$$-.707 \; T1 + .80 \; T2 = 0$$

Rearranging this equation yields:

$$T1 = 1.13 \; T2$$

Placing the solution for C from the ΣF_y equation and this relationship between $T1$ and $T2$ into the ΣF_x equation, we find:

$$\Sigma F_x = 0$$

$$-.707 \; T1 - .60T2 + .54C = 0$$

$$-.707 \; (1.13 \; T2) - .60T2 + .54 \; (23.81 \text{ KN}) = 0$$

Solving:

$$T2 = 9.2 \text{ KN}$$

Therefore, since $T1 = 1.13 \; T2$:

$$T1 = 1.13(9.2 \text{ KN}) = 10.4 \text{ KN}$$

Sometimes it is easier to deal with the coordinates of a vector when trying to calculate the direction angles of a vector which lies in space. The magnitude of the vector can simply be calculated from trigonometry as follows:

$$r^2 = r_x{}^2 + r_y{}^2 + r_z{}^2$$

where

$r_x = x$ coordinate at end of vector minus x coordinate at beginning of the vector

$r_y = y$ coordinate at end of vector minus y coordinate at beginning of the vector

$r_z = z$ coordinate at end of vector minus z coordinate at beginning of the vector

The direction cosine is simply the "length" of a component (x, y, or z) divided by the total "length" of the vector. Therefore the direction cosine necessary in solving for the length can easily be found by the following relationship:

$$\cos \theta_x = r_x / r$$

$$\cos \theta_y = r_y / r$$

$$\cos \theta_z = r_z / r$$

The following example will illustrate the coordinate use of direction angles for solving a non-coplanar type of problem.

EXAMPLE 3.11

For the tripod shown in this example, calculate the compression forces in each leg from the 1000 lb applied force.

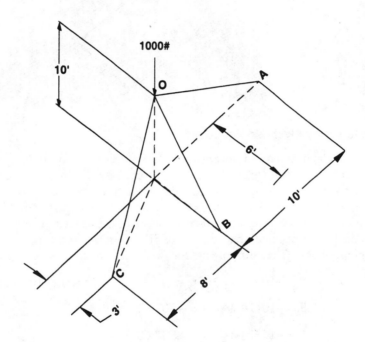

Problem Identification

The force system is concurrent through point O and is a non-coplanar or space structure. Use direction angles by coordinates and break down the three forces into their x, y, and z components.

Free-Body Diagram

The free-body diagram which follows consists of the compression in each leg of the tripod ($C1$, $C2$, $C3$) and the vertical load of 1000 lbs. Setup the x, y, and z axis as shown.

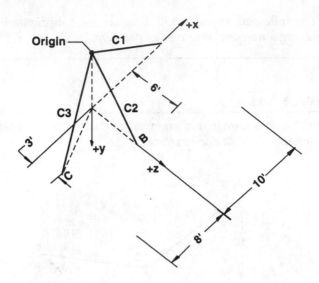

Solution by Algebraic Equations

Before algebraic equations are applied, it is necessary to calculate the components of each individual force. The 1000 lb force only has a y component since it lies strictly along the y axis (and perpendicular to the x and z axes), while the compression force in $C1$ has no z component since it is perpendicular to the z axis. Similarly, the compression force in the strut, $C2$, does not have an x component since it is perpendicular to the x axis. The compression force in $C3$ has x, y, and z components. The components for the forces are computed as follows:

For $C1$: $r_x = -10 - 0 = -10,\ r_y = -10 - 0 = -10,\ r = \sqrt{(-10)^2 + (-10)^2} = 14.14$

$F_x = -10/14.14\ C1 = -.707\ C1$

$F_y = -10/14.14\ C1 = -.707\ C1$

For $C2$: $r_y = -10 - 0 = -10,\ r_z = -6 - 0 = -6,\ r = \sqrt{(-10)^2 + (-6)^2} = 11.66$

$F_y = -10/11.66\ C2 = -.857\ C2$

$F_z = -6/11.66\ C2 = -.515\ C2$

For $C3$: $r_x = +8,\ r_y = -10,\ r_z = -3,\ r = \sqrt{(8)^2 + (-10)^2 + (-3)^2} = 13.15$

$F_x = 8/13.15\ C3 = .608\ C3$

$F_y = -10/13.15\ C3 = -.76\ C3$

$F_z = -3/13.15\ C3 = -.228\ C3$

Placing these in the three equations of force equilibrium we find the following:

$$\Sigma F_x = 0$$

$$-.707 \; C1 + .608 \; C3 = 0$$

$$C1 = .86 \; C3$$

$$\Sigma F_y = 0$$

$$.707 \; C1 + .857 \; C2 + .76 \; C3 = 1000$$

$$\Sigma F_z = 0$$

$$-.515 \; C2 - .228 \; C3 = 0$$

Solving the $\Sigma F_z = 0$ equation yields:

$$C2 = -.443 \; C3$$

Placing this relationship and the relationship found in the ΣF_x equation into the ΣF_y equation, we find:

$$\Sigma F_y = 0$$

$$.707 \; C1 + .857 \; C2 + .76 \; C3 = 1000$$

$$.707(.86 \; C3) + .857(-.443 \; C3) + .76 \; C3 = 1000$$

$$C3 = 1011.8 \; lbs$$

Solving for C1 and C2:

$$C1 = .86 \; C3$$

$$C1 = .86 \; (1011.8) = 870.1 \; lb$$

$$C2 = -.443 \; C3$$

$$C2 = -.443(1011.8) = -448.2 \; lb.$$

Note that the negative sign for C2 would mean that the force actually acts opposite from what was assumed, meaning that this member would actually be in tension. If this structure was truly a tripod, it would need to have all members in compression.

3.10   Summary

For bodies to remain at rest when loads are applied, counteracting forces must be developed for the body to remain in equilibrium. Such counteracting forces are referred to as reactions. When a body is in equilibrium, it must not move translationally or rotationally. This means that the sum of all the forces on the body must be equal to zero and the sum of all the moments on the body about a certain point must also equal zero.

Many types of force systems—such as parallel, concurrent, and nonconcurrent—can exist for a given problem. The identification of the system that is at hand in a problem will aid us in the potential solution process to be used. Many times, the use of algebraic equations of equilibrium will be employed in a solution process. However, before such equations can be developed, it is mandatory that a proper free-body diagram of the force system be drawn that correctly shows all the forces and potential forces that act on the body. Without a proper free-body diagram, the problem cannot be correctly worked.

Solution of nonplanar, concurrent systems involves the solution of algebraic equations of equilibrium in the x, y, and z directions. This is most easily accomplished by using the direction angles of each vector as it lies in space. Such direction angles can easily be solved by using the coordinates of each force vector.

EXERCISES

1. Draw the proper free-body diagram of the system shown.

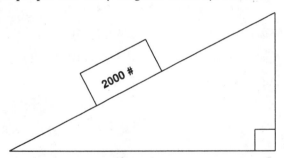

2. Draw the proper free-body diagram of the system shown.

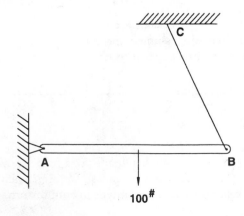

3. Draw the proper free-body diagram of the system shown.

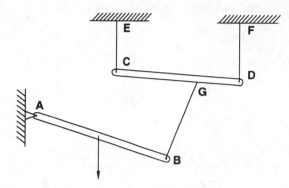

4. Draw the proper free-body diagram of the system shown.

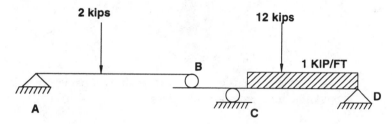

5. Draw the proper free-body diagram of the system shown.

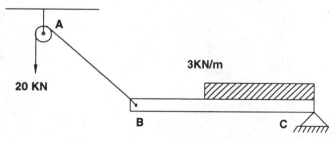

6. Determine the reactions at points *A* and *B* for the structure shown below. The reaction at *A* is pinned and the reaction at *B* is a roller.

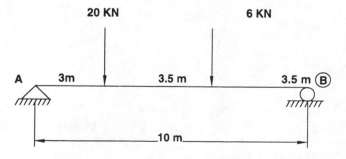

7. Determine the reactions for the beam shown below. The reaction at *A* is pinned and the reaction at *B* is a roller.

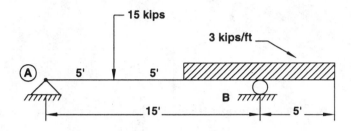

8. Determine the tension in the cables shown, using both the trigonometric and algebraic equation processes.

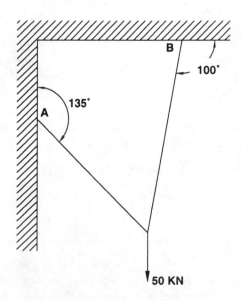

9. Determine the tension in cable *A* and cable *B* in the problem shown below.

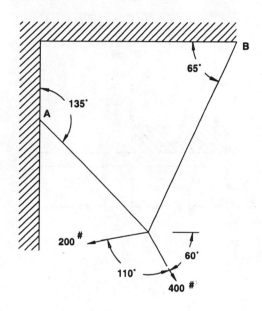

10. Solve the reactions for the beam shown below. The applied moment acts as shown in the clockwise direction.

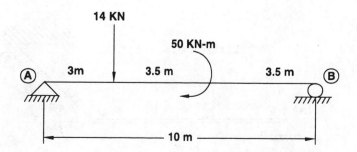

11. The cantilever beam supports the loads and applied moments as shown. Determine the reactions at the wall.

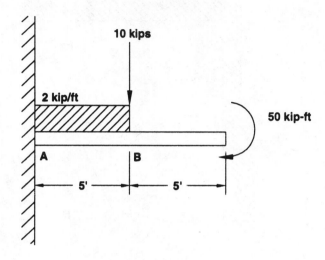

12. The member shown is supported at point *A* by a pin and by a cable at point *B*. Determine the pin reactions and the tension in the cable.

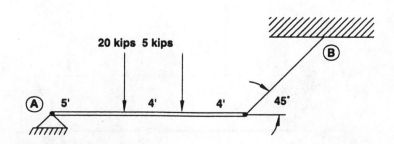

13. The beam shown is supported by a pin at C and by the cable AB. Determine the pin reaction and the tension in the cable.

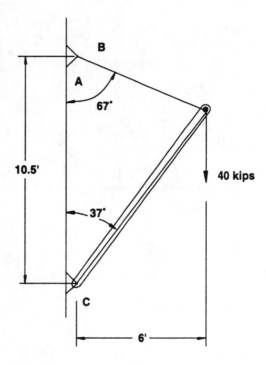

14. The A-frame shown has a pin at A and a roller at C. Determine the reactions at these supports.

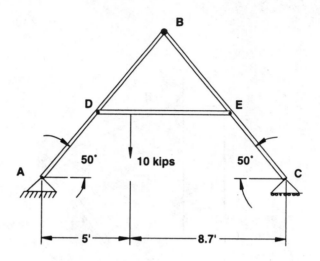

15. Determine the support reactions for the truss shown below.

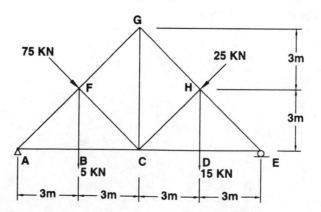

16. Determine the forces in members *OA*, *OB*, and *OC* in the concurrent space system shown below.

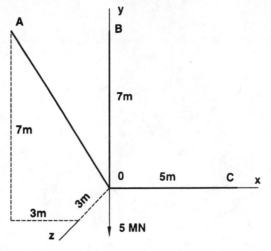

17. Determine the forces in the members of the derrick shown.

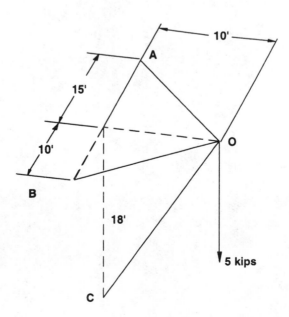

18. Determine the compressive forces which exist in each leg of the tripod shown below.

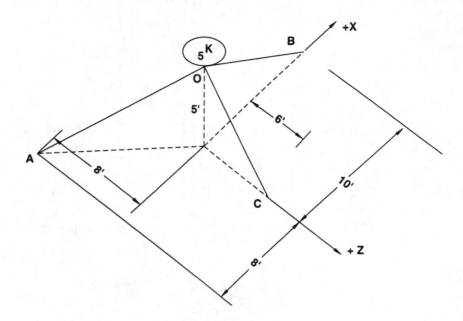

4

TRUSS AND FRAME ANALYSIS

4.1 Introduction

Many great structures of the world are composed of individual elements that react in unison when forces are applied to them. The reactions of the individual elements in these structures are governed by the laws of equilibrium, which have been discussed previously. In this chapter, we will use these principles of equilibrium to analyze the behavior of two very common structures—pin-connected trusses and pin-connected frames.

Trusses and frames make up large numbers of structures that are used throughout the world; from bridges and roof structures, to towers and cranes (Figure 4-1). The analysis of these structures is an exercise consisting of breaking down a large structure into smaller free-bodies. Prior to this chapter, much of the discussion focused on the solution of unknown support reactions. However, the focus in this chapter is on describing the forces that act on the individual elements of the full structure. The solution techniques previously used will be employed and modified to suit the particular application at hand.

The distinction between trusses and frames lies in the application of forces to the members that comprise each type of structure. A pin-connected truss has all of its members ideally in pure axial tension or pure axial compression, and as such, *all members are considered to be two-force members*. This means that all the members in a truss are subjected to colinear forces that push or pull that member directly along the line of the member (Figure 4-2). This contrasts with the

Figure 4-1 Truss Bridge over the Rappahannoch River, Virginia (Courtesy Bethlehem Steel Corporation)

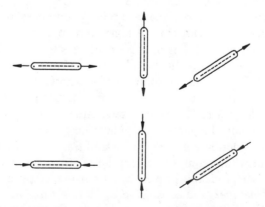

Figure 4-2 Illustration of Two-Force Members

forces which act on members that comprise a pin-connected frame. In a frame, some of the members are subjected to bending, which introduces a more complicated type of behavior. In a frame, it is important to realize that some of the members may actually be two-force or truss-type members. However, at least some of the members are subjected to bending.

The following sections describe analysis techniques used in the solution of pin-connected trusses and frames.

4.2   Truss Behavior

A truss is typically a framework of members arranged in a series of triangles that all lie in the same plane. The arrangement of members into triangular patterns is necessitated by the fact that this is a stable shape able to produce forces that will prevent "racking" under the application of lateral forces. The location where members are joined together is referred to as a joint, and these joints will be assumed to be frictionless pins. This will be explained shortly.

Trusses come in many standard configurations, as shown in Figure 4-3, and can be modified to adapt to a specific function. Trusses are made up of top chord members, bottom chord members, and web members being either diagonal or vertical. In a simply supported truss, the bottom chord members are generally in tension, while the top chord members are usually in compression. The interior web members can generally be placed in either tension or compression, depending on where the loads on the truss are applied. Trusses can be considered as large beams that have been constructed to economically span longer distances, while providing a substantial reduction in weight. Although beam behavior will not be discussed until Chapter 12, the student should realize that as loads are applied to a truss,

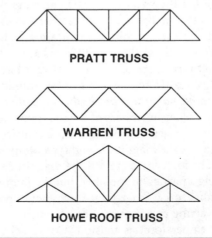

PRATT TRUSS

WARREN TRUSS

HOWE ROOF TRUSS

Figure 4-3   Common Types of Trusses

the tension and compression received by the bottom and top chords is analogous to the tension and compression felt in the flanges of a beam. The tension and compression distributed in the web members is also analogous to the shear forces developed by a standard beam.

In our analysis of a truss there are several assumptions which are necessary to make our job less tedious. These assumptions are as follows:

1. Truss members are connected by frictionless pins. Although many older trusses were pinned, today's trusses are generally bolted or welded into large gusset plates. Because of this rigidity at the joints, actual trusses have some bending moment associated with the joint. Even so, this does not provide a large discrepancy in the magnitude of force in the actual member and therefore the assumption is warranted.
2. All the members lie in the same plane.
3. All of the forces on the truss are applied at the joints. This ensures that any bending associated with the truss is not directly attributed to the loads and keeps the truss members primarily in axial tension or compression.
4. The selfweight of the members is negligible and is either neglected or applied equally at the joints as a concentrated load.

Before a discussion on the various analysis techniques, it is important to distinguish the types of forces which are commonly referred to in truss analysis. There are two types of forces which exist in the ideal truss—member forces and joint forces. **Member forces** are the tension and compression forces which actually exist in the members of the truss as loads are placed upon it. If the members forces are "pulling on" a member it is said to be in tension. Likewise, if the member forces are "pushing on" a member it is said to be in compression. These forces are necessary to design the size of a particular truss member and therefore are the forces needed in the analysis. These member forces also act on the pinned joints creating "reactions" at the joints which are equal and opposite of the member force. These "reactions" at the pinned joints are referred to as **joint forces.** Just as with the member forces, the joint forces always have to be in equilibrium (Figure 4-4). Notice that relative to a given member framing into a joint, *the joint forces are always the same magnitude but in the opposite direction of the member forces.*

Because the various members of a truss frame into, and are concurrent about, specific joints it is easier to work with the joint forces rather than the member forces, although it is the magnitude and type of member forces which the designer needs to find. The joint forces reveal the magnitude of the member force framing into a joint along a particular member, but the joint force acts in a direction opposite of the member force. Since joint forces act oppositely of the member force, if the joint forces acting at opposite ends of a member appear to be "pushing on" a member, the member force is tension. Conversely, if the joint forces acting on

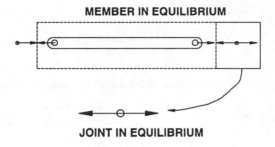

MEMBER IN EQUILIBRIUM

JOINT IN EQUILIBRIUM

Figure 4-4   Difference Between Joint Forces and Member Forces

opposite ends of a member appear to be "pulling on" a member, the member force is actually compression.

The following two sections discuss two common techniques used in the analysis of trusses. Both methods are based on the principles of equilibrium which were discussed at length in the previous chapter.

4.3   Method of Joints

The first method of truss analysis which is discussed here is the method of joints. This method draws a free-body diagram of each joint and, using the principles of equilibrium, solves for the unknown joint forces at that particular joint. Because all of the forces in the free-body of any particular joint act through the joint, the system is a concurrent system. As such, the following equations of equilibrium will typically be utilized:

$$\Sigma F_x = 0$$

$$\Sigma F_y = 0$$

Each joint must be in equilibrium, with the total forces acting in the x direction being equal to zero and similarly the total forces acting in the y direction also equal to zero. It is also very important to remember that the *joint forces caused by a member are in the same line as the member*. This means that a vertical member, since it is a two-force member, creates only a vertical joint force, a horizontal member creates only a horizontal joint force, and a diagonal member creates a joint force that acts along the same line (at the same angle) as the diagonal member.

When using the method of joints, the analysis proceeds from joint to joint throughout the truss. Typically, a joint with a small number of unknowns is the best place to start. The joint forces found at one end of a member will systematically be transferred to the joint at the opposite end of the member and in the opposite direction. The author refers to this as "flip-flopping". Flip-flopping describes the procedure whereby the joint forces on one end of the member have to be equal but opposite of those on the other end of the member (Figure 4-5).

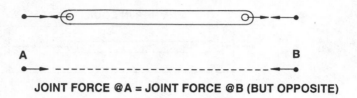

Figure 4-5   Relationship of Joint Forces at Opposite Ends of a Member

The procedure for method of joints analysis can be described in detail as follows:

1. Solve the external support reactions of the truss. This provides more "known" forces to help solve the free-body diagrams of the joints located at the supports.

2. Draw free-body diagrams of each joint in the truss. Remember to place all external loads and reactions that act on a joint *and* all potential joint forces caused by members framing into that joint. Break down diagonal joint forces into x and y components. Also draw a "big picture" of the truss. After finding joint forces on the free-body diagrams of the joints, we will transfer these forces onto the "big picture". This will help in keeping track of the joint forces and also provide a convenient place to label the member forces.

3. Begin at a joint which does not have many potential unknown joint forces. Typically, a good place to start may be the joint at the reaction (which has been found per step 1). Apply the equations of equilibrium ($\Sigma F_x = 0$ and $\Sigma F_y = 0$) at that particular joint. Solve for the unknown joint forces.

4. Flip-flop the joint forces that act at one end of a member to the opposite end of that member. Make sure they act in the opposite direction. Member forces can be solved as the joint forces are found at each end of a member. *The member force has the same magnitude as the joint forces for a member, but act in the opposite direction.* Therefore, if the joint forces "look like tension", the member is in compression, and vice versa.

5. Go on to another joint until the truss is completed.

The method of joints is a long process that requires the systematic utilization of some standard procedure, such as that outlined above. It is also easy to make a mistake due to the drawing of joint forces in the wrong direction or transposing numbers. Therefore, it is imperative that the student learn a specific system of solving by the method of joints and stick to it. Repeating the same steps again and again is the surest way to ensure that mistakes are minimized. Because of this, the author leans towards the big-picture approach, for it is easier to track the results as the problem progresses.

The following example illustrates the use of the aforementioned solution process.

EXAMPLE 4.1

In the truss shown below, solve for all the member forces using the method of joints.

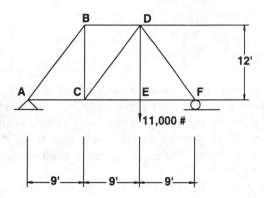

Problem Identification

Solve all members using method of joints; state the magnitude of force on each, and whether they are in tension (T) or compression (C).

Free- Body Diagram

A free-body diagram of the full truss must be drawn such that support reactions can be solved. This is done as follows:

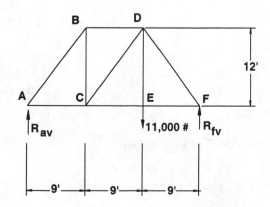

Solving the reactions, we find:

$$\Sigma F_y = 0$$

$$R_{av} + R_{fv} - 11000 \text{ lb} = 0$$

$$\Sigma M_a = 0$$

$$+11000 \text{ lb}(18 \text{ ft.}) - R_{fv}(27 \text{ ft.}) = 0$$

$$R_{fv} = 7333 \text{ lb}$$

$$R_{av} = 3667 \text{ lb}$$

After support reactions have been solved, a free-body diagram of each joint should likewise be drawn. The free-body diagram of a joint should show all the loads and reactions on that particular joint, as well as all the potential joint forces applied from each member framing into that joint. A big picture of the truss should also be drawn to help us keep track of all the work that will be completed on successive joints.

Beginning with joint A, we would place the reaction force of 3667 lbs on the joint, in addition to the joint force applied from member AC and AB. Notice the joint force from member AB has been represented by its x and y components.

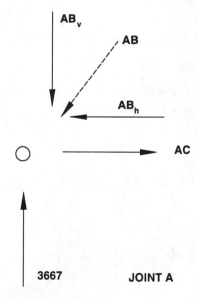

JOINT A

Algebraic Equations of Equilibrium

Joint A
Applying the equations of equilibrium:

$$\Sigma F_y = 0$$

$$+3667 \text{ lbs} - AB_v = 0$$

Solving:

$AB_v = +3667$ (positive indicates correct direction, down, as assumed)

Because the resultant force of AB has to lie along the line of the member, AB_h can be found by trigonometry. Since AB lies at a 53.1° angle with the horizontal, its x and y components have to vectorially add to produce a resultant along that line. As such, AB_h can be calculated as follows:

$$AB_h = (AB_v/\tan 53.1°) = (3667 \text{ lb})/\tan 53.1° = 2750 \text{ lb}$$

(This can just as easily be done using the ratios of the sides making up member AB, which is in space at 9 horizontally to 12 vertically. Therefore, the x component of a force that lies along the same line must be 9/12 of the vertical component).

$$\Sigma F_x = 0$$

$$+AC - AB_h = 0$$

$$+AC - 2750 \text{ lbs} = 0$$

Solving:

$$AC = 2750 \text{ lb}$$

3667

2750

2750

3667

JOINT A

Joint A is now complete and the results can be transferred to the big picture as shown below. Notice that the arrowheads on the diagonal (member AB) act to form a resultant that is in the line of the member. This always has to be the case. Either the arrowheads are together or the arrow tails are together on the line, because these are the only ways that the resultant force can lie along a given line.

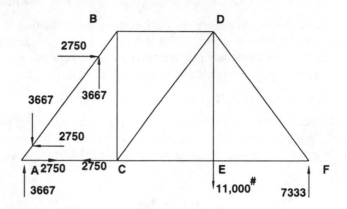

Flip-flopping the joint's forces from joint *A* to the other end of members *AB* and *AC* is the next step before moving on. From the big picture, it can easily be seen that the *member forces* for *AC* and *AB* are as follows:

$AC = 2750$ lb tension (joint forces appear to be in compression)

$AB = \sqrt{(3667)^2 + (2750)^2} = 4583.6$ lb compression

(joint forces appear to be in tension)

Next proceed to joint *B*.

Joint B

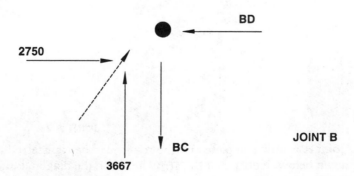

At joint B, we have no external forces and three members that may bring joint forces into joint B. The joint forces from member AB are known from the flip-flopping procedure, placing an upwards y component of 3667 lb on joint B and an x component of 2750 lb to the right. Members BC and BD only have the potential of bringing a vertical and horizontal component respectively to this joint. Solve the joint as follows:

Algebraic Equations of Equilibrium

Applying the equations of equilibrium,

$$\Sigma F_x = 0$$

$$+2750 \text{ lbs} - BD = 0$$

$$BD = 2750 \text{ lbs (in the direction shown on the free-body diagram)}$$

$$\Sigma F_y = 0$$

$$+3667 - BC = 0$$

$$BC = 3667 \text{ lbs (in the direction shown on the free-body diagram)}$$

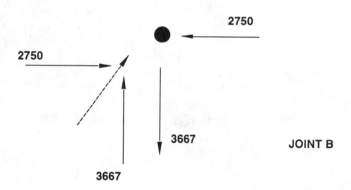

Joint B is now complete and the results can be transferred to the big picture as shown.

Flip-flopping the joint's forces from joint B to the other end of member BC and BD is the next step before moving on. From the big picture it can easily be seen that the member forces for members BC and BD are:

$BC = 3667$ lb tension (looks like compression)

$BD = 2750$ lb compression (looks like tension)

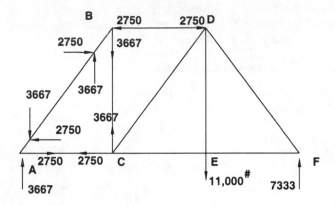

Next proceed to joint C.

Joint C

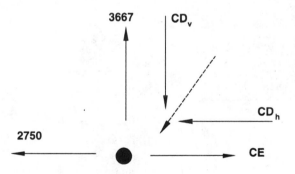

JOINT C

At joint C, we have no external forces and four members that may bring joint forces into joint C. The joint forces from member AC and BC are known from the previous flip-flopping procedures, placing the known joint forces on joint C as shown. Members CE and CD have the potential of bringing both vertical and horizontal forces to this joint. Remember member CD is a diagonal, and therefore we will break its joint force into x and y components. Solve the joint as follows:

Algebraic Equations of Equilibrium

Applying the equations of equilibrium:

$$\Sigma F_x = 0$$

$$-2750 \text{ lbs} - CD_h + CE = 0$$

$$\Sigma F_y = 0$$

$$+3667 - CD_v = 0$$

$$CD_v = 3667 \text{ lbs (in the direction shown on the previous}$$
free-body diagram)

Since CD_v has been found, the magnitude of CD_h can likewise be found. The resultant of these two forces must lie along the line of the member.

$$(CD_h/CD_v) = (9/12)$$

$$CD_h = (9/12) \ CD_v = (9/12)(3667 \text{ lb}) = 2750 \text{ lb (in the direction assumed)}$$

Since CD_h has now been found, the ΣF_x equation can be reentered and CE solved for, as follows:

$$-2750 \text{ lb} - CD_h + CE = 0$$

$$-2750 \text{ lb} - (2750) + CE = 0$$

Solving:

$$CE = 5500 \text{ lb in the assumed direction}$$

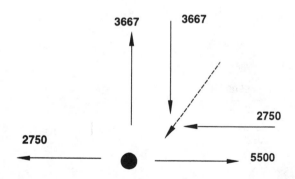

JOINT C

Joint C is now complete and the results can be transferred to the big picture as shown. Flip-flopping the joint's forces from joint C to the other end of member CE and CD is the next step before moving on. From the big picture, it can easily be seen that the member forces for members CD and CE are:

$$CD = \sqrt{(3667)^2 + (2750)^2} = 4583.6 \text{ lb compression}$$

(joint forces appear to be in tension)

$CE = 5500$ lb tension (looks like compression)

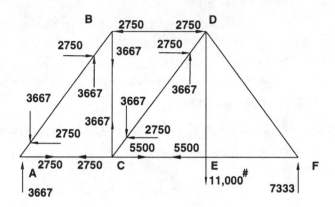

The next place that one might proceed is to joint E.

Joint E

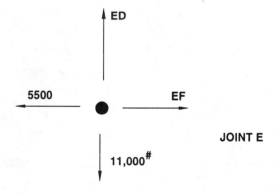

At joint E, we have one external forces and three members that may bring joint forces into joint E. The joint force from member CE is known from the previous flip-flopping procedures, placing the known joint force on joint E, as shown. Members ED and EF have the potential of bringing a vertical and horizontal force respectively to this joint. Remember that member ED is a vertical, and therefore can only have a vertical force in it. Member EF is horizontal, and therefore can only carry a horizontal force into joint E. Solve the joint as follows:

Algebraic Equations of Equilibrium

Applying the equations of equilibrium:

$$\Sigma F_x = 0$$

$$-5500 \text{ lbs} + EF = 0$$

Solving:

$$EF = 5500 \text{ lb}$$

$$\Sigma F_y = 0$$

$$-11000 \text{ lb} + ED = 0$$

Solving:

$$ED = 11000 \text{ lb}$$

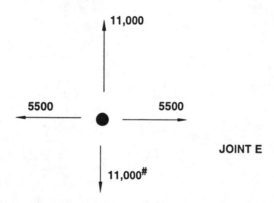

JOINT E

Joint E is now complete and the results can be transferred to the big picture as shown. Flip-flopping the joint's forces from joint E to the other end of members ED and EF is the next step. From the big picture, it can easily be seen that the member forces for members ED and EF are:

ED = 11000 lb tension (joint forces appear to be in compression)

EF = 5500 lb tension (joint forces appear to be in compression)

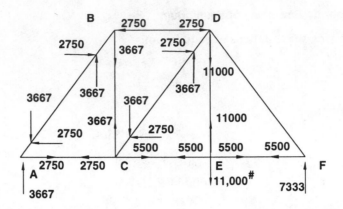

The next place that one might proceed is to joint D.

Joint D

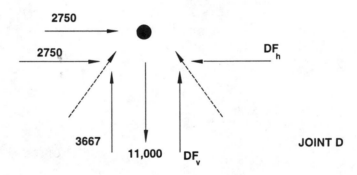

JOINT D

At joint D, we have no external forces and four members that may bring joint forces into joint D. The joint forces from member ED, CD, and BD are known from the previous flip-flopping procedures, placing the known joint forces on joint D, as shown above. Member DF brings in the only unknown joint forces and has the potential of bringing a vertical and horizontal component to this joint. Solve the joint as follows:

Algebraic Equations of Equilibrium

Applying the equations of equilibrium:

$$\Sigma F_x = 0$$

$$+2750 + 2750 - DF_h = 0$$

Solving:

$$DF_h = 5500 \text{ lb}$$

$$\Sigma F_y = 0$$

$$-11{,}000 \text{ lb} + 3667 + DF_v = 0$$

Solving:

$$DFv = 7333 \text{ lb}$$

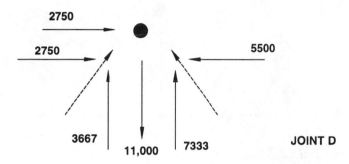

JOINT D

Notice that the components of DF that were solved independently of each other are still in a 9:12 relationship. This is a good confirmation that the process has been error-free because the resultant force in member DF lies along the line of the member. Joint D is now complete and the results can be transferred to the big picture. Flip-flopping the joint's forces from joint D to the other end of members DF is the next step. From the big picture, it can easily be seen that the member forces for DF is:

$$DF = \sqrt{5500^2 + 7333^2} = 9167 \text{ lb compression (looks like tension)}$$

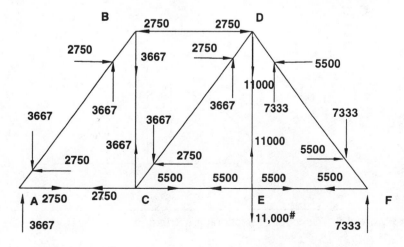

Moving onto to the last joint (joint *F*), notice that there are no unknowns, and as such, the joint should be in equilibrium. From the diagram shown below, this is indeed the case.

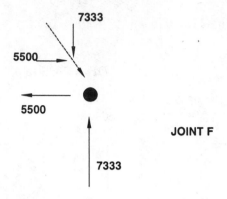

4.4   Method of Sections

In this section we will discuss an alternate procedure to analyze the member forces that exist in a determinate truss. This procedure is referred to as the method of sections and involves "cutting" the truss through the members whose forces we wish to find (Figure 4-6).

After the "cut" is made, equilibrium equations are used to determine the joint forces that act on the members that have been cut. The equilibrium equations that can be used are the standard equations as follows:

$$\Sigma F_x = 0$$

$$\Sigma F_y = 0$$

$$\Sigma M = 0$$

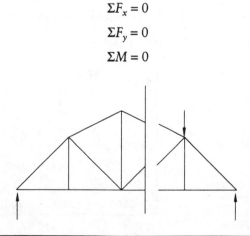

Figure 4-6   Concept of Sectioning a Truss

The sectioning of the truss creates a new free-body of the original truss, which also must be in equilibrium. This section must develop external forces at the "cuts" to keep it from falling apart. This section of the truss must develop reacting joint forces at the locations of the cut to stabilize the section. It is very convenient to view the joint forces at the section line as unknown support reactions. Therefore, it is important to realize that the more members cut in a particular section, the more unknowns will be present in the equations of equilibrium (Figure 4-7)

The steps to follow in using the method of sections can be listed as follows:

1. Solve for the support reactions of the complete truss. This provides additional known forces that aid in completing the equilibrium equations.
2. Section the truss through the members that are to be found. Remember to "cut" the truss completely in two and that every cut member becomes an unknown force on that section. More than one section can be cut on a given truss and the cuts do not necessarily have to be vertical.
3. Work with only one piece of the truss as the new free-body diagram. Remember that the side chosen does not matter, but that all loads, reactions, and joint forces at the cut members have to be included.
4. Write and solve the appropriate equilibrium equations. Similar to the method of joints, the answers are joint forces that act at the joint of the cut members. Therefore, if the force appears to be placing the member in compression, the member is actually in tension (and vice versa).

These steps can be repeated as needed throughout the truss. As the student becomes comfortable with this procedure, he or she will find that it can be used to provide very fast information on only the chosen members. This makes it much less tedious than the method of joints.

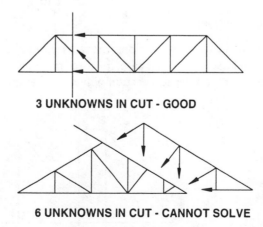

3 UNKNOWNS IN CUT - GOOD

6 UNKNOWNS IN CUT - CANNOT SOLVE

Figure 4-7   Number of "Cut" Members is Equal to Number of Unknowns

The following example is a repeat of Example 4.1 and will illustrate this technique.

EXAMPLE 4.2

For the truss shown below find the member forces in *BD*, *CD*, and *CE* using the method of sections.

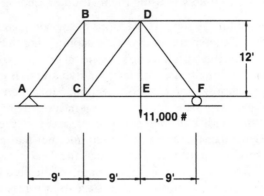

Problem Identification

Using the method of sections, find the member forces in *BD*, *CD*, and *CE*. To begin with, the support reactions for the truss should be found. This was previously done in Example 4.1 with the following results:

$$R_{fv} = 7333 \text{ lb}$$
$$R_{av} = 3667 \text{ lb}$$

Free-Body Diagram

After finding the support reactions, the truss should be "cut" through the members which are to be solved. In this case, the three members specified are directly beneath each other so that a vertical section will easily contain all members to be found.

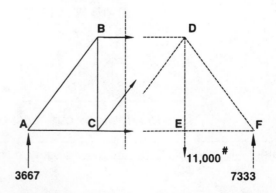

In this case, choose the left side of the truss to work with as the free–body diagram. Notice that all loads, reactions, and joint forces of the cut members are included. The direction of the joint forces for the cut members are assumed.

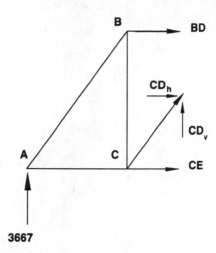

Algebraic Equations of Equilibrium

The algebraic equations of equilibrium may be applied in several different fashions to find the desired members. However, this example will apply the equation in their most standard form as follows:

$$\Sigma F_x = 0$$

$$\Sigma F_y = 0$$

$$\Sigma M = 0$$

For the free-body diagram shown, the algebraic equations yield the following:

$$\Sigma F_x = 0$$

$$+BD + CD_h + CE = 0 \text{ (At this moment it cannot be solved)}$$

$$\Sigma F_y = 0$$

$$+3667 \text{ lb} + CD_v = 0$$

Solving:

$$CD_v = -3667 \text{ lb} \quad \text{(the negative indicates the force acts in the opposite direction from the way it was assumed).}$$

Since CD_v is a component of CD, and knowing that truss members have resultant forces that lie along the line of the member, CD_h can be found using the geometric layout of the member as follows:

$$(CD_h / CD_v) = (9/12)$$

$CD_h = (9/12)\,CD_v = (9/12)(-3667\text{ lb}) = -2750\text{ lb (direction opposite of assumed)}$

The member force in CD can now be found by:

$$CD = \sqrt{-(2750)^2 + -(3667)^2} = 4583.6 \text{ lbs compression (looks like tension)}$$

$$\Sigma M = 0$$

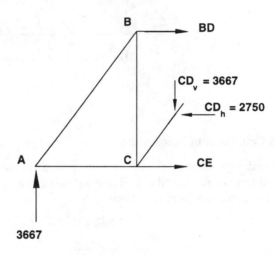

Remember that it is possible to sum moments anywhere, but it is most advantageous to sum moments about a point where several of the unknown forces cancel out. This can be done by choosing a point where several of the members (or their lines of action) intersect. In this example, the two remaining unknown member forces are BD and CE. If point C is chosen as the location to sum moments, the unknown member force through CE and the member force through CD (which has previously been found) cancel out. Summing moments about C we find:

$$\Sigma M_c = 0$$

$$+3667\text{ lbs}(9\text{ ft.}) + BD(12\text{ ft.}) = 0$$

$BD = -2750\text{ lb (negative indicates direction opposite of assumed)}$

$BD = 2750\text{ lbs compression}$

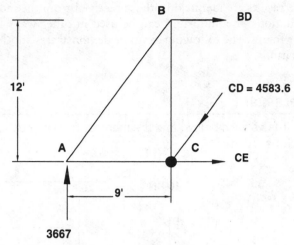

Finally CE can be solved for by reentering the $\Sigma F_x = 0$ equation with the now known values for BD and CD_h. This is as shown:

$$+BD + CD_h + CE = 0$$

$$(-2750 \text{ lb}) + (-2750 \text{ lb}) + CE = 0$$

$$CE = 5500 \text{ lb tension (in the direction assumed)}$$

Summarizing the results:

$$CE = 5500 \text{ lb tension}$$

$$CD = 4583.6 \text{ lb compression}$$

$$BD = 2750 \text{ lb compression}$$

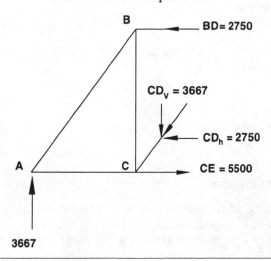

The results are identical to those received from the method of joint solution. The equations of equilibrium can be used in other ways to find the solution to member forces. The following example demonstrates another method of applying the equations.

EXAMPLE 4.3

For the truss shown below, find the member forces in *BC*, *BG*, and *FG,* using the method of sections.

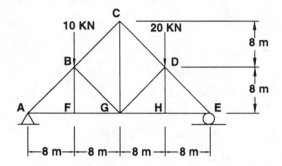

Problem Identification

Using the method of sections, find the member forces in *BC, BG,* and *FG.* To begin with, the support reactions for the truss should be found. This is done as follows:

$$\Sigma F_y = 0$$

$$+R_{av} + R_{ev} - 10 \text{ KN} - 20 \text{ KN} = 0$$

$$\Sigma M_a = 0$$

$$+10 \text{ KN}(8 \text{ m}) + 20 \text{ KN} (24 \text{ m}) - R_{ev}(32 \text{ m}) = 0$$

Solving:

$$R_{ev} = 17.5 \text{ KN}$$

Substituting back into the ΣF_y equation:

$$R_{av} = 12.5 \text{ KN}$$

Free-Body Diagram

After finding the support reactions, the truss should be "cut" through the members which are to be solved. In this case, the three members specified are directly beneath each other so that a vertical section will easily contain all members to be found.

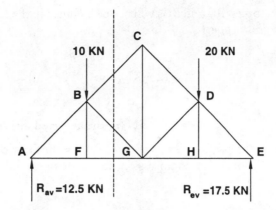

Let us choose the left side of the truss to work with as the free body diagram. Notice that all loads, reactions, and joint forces of the cut members are included. The direction of the joint forces for the cut members are assumed.

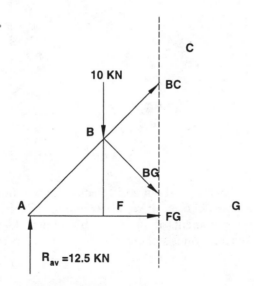

Algebraic Equations of Equilibrium

The algebraic equations of equilibrium can be applied in several different manners to find the desired members. In this example, several moment equations will be used to solve for the unknown forces. When applying moment equations, the choice of what point to sum moments about is very important. Choose a point which tends to cancel out other unknown forces. In this example, if member FG is to be found, the place to sum moments is where the other two unknown member forces cancel out. This would be point B. By summing moments at point B, two unknown forces (BC and BG) will not be included in the equation, *since their forces act through the point.*

$$\Sigma M_B = 0$$

$$+12.5 \text{ KN}(8 \text{ m}) - FG(8 \text{ m}) = 0$$

Solving:

$$FG = 12.5 \text{ KN (in the assumed direction)}$$

Since the joint force for FG acts at point F, from the flip-flopping procedure the joint forces in FG looks like compression, therefore the member force in FG is tension.

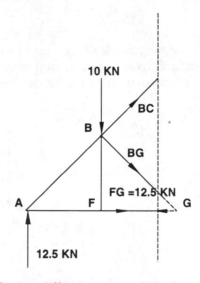

Summing moments about a different point will find other unknown forces if the point is wisely chosen. If we are trying to find the member force in BC, the location to sum moments about would be point G. Even though point G is "off" the section, its location is still known and is where the forces in BG and FG cancel out. This is done as follows:

$$\Sigma M_G = 0$$

$$+12.5 \text{ KN}(16 \text{ m}) - 10 \text{ KN}(8 \text{ m}) + BC(11.3 \text{ m}) = 0$$

Solving:

$$BC = -10.62 \text{ KN (the negative indicates the direction is opposite of the assumed)}$$

Since the joint force for BC acts at point B, from the flip-flopping procedure the joint forces appear to be in tension, therefore the member force of BC is compression.

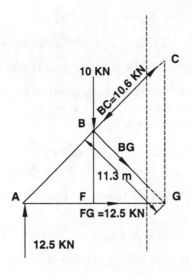

It should also be noted that the resultant joint force in *BC* was used in the moment equation because the perpendicular distance from the point of summation (point *G*) to the force is 11.3 m. Sometimes the perpendicular distance may not be as easy to calculate and the member may be broken down into its rectangular components. To solve for the force in member *BC* in this fashion, we would solve for one of the components (either BC_v or BC_h). This is done by sliding the components to a point where one or the other acts through the point where moments are being summed, thus canceling that component force out. In this problem, if the components of *BC* are slid up to point *C*, BC_v acts through point *G* and therefore cancels out (see figure). The moment equation would then be:

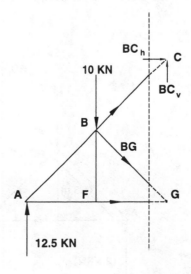

$$\Sigma M_G = 0$$

$$+12.5 \text{ KN}(16 \text{ m}) -10 \text{ KN}(8 \text{ m}) + BC_h(16 \text{ m}) = 0$$

$$BC_h = -7.5 \text{ KN (acting opposite from direction assumed)}$$

Knowing that the other component, BC_v, must be of such magnitude and direction to make the resultant lie along the line of the member (45° line), BC_v would be found as follows:

$$Bc_v/Bc_h = 8/8$$

$$BC_v = (8/8)\ (-7.5 \text{ KN}) = -7.5 \text{ KN}$$

The resultant force BC can be found by the Pythagorean relationship as follows:

$$BC = \sqrt{(-7.5^2) + (-7.5)^2} = 10.62 \text{ KN (same as before)}$$

The final member, BG, can be found in similar manner. If the member force in BG is desired, the location to sum moments about would be point A. This location is nice, since the forces in BC and FG act through point A and thus cancel out. This is done as follows:

$$\Sigma M_A = 0$$

$$+10 \text{ KN}(8 \text{ m}) + BG(11.3 \text{ m}) = 0$$

Solving:

$$BG = -7.08 \text{ KN (the negative indicating the direction is opposite)}$$

Since the joint force for BG acts at point B, from the flip-flopping procedure the joint force in BG looks like it is in tension, therefore the member force is in compression.

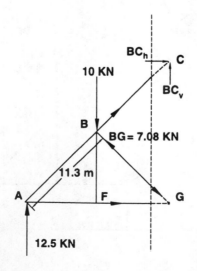

Again notice that the resultant force in *BG* was solved for, since the perpendiclar distance was used in the moment equation.

The method of sections is a very flexible solution technique for trusses. The typical equations of equilibrium can be applied in many different manners. The basic procedure outlined in the previous examples should be followed, realizing that each member cut represented an unknown joint force. Therefore it is best to choose a "section" which minimizes the number of members "cut", usually to a maximum of three.

4.5   Introduction to Frames

As mentioned previously, frames differ from trusses by the fact that they are composed of members on which more than two forces are acting. Frames have members which have forces acting somewhere along the member or are under the influence of significant selfweight. This being the case, such members are subjected to bending, and therefore the resultant of forces on a member will not act along the line of the member (Figure 4-8).

Frames are assumed to be pin-connected. The members composing the frame are considered to have the ability to develop reaction forces while being free to rotate. The initial analysis of frames will focus on finding the reactions that the members place upon the pinned connections. These forces are developed at a member's end and are referred to as **pin reactions**. Finding these pin reactions will be an important first step to learning more about shear and bending forces which act on such members.

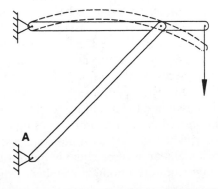

Figure 4-8   Frame Member Subjected to Bending

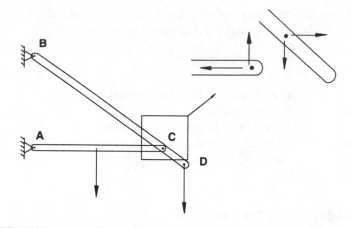

Figure 4-9   Pin Reactions Bring Equal and Opposite at Adjoining Joints

A key to understanding the analysis of frames is the Newtonian concept of action and reaction. As a member in a frame is connected to another, it follows that the forces acting on one end of a member (termed pin reactions) must be resisted by equal and opposite forces from the other member's end (Figure 4-9). This concept will be used extensively in the solution of frames in the upcoming section.

4.6   Frame Analysis

The analysis of pin-connected frames is based on the concepts of isolating a free-body and writing algebraic equations that have already been used extensively throughout Chapters 3 and 4. The isolation of free-bodies for the frame entails separating the members of the frame and proceding to use equilibrium equations on the individual pieces of the frame. This technique may be thought of as "exploding" the frame. The equations of equilibrium are used on each individual free-body that is drawn and the resulting solutions for unknown pin reactions are subsequently transferred to other free-bodies using the concept of action and reaction. This concept, when applied to frame analysis, states that the pin reaction from one end of a member must be equal in magnitude and opposite in direction to the pin reactions from the adjoining member at the same joint (Figure 4-10).

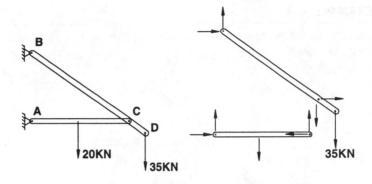

Figure 4-10   Exploding of a Frame

This analysis is not intended to find the member forces, as was the case in truss analysis. Rather, it is only to determine the end forces on the members that we refer to as pin reactions. The steps to frame analysis may be outlined as follows:

1. Determine the external support reactions (pins, rollers, etc.) for the complete structure. If all supports reactions cannot be determined, find as many as possible. Do not assume that particular members are two-force members (truss members) unless absolutely certain.
2. "Explode" the frame into separate free-body diagrams of each member. Remember to include all forces that act on each member and all pin reactions (typically the x and y components) for each pin along the member.
3. Use the three equilibrium equations to solve for unknown pin reactions. If the process gets bogged down or if all the unknowns are solved on a particular member, move on to another. As the process proceeds to another member, remember to transfer the solved pin reactions to the adjoining members, making sure that they are of equal magnitude and opposite direction.
4. Continue until all pin reactions are solved.

While this method of frame analysis is rather straight forward, keeping track of the pin reactions is very important. Remember to transfer pin reactions to the adjoining members as the process leaves a particular member and double check to make sure the magnitudes and directions of the transferred pin reactions are accurate. Also, visually check your answers on an individual free-body to make sure there are no obvious discrepancies. The following examples will illustrate the analysis of simple frames.

EXAMPLE 4.4

Determine the pin reactions for the frame shown. The surface at *A* and *B* are to be considered smooth.

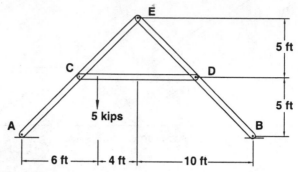

Problem Identification

Solve the pin reactions for the frame by first trying to solve the external support reactions and then exploding the frame into individual free-bodies.

Free-Body Diagram

The initial free-body diagram is of the whole frame and is used to solve for the external support reactions as follows:

$$\Sigma F_x = 0$$

$$-R_{ah} + R_{bh} = 0$$

Solving:

$$R_{ah} = R_{bh} \text{ (equal magnitude and opposite direction)}$$

$$\Sigma F_y = 0$$

$$+R_{av} + R_{bv} - 5 \text{ kips} = 0$$

$$+R_{av} + R_{bv} = 5 \text{ kips}$$

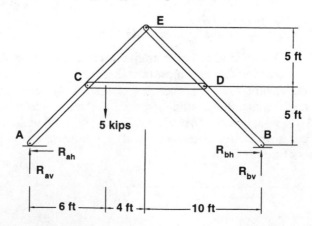

$$\Sigma M_a = 0$$

$$\overline{+5} \text{ kips}(6 \text{ ft.}) - R_{bv}(20 \text{ ft.}) = 0$$

Solving:

$R_{bv} = 1.5$ kips (the positive sign indicates the correct direction assumed)

Substituting $R_{bv} = 1.5$ kips into the $F_y = 0$ equation yields:

$$+R_{av} +(1.5 \text{ kips}) = 5 \text{ kips}$$

$$R_{av} = 3.5 \text{ kips}$$

Because the problem stated that the floor was a smooth surface, the supports can be assumed to act as rollers, and therefore develop no horizontal reactions. (R_{ah} and $R_{bh} = 0$) Continue by exploding the frame into separate free-bodies of each member. This can be done as follows:

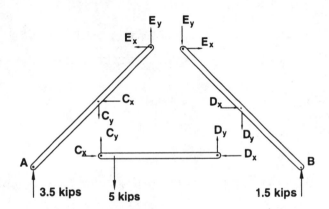

For each of these free-bodies, the equilibrium equations can now be applied.

Algebraic Equations of Equilibrium

Member CD

$$\Sigma F_x = 0$$

$+C_x - D_x = 0$ (must be equal in magnitude and in opposite directions)

$$\Sigma F_y = 0$$

$$+C_y - 5 \text{ kips} +D_y = 0$$

$$\Sigma M_c = 0$$

$$+5 \text{ kips}(1 \text{ ft.}) -D_y(10 \text{ ft.}) = 0$$

Solving:

$$D_y = 0.5 \text{ kip (in the direction which was assumed)}$$

Substituting $D_y = .5$ kip back into the $\Sigma F_y = 0$ equation and solving:

$$+C_y - 5 \text{ kips} + (0.5 \text{ kips}) = 0$$

$$C_y = 4.5 \text{ kips (in the assumed direction)}$$

C_x and D_x cannot be solved at this moment, so it is time to transfer the C_y and D_y pin reactions to the other free-bodies and move on. Make sure that Cy and Dy are transferred in the opposite direction as shown.

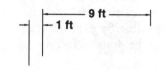

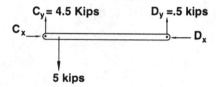

Member ACE

$$\Sigma F_x = 0$$

$-C_x - E_x = 0$ (must be equal in magnitude and in opposite directions)

$$\Sigma F_y = 0$$

$$-4.5 \text{ kips} + 3.5 \text{ kips} + E_y = 0$$

Solving:

$$E_y = 1 \text{ kip (in the assumed direction).}$$

The last two unknowns on this member are E_x and C_x, which we know must be equal and opposite. The application of a moment equation will isolate one of these unknowns.

$$\Sigma M_c = 0$$

$$+3.5 \text{ kips}(5 \text{ ft.}) - 1 \text{ kip}(5 \text{ ft.}) - E_x(5 \text{ ft.}) = 0$$

Solving:

$$E_x = 2.5 \text{ kips (the positive indicating the correct direction assumed)}$$

Substituting this back into the $\Sigma F_x = 0$ equation:

$$-C_x - (2.5 \text{ kips}) = 0$$

$$C_x = -2.5 \text{ kips (indicating the direction opposite}$$
$$\text{from what was assumed).}$$

Member ACE is now complete as shown and its pin reaction can be transferred.

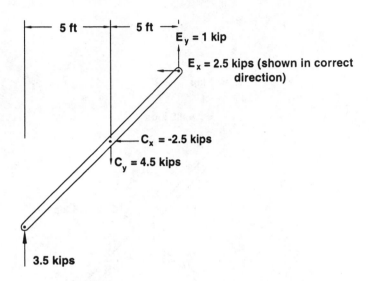

Member CD

With the transfer of the C_x pin reaction from member ACE, member CD is now finished off very easily. Applying the horizontal equilibrium equation as follows:

$$+C_x - D_x = 0$$

With C_x being found previously as –2.5 kips:

$$+(-2.5 \text{ kips}) - D_x = 0$$

$$D_x = -2.5 \text{ kips}$$

The finished free-body for member CD is as follows and the pin reactions transferred to member BDE:

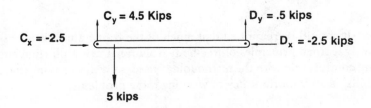

Member BDE

Member *BDE* is already solved, due to the transferring of pin reaction from the other two members. A quick check of the free-body shown below verifies equilibrium; since we have 2.5 kips to the left and to the right, as well as having 2.5 kips upwards and downwards.

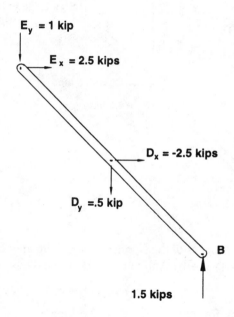

It should be reiterated that much of the frame analysis is based on the previously learned principles of free-body isolation and equilibrium equations. Becoming comfortable with the methodology used, as well as having good organizational skills, is an important key to working these problems.

EXAMPLE 4.5

Determine the pin reactions for the frame shown.

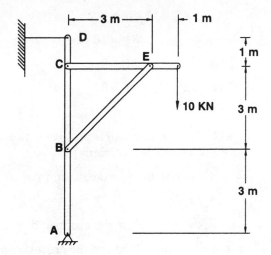

Problem Identification

Solve the pin reactions for the frame. First try to solve the external support reactions and then explode the frame into individual free-bodies.

Free-Body Diagram

The initial free-body diagram is of the whole frame and is used to solve for the external support reactions.

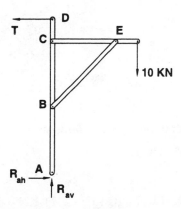

This can be done as follows:

$$\Sigma F_x = 0$$

$$+R_{ah} - T = 0$$

$$R_{ah} = T \text{ (equal magnitude and opposite direction)}$$

$$\Sigma F_y = 0$$

$$+R_{av} - 10 \text{ KN} = 0$$

$$R_{av} = 10 \text{ KN}$$

$$\Sigma M_a = 0$$

$$+10 \text{ KN (4 m)} - T \text{ (7 m)} = 0$$

Solving:

$$T = 5.71 \text{ KN (the algebraic sign is positive indicating the direction}$$
$$\text{was assumed correctly)}$$

Substituting $T = 5.71$ KN into the $F_x = 0$ equation yields:

$$R_{ah} = T$$

$$R_{ah} = 5.71 \text{ KN (in the assumed direction)}$$

The external reactions are now solved and the process continues by exploding the frame into separate free-bodies of each member. This can be done as follows:

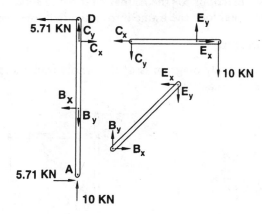

Algebraic Equations of Equilibrium

Member ABCD

$$\Sigma F_x = 0$$

$$5.71 \text{ KN} - B_x + C_x - 5.71 \text{ KN} = 0$$

Simplifying:

$$C_x = B_x \text{ (must be equal in magnitude and in opposite directions)}$$

$$\Sigma F_y = 0$$

$$+C_y + 10 \text{ KN} - B_y = 0$$

$$\Sigma M_c = 0$$

$$-5.71(1 \text{ m}) - 5.71 \text{ KN}(6 \text{ m}) + B_x(3 \text{ m}) = 0$$

Solving:

$$B_x = 13.3 \text{ KN (in the direction which was assumed)}$$

Substituting $B_x = 13.3$ KN back into the $F_x = 0$ equation:

$$C_x = B_x$$

$$C_x = 13.3 \text{ KN}$$

C_y and B_y cannot be solved at this moment, so it is time to transfer the C_x and B_x pin reactions to the other free-bodies and move on. Make sure that C_x and B_x are transferred in the opposite direction as shown on free-body $ABCD$.

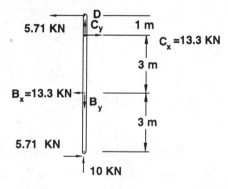

Member CE

$$\Sigma F_x = 0$$

$$-13.3 \text{ KN} + E_x = 0$$

Solving:

$$E_x = 13.3 \text{ KN (in the assumed direction)}$$

$$\Sigma M_c = 0$$

$$-E_y(3 \text{ m}) + 10 \text{ KN}(4 \text{ m}) = 0$$

Solving:

$$E_y = 13.3 \text{ KN (in the assumed direction)}$$

$$\Sigma F_y = 0$$

$$-C_y + 13.3 \text{ KN} - 10 \text{ KN} = 0$$

Solving:

$$C_y = 3.3 \text{ KN (in the assumed direction)}$$

Member *CE* is now completed as shown below and its pin reactions for adjoining joints transferred to the other free-bodies. The process can continue by moving back to member *ABCD* and finishing this free-body.

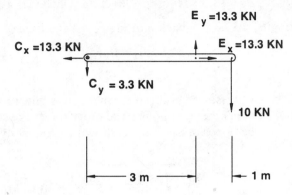

Member ABCD

$$\Sigma F_y = 0$$
$$+3.3 \text{ KN} + 10 \text{ KN} - B_y = 0$$

Solving:

$$B_y = 13.3 \text{ KN (in the assumed direction)}.$$

This member is now complete as the next figure shows and its pin reactions for adjoining joints are transferred.

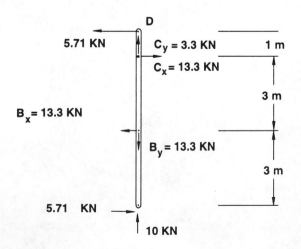

Member BE
The final member, member *BE*, is already solved due to the transferring of pin reaction from the other two members. A quick check of the free-body verifies equilibrium, since we have 13.3 KN to the left and to the right, as well as having 13.3 KN upwards and downwards. A closer look will reveal that this member is a two-force member, since the components form a resultant that lies along the line of the member.

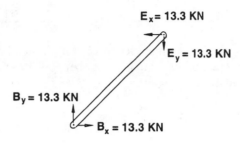

4.7   Summary

Trusses and frames are common structures that are encountered throughout the engineering world. The fundamental analysis and knowledge about the behavior of these structures is a very beneficial skill to all engineers and technologists.

Trusses are composed of members that are joined together at a location referred to as a joint. These truss joints are assumed as pinned connections, to simplify the analysis procedure. Truss members are all two-force members—meaning that they are either axially loaded tension or compression members. The two methods of truss analysis are the method of joints and the method of sections.

Frames differ from trusses by not being composed of all two-force members. Frames have members that are loaded in the middle of the member or are subject to bending. Due to this more complicated behavior, frame analysis solves for the forces at the joints along the members that are referred to as pin reactions. The solution technique used on frames uses separate free-bodies of the pieces of the frame and solves those pieces using equilibrium equations on each.

EXERCISES

1. Determine the member forces for all members in each truss shown using the method on joints. Remember to list each member force as either tension or compression.

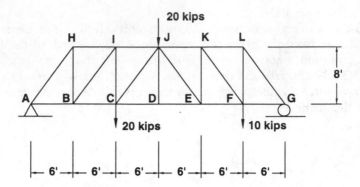

2. Determine the member forces for all members in each truss shown using the method of joints. Remember to list each member force as either tension or compression.

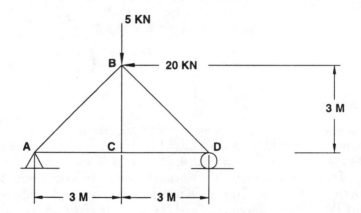

3. Determine the member forces for all members in each truss shown using the method of joints. Remember to list each member force as either tension or compression.

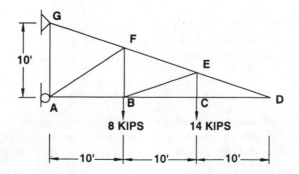

4. Determine the member forces for all members in each truss shown using the method of joints. Remember to list each member force as either tension or compression.

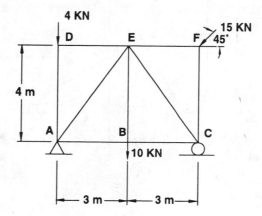

5. Determine the member forces for all members in each truss shown using the method of joints. Remember to list each member force as either tension or compression.

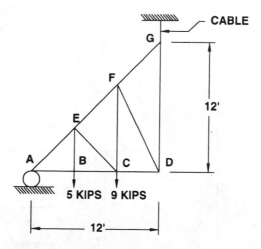

6. Determine the member forces for all members in each truss shown using the method of joints. Remember to list each member force as either tension or compression.

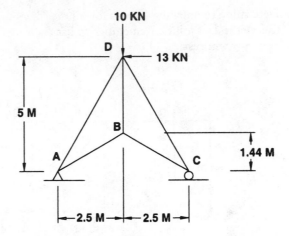

7. Determine the member forces for all members in each truss shown using the method of joints. Remember to list each member force as either tension or compression.

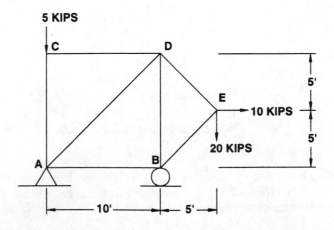

8. Using the method of sections, determine the member forces in members *BC, BI,* and *KF* of the truss shown below.

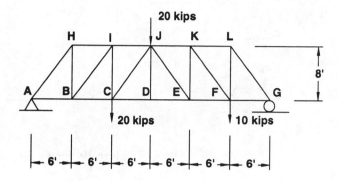

9. Using the method of sections, determine the member forces in members *BC, FC,* and *FG* of the truss shown below.

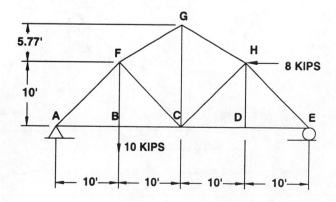

10. Using the method of sections, determine the member forces in members *GH*, *CH*, and *CD* of the truss shown below.

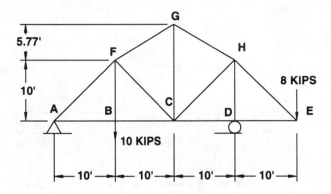

11. Using the method of sections, determine the member forces in members *BC*, *CE*, and *ED* of the truss shown below.

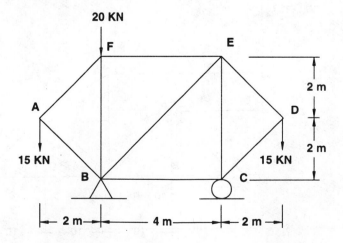

12. Using the method of sections, determine the member forces in members *BC*, *GC*, and *DF* of the truss shown below.

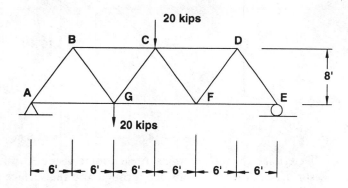

13. Using the method of sections, determine the member forces in members *BC*, *BG*, and *FG* of the truss shown below.

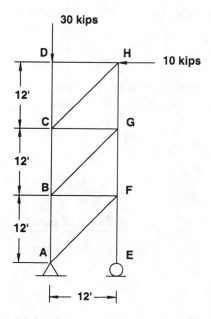

14. Determine the pin reactions at each joint of the simple frame as shown. Neglect selfweight of the members.

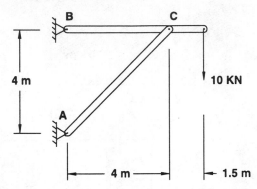

15. Determine the pin reactions at each joint of the A-frame as shown. The surface at *A* is to be considered rough and the surface at *B* is to be considered smooth. Neglect selfweight of the members.

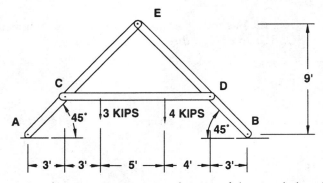

16. Determine the pin reactions at each joint of the simple bracket as shown. Neglect selfweight of the members.

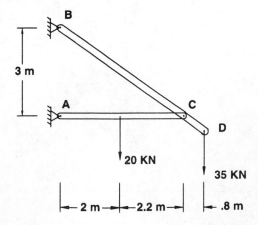

17. A folding chair diagram is shown in the following. Determine the pin reactions at each joint, neglecting selfweight of all components. *A* and *B* are both to be considered as smooth surfaces.

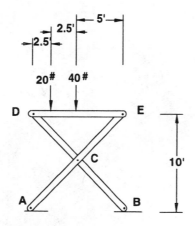

CHAPTER

5

FRICTION

5.1 Introduction

Friction is a force developed between two surfaces as they move relative to each other (Figure 5-1). This force tries to restrain or decrease motion and acts in a direction opposite of the movement or impending movement of the body. Friction is thought to be primarily a function of interatomic attraction[1] and the roughness between the two surfaces. Therefore, high amounts of friction force are developed if the atoms of each surface bond strongly to each other, or if the surfaces are rough. Materials which do not develop strong atomic bonds and are smooth will be seen to have a lower value of frictional forces.

There are two types of friction that can act on objects. These are commonly referred to as **static friction** and **kinetic friction**. Static friction refers to the frictional force developed on the object prior to motion. Kinetic friction refers to the

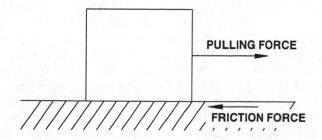

Figure 5-1 Friction Developed along a Sliding Surface

frictional force that acts on an object while motion is occurring. A good example of this may be pushing a broken-down car on a level street. The car does not move until a certain amount of force is applied. At this time, static friction forces are actively restraining the car from moving. As the car finally begins to move, frictional forces still affect the object and these are referred to as kinetic friction forces. These kinetic friction forces are less than the static forces, and hence, the car feels easier to keep moving once it is already in motion. However, if the pushing is stopped, the car will at some point stop coasting due to the kinetic frictional forces.

This text will focus primarily on static friction, since the behavior of rigid bodies in equilibrium is such that the bodies are normally at rest. The following sections will further explore the fundamentals of static behavior and the application of such fundamentals in the solution of various types of static problems.

5.2   Static Friction

As mentioned previously, static friction is the force developed prior to movement of one body relative to another. It is the force that needs to be overcome such that movement finally occurs. Static friction can simply be thought of as the resistance to sliding. In many friction-type problems, we will need to calculate the maximum force that is needed to overcome static friction so that motion will occur. Although many problems are typically concerned with the maximum static frictional force, it must be realized that static frictional forces develop in response to any magnitude of "push" on the object. If there is no "push" on the object, the frictional force is latent or not developed. As a "push" or sliding force is applied and increases in magnitude, the frictional force must likewise increase if the object is to remain stationary. As the applied sliding force exceeds the static frictional force at its maximum value, motion occurs. This can be shown in Figure 5-2.

What determines the maximum value of static frictional force that can be developed? The maximum value of static frictional force is dependent on two variables; the **normal force** and the **coefficient of friction**. This normal force is the reaction which occurs perpendicular to the moving object and the surface over which it moves (Figure 5-3). The normal force is related to the weight of the moving object; the heavier the object is, the harder it will be to move. An example of this can be shown very easily. Close your text and place it on the table. Take your finger and begin to gradually apply a force which tends to slide the book. Very soon the book begins to slide across the table. Now place your other hand on top of the book and apply a downward force on the book, and with your other hand, begin to apply a lateral push trying to slide the book. Notice how much harder it is to slide the book? This is due to the fact that in the second case the normal (reaction) force from the table onto the book was much higher, thereby increasing the static frictional force needed to make the book slide. Therefore, it can be stated that the frictional resistance acting on an object is proportional to the normal force.

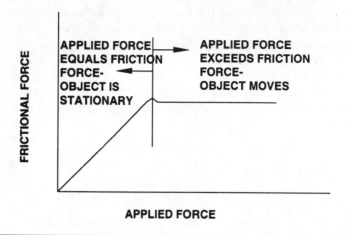

Figure 5-2   Static Friction versus Kinetic Friction

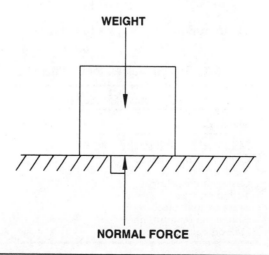

Figure 5-3   Normal Force Developed Perpendicular to Surface

The value of the maximum static frictional force which can be generated is also dependent upon the types of surfaces (on the object and sliding surface) that are present. Because of the roughness and atomic structure, some surfaces can develop higher resistances and seem "rougher" than others. This value of roughness is referred to as the **coefficient of static friction, μ,** and is approximately constant for a given material. Typical values of the coefficient of static friction are shown in Table 5-1.

The maximum amount of frictional force, F_f, is dependent on both the value of normal force and the coefficient of static friction, μ. This can be expressed as follows:

$$F_f = \mu \times N \qquad \text{(Eq. 5-1)}$$

where

$$F_f = \text{maximum friction force}$$

$$\mu = \text{coefficient of static friction}$$

$$N = \text{normal force}$$

The frictional force and the normal force are always perpendicular to each other and it has been shown mathematically that the frictional force is a portion of the normal force. Being perpendicular forces, the normal and frictional force can be viewed as rectangular components which are actually part of a resultant force applied at some angle (α) to the sliding surface. This angle is sometimes referred to as the **angle of friction** and is simply defined as follows (Figure 5-4):

$$\tan \alpha = F_f / N \qquad \text{(Eq. 5-2)}$$

Previously from Equation 5-1, we found that:

Table 5-1 Typical Values for Coefficients of Static Friction

Coefficients of Static Friction (Dry Surface)	
Materials	
Mild steel on mild steel	0.74
Copper on mild steel	0.53
Mild steel on lead	0.95
Cast iron on cast iron	1.10
Teflon on steel	0.04
Brass on mild steel	0.51
Aluminum on aluminum	1.05
Oak on oak (perpend.)	0.54

From *"Marks' Standard Handbook for Mechanical Engineers"*, McGraw-Hill, 8th ed., 1978.

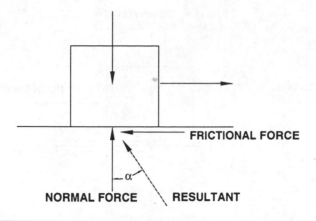

Figure 5-4   Angle of Friction

$$F_f = \mu \times N \qquad\qquad \text{(Eq. 5-1)}$$

Rearranging Eq. 5-1 in terms of μ we find:

$$\mu = F_f / N$$

Comparing Equations 5-1 and 5-2 we can demonstrate the following:

$$\mu = \tan \alpha$$

This is a very useful relationship when solving frictional type problems. Typical friction-related problems are discussed at length in the remaining sections of this chapter.

5.3   Forces Needed to Slide an Object

The most basic problem that involves frictional forces is the force necessary to cause an object to slide across a surface. In essence, this problem is to calculate the force necessary to break equilibrium and cause the object to move. Typical approaches to this problem consider objects on level and inclined surfaces (sometimes called inclined planes). This section will explore both of these cases.

When an object is to slide across a level surface, the forces that typically act on the object are as follows (Figure 5-5):

- the weight of the object
- the normal (or reaction) force
- the push or pulling force on the object
- the frictional force developed along the surfaces

A common solution of the problem is to draw the free-body diagram and write equilibrium equations ($\Sigma F_x = 0$, $\Sigma F_y = 0$). As movement is about to occur, the

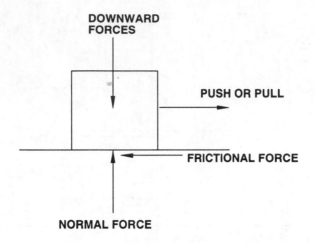

Figure 5-5 Forces Acting on an Object Subject to Sliding

frictional force will reach its maximum value. Any pushing or pulling force that exceeds this value will cause movement. The following examples demonstrate this type of problem.

EXAMPLE 5.1

A block weighing 50 KN rests on a level surface as shown. If the coefficient of static friction between the surfaces is 0.30, determine the force needed to start the block in motion.

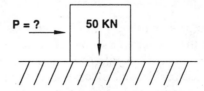

Problem Identification

Determine the force needed to overcome the frictional force, F_f.

Free-Body Diagram

The free-body diagram should include the weight of the block, the normal force from the surface onto the block, the unknown force that will push the block, and the frictional force that has to be overcome. (If the normal and frictional forces are resolved into a single resultant, the system easily can be viewed as a concurrent system. Such a system could also be solved using the graphical procedures that were discussed in Chapter 3.)

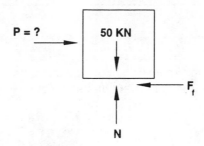

Algebraic Equations of Equilibrium

Since the force system is concurrent, the equations of equilibrium that are needed are $\Sigma F_x = 0$ and $\Sigma F_y = 0$. These can be written as follows:

$$\Sigma F_x = 0$$

$$+P - F_f = 0$$

Rearranging this equation:

$$P = F_f$$

$$\Sigma F_y = 0$$

$$-50 \text{ KN} + N = 0$$

Solving:

$$N = 50 \text{ KN}$$

Recognizing that the maximum value of static frictional force, F_f, is dependent upon the value of normal force and the coefficient of static friction, μ, it was stated that:

$$F_f = \mu \times N \qquad \text{(Eq. 5-1)}$$

Therefore, in this case:

$$F_f = 0.30 \times 50 \text{ KN} = 15 \text{ KN}$$

Substituting this value back into the $\Sigma F_x = 0$ we find:

$$+P - (15 \text{ KN}) = 0$$

$$P = 15 \text{ KN}$$

Therefore, any force greater than 15 KN will begin to move the block across the surface. Notice the sign convention used up and to the right as positive, and down and to the left as negative. While the use of any sign convention is arbitrary, it has to remain constant within a given problem.

EXAMPLE 5.2

Determine the minimum weight of a block that will not move when acted on by a pull of 200 lbs. The coefficient of static friction is 0.33.

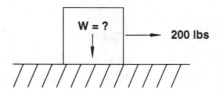

Problem Identification

Determine the minimum weight of the block, such that it will not move. This means that the block will have to generate a minimum frictional force, F_f, of 200 lbs.

Free-Body Diagram

The free-body diagram includes the unknown weight of the block, the unknown normal or reaction force from the surface onto the block, the 200 lb force that is pushing the block, and the frictional force that has to be overcome. (If the normal and frictional forces are resolved into a single resultant, the system can be viewed as a concurrent system. Such a system could also be solved using the graphical procedures discussed in Chapter 3).

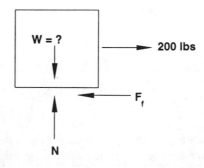

Algebraic Equations of Equilibrium

Since the force system is concurrent, the equations of equilibrium that will be needed are $\Sigma F_x = 0$ and $\Sigma F_y = 0$. These can be written as follows:

$$\Sigma F_x = 0$$

$$+P - F_f = 0$$

$$200\text{lb} - F_f = 0$$

Solving this equation:

$$F_f = 200 \text{ lb}$$

$$\Sigma F_y = 0$$

$$+N - W = 0$$

Solving this equation:

$$W = N$$

Recognizing that the maximum value of static frictional force, F_f, is dependent upon the value of normal force and the coefficient of static friction, μ, it was stated that:

$$F_f = \mu \times N \qquad\qquad \text{(Eq. 5-1)}$$

With Ff = 200 lb and μ = 0.33, we can solve for the normal force as follows:

$$200 \text{ lb} = 0.33(N)$$

$$N = 200 \text{ lb}/0.33 = 600 \text{ lb}$$

Substituting this value for N back into the $\Sigma F_y = 0$ we find:

$$+N - W = 0$$

$$W = N = 600 \text{ lb}$$

Therefore, the minimum weight to keep the block stationary would be 600 lbs. Anything less than this amount will cause the block to move.

Many times a friction problem is associated with the object having to be moved over an inclined surface. This complicates the situation slightly, since the normal force still acts perpendicular to the inclined surface, but the weight of the object still acts directly downwards. This is illustrated in Figure 5-6. Notice that the angle between the normal force and the weight force is exactly the same as the angle of the incline. We will refer to this angle as φ.

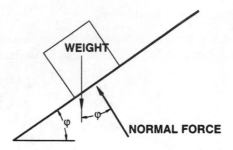

Figure 5-6   Angle of Inclination

This problem is typically handled by writing and solving the algebraic force equations in the x and y directions ($\Sigma F_x = 0$ and $\Sigma F_y = 0$). However, a problem arises since all the forces are not in the horizontal and vertical direction. This problem is overcome by orienting the x-axis to act in the direction of the incline, with the y-axis perpendicular to it (Figure 5-7). All the applicable forces will then be resolved to lie in either the modified x-axis (up the incline) or the modified y-axis (perpendicular to the incline). The following examples will illustrate the use of this principle.

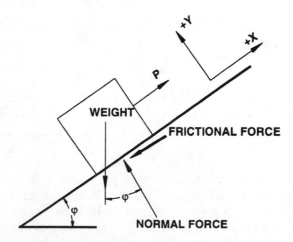

Figure 5-7   Force Acting on an Object Subject to Sliding on an Incline

EXAMPLE 5.3

A block weighing 50 KN rests on an inclined surface as shown. If the coefficient of static friction between the surfaces is 0.30, determine the force needed to start the block in motion up the incline.

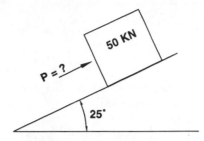

Problem Identification

Determine the force needed to overcome the frictional force, F_f and a component of the weight of the block. Realize that the weight of block will have a component that will act to restrain this move up the plane.

Free-Body Diagram

The free-body diagram includes the weight of the block, the normal or reaction force from the surface onto the block, the unknown force that will push the block, and the frictional force. Notice that the weight force is broken into components that lie in the assumed x and y axes. The components of the weight are found as follows:

$$X = 50 \text{ KN sin } 25° = 21.1 \text{ KN}$$

$$Y = 50 \text{ KN cos } 25° = 45.3 \text{ KN}$$

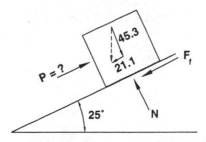

Algebraic Equations of Equilibrium

Since the force system is concurrent, the equations of equilibrium that will be needed are $\Sigma F_x = 0$ and $\Sigma F_y = 0$. The positive x-axis is assumed to be up the incline and the y-axis is perpendicular to the incline surface. The equations can be simply written as follows:

$$\Sigma F_x = 0$$

$$+P - 21.1 \text{ KN} - F_f = 0$$

$$+P - F_f = 21.1 \text{ KN}$$

$$\Sigma F_y = 0$$

$$+N - 45.3 \text{ KN} = 0$$

Solving:

$$N = 45.3 \text{ KN}$$

Recognizing that the maximum value of static frictional force, F_f, is dependent upon the value of normal force and the coefficient of static friction, μ, it was stated in Equation 5-1 that:

$$F_f = \mu \times N \qquad \text{(Eq. 5-1)}$$

Therefore, the F_f in this case is calculated as follows:

$$F_f = 0.30 \ (45.3 \text{ KN}) = 13.6 \text{ KN}$$

Substituting this value back into the $\Sigma F_x = 0$ we find:

$$+P - 21.1 \text{ KN} - F_f = 0$$

$$P - 21.1 \text{ KN} - (13.6 \text{ KN}) = 0$$

$$P = 34.7 \text{ KN}$$

Therefore, the force necessary to place the block on the verge of moving up incline is 34.7 KN. Pay attention to the sign convention that is used during the solution process.

EXAMPLE 5.4

Given the same information as Example 5.3, determine the force needed to start the block in motion *down* the incline.

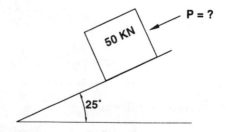

Problem Identification

Determine the force needed to overcome the frictional force, F_f, and move the block down the incline. Realize that the weight component of the block now helps in moving the block in this direction.

Free-Body Diagram

The free-body diagram includes the weight of the block, the normal force from the surface onto the block, the unknown force that will push the block, and the frictional force (which now acts up the incline). Notice that the weight force is broken into components that lie in our assumed x and y axes. The components of the weight are found as follows:

$$X = 50 \text{ KN sin } 25° = 21.1 \text{ KN}$$

$$Y = 50 \text{ KN cos } 25° = 45.3 \text{ KN}$$

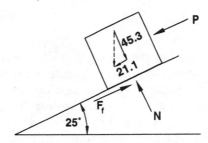

Algebraic Equations of Equilibrium

Since the force system is concurrent, the equations of equilibrium which will be needed are $\Sigma F_x = 0$ and $\Sigma F_y = 0$. The positive x-axis is assumed to be up the incline and the y-axis is perpendicular to the surface of the incline. The frictional force is now in the positive direction while the pushing force is in the negative direction. The equations can be simply written as follows:

$\Sigma F_x = 0$

$$-P - 21.1 \text{ KN} + F_f = 0$$

$$-P + F_f = 21.1 \text{ KN}$$

$\Sigma F_y = 0$

$$+N - 45.3 \text{ KN} = 0$$

Solving:

$$N = 45.3 \text{ KN}$$

Recognizing that the maximum value of static frictional force, F_f, is dependent upon the the value of normal force and the coefficient of static friction, μ, it was stated in Equation 5-1 that:

$$F_f = \mu \times N \qquad \text{(Eq. 5-1)}$$

Therefore the F_f in this case is calculated as:

$$F_f = 0.30(45.3 \text{ KN}) = 13.6 \text{ KN}$$

Substituting this value back into the $\Sigma F_x = 0$ we find:

$$-P - 21.1 \text{ KN} + F_f = 0$$

$$-P - 21.1 \text{ KN} + 13.6 = 0$$

$$P = -7.5 \text{ KN}$$

The force needed to push the block down the incline is –7.5 KN. The negative sign indicates that the block will move down the incline without any push at all! In fact, because the frictional force is 13.6 KN and the weight component in the x-direction is 21.1 KN, any *pull* less than 7.5 KN will still allow the block to slide down the incline.

5.4   Wedges

Practically everyone has used a wedge to lift an object into a higher or more advantageous position. Wedges are a simple and very useful application of friction where the frictional forces are used to develop normal forces that move an object. The most common use of a wedge involves driving the wedge under another object thereby creating frictional forces along three planes.

The method of solution will involve writing equations of equilibrium for the two objects and transferring the frictional and normal forces across the plane making contact between the two objects. Some substitution of mathematical relationships will be required, although the basic solution process will remain consistant with previous procedures. The following example will illustrate this technique on such a problem.

EXAMPLE 5.5

Determine the applied force necessary to act on the wedge such that object #2 moves upward. Object #2 weighs 100 lbs and the weight of the wedge is negligible. The coefficient of friction for all surfaces, μ, is 0.50.

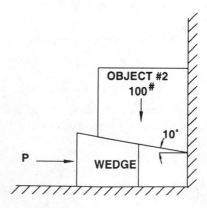

Problem Identification

Determine the force needed, P, on the wedge to overcome the frictional forces between the wall and object #2 such that object #2 moves up the wall. Realize that the two objects will be broken down in separate free-body diagrams and that the forces acting on each will then be used in the solution.

Free-Body Diagram

Object #2

The free-body diagram includes the weight of the object, the normal and frictional forces along the wall, and the normal and frictional forces along the wedge. Notice that the forces along the wedge are broken into their rectangular components such that the standard equations of equilibrium will be utilized. The forces on object #2 are as follows:

OBJECT #2

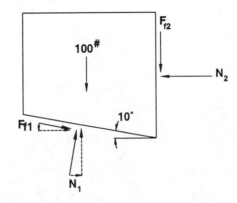

Algebraic Equations of Equilibrium, Object #2

The force system is concurrent and the equations of equilibrium needed are $\Sigma F_x = 0$ and $\Sigma F_y = 0$. The positive y-axis is assumed to act up the wall and the positive x-axis acts to the right. The frictional forces and normal forces act as shown. The equations can be simply written as follows:

$$\Sigma F_x = 0$$

$$+N_1 \sin 10° + F_{f1} \cos 10° - N_2 = 0$$

Since $F_{f1} = \mu\, N_1$, this equation can be written as:

$$+N_1 \sin 10° + (0.5\, N_1) \cos 10° - N_2 = 0$$

Simplifying:

$$0.666\ N_1 - N_2 = 0$$

$$N_2 = 0.666\ N_1 \qquad \text{(Eq. \#1)}$$

$$\Sigma F_y = 0$$

$$+N_1 \cos 10° - F_{f1} \sin 10° - F_{f2} - 100 = 0$$

Recognizing $F_{f2} = \mu\ N_2$ and $0.666\ N_1 = N_2$ from Equation #1:

$$F_{f2} = \mu\ (0.666\ N_1)$$

Substituting this into the ΣF_y equation:

$$+N_1 \cos 10° - F_{f1} \sin 10° - (0.50)(0.666\ N_1) - 100 = 0$$

Simplifying:

$$+N_1 \cos 10° - (0.50\ N_1) \sin 10° - (0.50)(0.666\ N_1) - 100 = 0$$

Solving:

$$.565\ N_1 = 100$$

$$N_1 = 177\ \text{lb}$$

Therefore, since $N_1 = 177$ lb:

$$N_2 = 0.666\ N_1 = 0.666(177\ \text{lb}) = 117.8\ \text{lb} \qquad \text{(Eq. \#1)}$$

$$F_{f1} = \mu\ N_1 = (0.50)(177\ \text{lb}) = 88.5\ \text{lb}$$

$$F_{f2} = \mu\ (0.66\ N_1) = (0.50)(0.666 \times 177\ \text{lb}) = 58.9\ \text{lb}$$

Free-Body Diagram

The Wedge

The free-body diagram for the wedge should include the force on the wedge, the normal and frictional forces from the floor, and from object #2. Realize that the frictional and normal forces from object #2 have already been found from the free-body diagram of object #2 and are transferred to the wedge. Notice that the transferred forces acting on the wedge are in the opposite direction as compared to object #2. The forces along the wedge are broken into their rectangular components such that the standard equations of equilibrium will be utilized. The forces on this object are as follows:

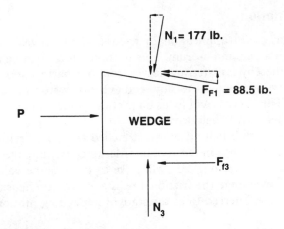

Algebraic Equations of Equilibrium, The Wedge

The equations needed are $\Sigma F_x = 0$ and $\Sigma F_y = 0$. The positive y-axis is assumed to act up the wall and the positive x-axis acts to the right. The frictional forces and normal forces act as shown. The equations can be simply written as follows:

$$\Sigma F_x = 0$$

$$-N_1 \sin 10° - F_{f1} \cos 10° - F_{f3} + P = 0$$

$$\Sigma F_y = 0$$

$$-N_1 \cos 10° + F_{f1} \sin 10° + N_3 = 0$$

Since N_1 and F_{f1} are known from the previous work on object #2, the $\Sigma F_y = 0$ equation is solved as:

$$-(177 \text{ lb})\cos 10° + (88.5 \text{ lb})\sin 10° + N_3 = 0$$

$$N_3 = 158.9 \text{ lb}$$

Knowing that $F_{f3} = \mu\, N_3$ then:

$$F_{f3} = \mu\, N_3$$

$$F_{f3} = (0.50)(158.9 \text{ lb}) = 79.4 \text{ lb}$$

Substituting this value of F_{f3} into the $\Sigma F_x = 0$ equation we find:

$$-N_1 \sin 10° - F_{f1} \cos 10° - F_{f3} + P = 0$$

$$-(177 \text{ lb})\sin 10° - (88.5 \text{ lb})\cos 10° - 79.4 \text{ lb} + P = 0$$

$$P = 197.3 \text{ lb}$$

The force needed to slide object #2 up the wall would have to exceed 197.3 lb.

5.5   Belt Friction

Another useful application of friction is the development of frictional forces between a belt and a drum. This frictional force can be used to start rotating the drum, thereby transferring power from one part to another. Conversely, the friction may also be used to slow and eventually stop the drum, thereby acting as a brake. This section will focus on friction forces and how they affect the operation of flat belt and V-belt drives.

As a flexible belt or cable is stretched around a circular drum, the belt on each side of the drum is in tension. In order to cause slippage of the belt relative to the drum, a force on one side will have to reach some value of maximum tension, T_{max}, to overcome the frictional resistance and the tension on the other side, T_0. This state is referred to as the **state of impending motion**. This state is shown in Figure 5-8.

The factors influencing the magnitude of the maximum tension force, T_{max}, are the tension force on the other side, T_0, the coefficient of friction between the belt and the drum, μ, and the angle of contact between the belt and the drum, β. By using equilibrium equations, the magnitude of the maximum tension force has been derived. Although the derivation is beyond the scope of this text, many references are available to show the derivation process.[2] The following relationship between these factors has been determined as:

$$\ln (T_{max}/T_0) = \mu\beta$$

Recognizing that the natural logarithm is simply $\log_e$ (where e is approximately 2.718), this would become as follows:

$$\log_e (T_{max}/T_0) = \mu\beta \qquad \text{(Eq.5-3)}$$

The definition of a logarithm states that "x is the logarithm of y to the base b" and mathematically states the following:

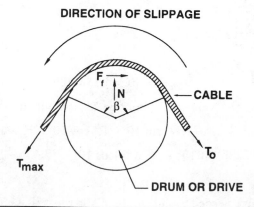

DIRECTION OF SLIPPAGE

CABLE

T_{max}

T_0

DRUM OR DRIVE

Figure 5-8   Forces Acting on a Flat Belt Pulley

$$y = b^x \text{ is equivalent to } x = log_b y$$

In this case with the base equal to e and x equal to $\mu\beta$, Equation 5-3 becomes:

$$e^{\mu\beta} = (T_{max}/T_0)$$

or as more typically expressed:

$$T_{max} = T_0 \, e^{\mu\beta} \qquad\qquad (Eq.\ 5\text{-}4)$$

where

T_{max} = the maximum tension needed to get the drum on the verge of rotating

T_0 = the tension on the other side

e = base of natural logarithms (2.718)

μ = coefficient of static friction

β = angle of contact between the belt and the drum, in radians

This equation can be used to solve many interesting problems with regard to flat belt drives. The following example will demonstrate its use.

EXAMPLE 5.5

On the flat belt drive shown below, calculate the tension force necessary to place the belt in a state of impending motion. The angle of contact is 180°, the coefficient of friction is 0.35, and T_o = 20 KN.

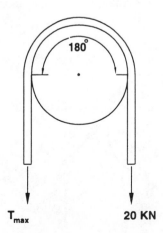

T_{max} 20 KN

Problem Identification

Solve for the force, T_{max}, which will place the drive on the verge of moving.

Solution

Using Equation 5-4, the force T_{max} can be calculated as follows:

$$T_{max} = T_0 e^{\mu\beta} \qquad \text{(Eq. 5-4)}$$

with

$$T_0 = 20 \text{ KN}$$

$$\mu = 0.35$$

$$\beta = 180/57.3° \text{ per radian} = 3.14 \text{ radians}$$

$$T_{max} = (20 \text{ KN})e^{(0.35)(3.14)} = 60 \text{ KN}$$

A V-belt drive is more popular than the flat belt drive for many types of machinery and consists of a belt that sits down in a groove. This type of drive forms a wedging action, due to the normal reaction on the sides of the groove (Figure 5-9). Because of this normal reaction, such a drive develops more frictional forces, which can be beneficial in mechanical applications. The maximum tension force, T_{max}, necessary to cause impending motion is found to be dependent on the same items, as was the case with the flat belt drive. By using equilibrium equations, the magnitude of the maximum tension force has been derived by modifying the flat-belt equation, which is explained in many references[3] and will not be repeated here. The formula for this maximum tension force is as follows:

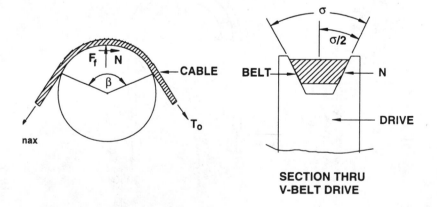

Figure 5-9 Forces Acting on a V-Belt Pulley

$$T_{max} = T_0 \, e^{(\mu\beta/\sin(\varphi/2))} \qquad \text{(Eq. 5-5)}$$

where

T_{max} = the maximum tension needed to get the drum on the verge of rotating

T_0 = the tension on the other side

e = base of natural logarithms (2.718)

μ = coefficient of static friction

β = angle of contact between the belt and the drum, radians

φ = angle of the groove

The following example will demonstrate the use of this formula.

EXAMPLE 5.6

For the V-belt shown below, calculate the maximum tension force to cause the belt to be on the verge of slipping. The coefficient of friction is 0.30, the angle of contact is 180°, and T_o is 200 lbs.

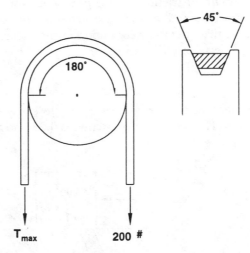

Problem Identification

Solve for the force T_{max} that will place the drive on the verge of moving.

Solution

Using Equation 5-5, the force T_{max} can be calculated as follows:

$$T_{max} = T_0 \, e^{\left(\mu\beta/\sin(\varphi/2)\right)}$$ (Eq. 5-5)

with

$T_0 = 200$ lbs

$\mu = 0.30$

$\beta = 180°/57.3°$ per radian $= 3.14$ radians

$\varphi = 45°$

$T_{max} = (200 \text{ lbs})e^{[(0.30)(3.14)/\sin 22.5°]} = 2347.6$ lbs

5.6   Summary

Friction is an extremely important force that governs many items in our day-to-day world. Such things as walking or the brakes on an automobile would not be possible if friction forces were not present. Friction forces can be classified as either static or kinetic. Static friction forces are those which are built up between an object and the sliding surface, keeping the object stationary. Kinetic frictional forces act once the object is sliding and tend to slow the object down and finally stop it.

The amount of frictional force generated is a function of the normal force between the object and the sliding surface and the coefficient of static friction (μ). The effect of these items is very important in understanding the effect of friction on sliding objects and belt drives.

EXERCISES

1. Determine the force, P, necessary to cause the block shown to begin to slide. The coefficient of friction is 0.33 and the block weighs 200 lbs.

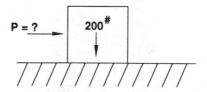

2. If the block in the previous problem weighed 500 lbs, what force would be necessary?

3. The block shown rests on an incline of 32°. Determine the force necessary to push the block up the incline, if the block weighs 30 KN. The coefficient of friction is 0.24.

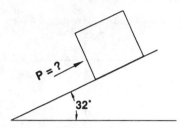

4. For the block in the previous example, determine the force necessary to push the block down the plane.
5. Determine the *horizontal* force needed to push the block up the incline shown. The block weighs 1000 lbs and the coefficient of friction is 0.35.

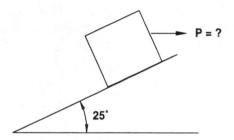

6. A 50 KN block sits on an inclined plane as shown. Determine the angle of the incline needed to have the block begin to slip down the plane. The coefficient of friction is 0.17.

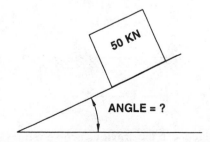

7. If the coefficient in the previous problem were doubled, what would be the angle necessary to cause slippage?
8. A block weighing 400 lbs sits on a level surface as shown. Calculate the force, P, applied as shown that would cause slippage if the coefficient of

friction is 0.25. Would the block begin to tip about its forward edge before or after this slippage would occur?

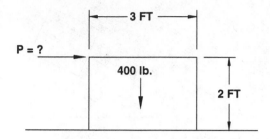

9. For the block in the preceding problem, calculate the additional distance above the point of application that would cause the block to tip before it would slide.

10. The ladder shown below is considered to rest on a frictionless surface at point *B* and rough surface at point *A*. If the coefficient of static friction at end *A* is 0.30, will the ladder slip under the loads shown?

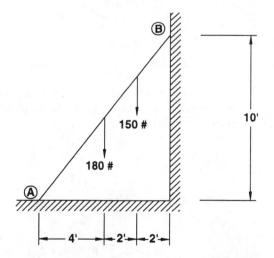

11. If the ladder in the previous problem does slip, what increased value of μ would prevent this from occurring?

12. A ladder rests on a wall that is considered frictionless at point *B* and a rough surface at point *A*. How far up the ladder can a 150 lb person climb before it begins to slip? Consider $\mu = 0.34$.

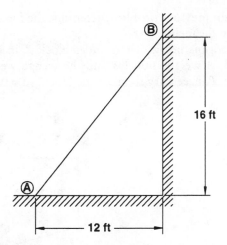

13. A 50 KN block rests upon a 100 KN block as shown. Determine the force, *P*, needed to make the bottom block begin to move if the coefficient of static friction between all surfaces (between the blocks and between the bottom block and the floor) is 0.25?

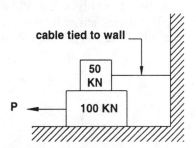

14. In the previous problem, what force would be needed to cause the bottom block to move if the coefficient between the blocks dropped to 0.20?

15. A block weighing 100 KN is attached to a 40 KN weight by a cable and frictionless pulley system, as shown. Calculate the force, *P*, needed to start the block moving up the incline. The coefficient of friction is 0.28.

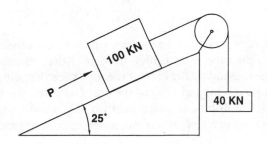

16. For the previous problem, determine the force necessary to push the block down the incline.

17. A wedge is being driven under block A in order to move it upward. If block A weighs 500 lbs and the wedge weighs 100 lbs, determine the force, P, necessary to move the block. ($\mu = 0.24$ between all surfaces)

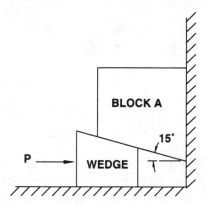

18. Determine the force needed on the wedge to move block A in the situation shown. The weight of block A is 300 KN and the weight of the wedge is neglected. The coefficient of friction for the wall and the floor is 0.30 and the coefficient of friction between the wedge and the block is 0.14.

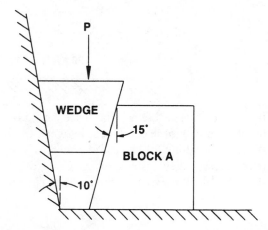

19. A rope is wrapped around a circular post to secure a ship that is pulling on the rope with a 5000 lb force. What minimum force is necessary on the free end (other end) of the rope such that slippage does not occur? The rope is wrapped around the post twice and $\mu = 0.22$.

20. A 300 lb box is being held off the ground by a rope looped over the branch of a tree. What minimum amount of force is to keep the box from

slipping if $\beta = 180°$ and $\mu = 0.30$? What minimum amount of force is to keep the box from slipping if $\beta = 135°$ and $\mu = 0.30$?

21. Two weights are suspended over a circular post as shown. Determine the minimum coefficient of static friction such that the weights remain motionless if $\beta = 180°$.

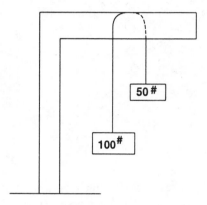

22. A flat belt passes over a drum, as shown. Determine the force needed to cause the 200 KN weight to be on the verge of being lifted if $\mu = 0.40$.

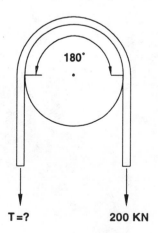

23. For the V-belt drive shown below, determine the maximum tension force needed to place the drive in a state of impending motion if β = 135° and μ = 0.20.

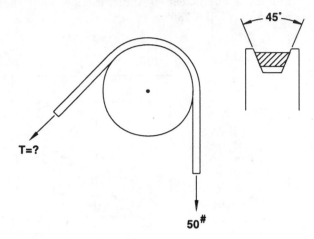

T=?

50#

24. If φ increases to 90°, what would the maximum tension force then need to be? Why does the force decrease as φ increases?

REFERENCES

1. Halliday, David and Resnick, Robert, *Fundamentals of Physics*, John Wiley and Sons, New York, 1974, p. 80.
2. Meriam, J. L., *Engineering Mechanics-Statics and Dynamics*, John Wiley and Sons, New York, 1978, p. 301.
3. Das, Braja M., Kassimali, Aslam, and Sami, Sedat, *Engineering Mechanics-Statics*, Richard D. Irwin Inc., Burr Ridge, Il., 1994, p.434.

6

CENTER OF GRAVITY AND CENTROIDS

6.1   Introduction

Previously, gravitational force has been discussed as one of the primary types of forces that act on a body. Viewing the gravitational force on a body from a microscopic perspective, it could be imagined that each atom of the body has its own discrete gravitational force that acts towards the earth's center. These individual forces on each atom can and will be viewed as a parallel force system and, as such, it may be advantageous to find a resultant of such a system (Figure 6-1). The location of such a resultant is referred to as the body's **center of gravity**.

The center of gravity of a three-dimensional body is important because it determines the location of the resultant force as that body acts on another. This location could be defined by x, y, and z coordinates relative to some set of predetermined axes (Figure 6-2). The determination of center of gravity is important to the field of machine design so that the designer can properly account for the weight of an object as it interacts within the machine assembly.

If the aforementioned body was assumed to be very thin, the location of the resultant would be located with respect to an area instead of a volume . This location of the resultant in a plane area is referred to as the **centroid** of the shape. This location is defined by using x, y coordinates relative to some predetermined set of axes (Figure 6-3). Centroids of an area are very useful in structural design, since the bending of homogeneous objects will tend to occur about an axis that locates the centroid . In essence, the axes which pass through the centroid are the primary axes of bending. This will be discussed further in Chapter 12. This chapter will

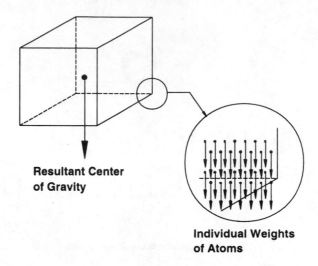

Figure 6-1 Concept of Center of Gravity

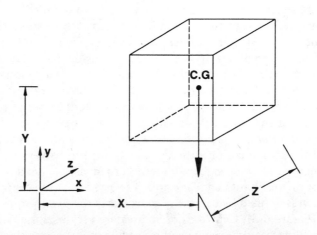

Figure 6-2 X, Y, and Z Coordinates Defining the Center of Gravity

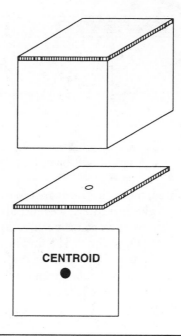

Figure 6-3   Concept of Centroid or Center of Gravity

focus on determining the centroids of plane areas, since this is of fundamental importance to many designers and technicians.

6.2   Centroid Fundamentals

As previously discussed, a centroid is the location where a gravitational resultant would act in a homogeneous plane area. Another way to think about the centroid of an area is to imagine it as the "balancing point" in a particular shape. Take your book, for example, which is virtually rectangular in shape and generally consistent in its makeup. Try to balance your book on the end of your finger. The first few attempts that you make will probably be unsuccessful because you have not located your finger on the centroid of the rectangular shape (and because some of you may be more steady than others). However, after a while, the point where you can balance the book will be found. This point (when viewing the book as a 2-D plane area) is the location of the centroid. The book balanced at this location because the resultant of the book's weight passed through the same point as the reaction caused by your finger (Figure 6-4). As discussed in Chapter 3, equilibrium will occur when two equal and opposite, collinear forces pass through a point resulting in no transitional or rotational movement. Hence, the book was balanced.

Finding the centroid of a shape is actually finding the *location* of the centroidal point. To find a location, it is common to given some type of "directions". To find

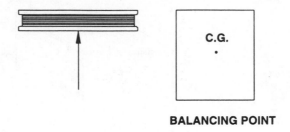

BALANCING POINT

Figure 6-4   Balancing Point of an Object Indicating Centroid

the centroid of an area, those directions are typically given by rectangular (x and y) distances. In order to give directions, we must always know from where the travel will begin. We would never give someone directions to our house without first asking "Where are you coming from?" The finding of centroids is no exception.

The centroids of common geometric shapes are shown in Table 6-1. It is very important to realize that the formulas given for the x and y distances (termed $\bar{x}$ and $\bar{y}$) to each centroid are relevant only for a particular point of origin or point of reference. For example, the centroid of a rectangle lies in the "center" of the rectangle and the $\bar{x}$ distance is given as $b/2$ (one-half the length of the base), while the $\bar{y}$ distance is given as $h/2$ (one-half the height). However, these formula values are only good if the point of origin is one of the rectangle's corners. If the point of origin was the center of the rectangle, the $\bar{x}$ distance and $\bar{y}$ distance would both be zero!

Table 6-1   Centroids of Standard Simple Shapes

Shape	Area	$\overline{X}$	$\overline{Y}$
Rectangle (ORIGIN, b, h)	bh	b/2	h/2
Triangle (ORIGIN, b, h)	1/2(bh)	2/3b	1/3h
Circle (ORIGIN, r)	πr^2	r	r
Semicircle (ORIGIN, r)	π r 2/2	r/2	.4244r

There are some key concepts that must be stated before continuing with this discussion of centroids:

KEY POINTS ABOUT CENTROIDS

- Centroids occur at a specific location in an area. This location is defined typically by an x and y axis. Such axes are referred to as **centroidal axes**.
- To located a centroid, one typically refers to an x and y distance from a point (or axes) of origin. The x and y distance are referred to as $\bar{x}$ and $\bar{y}$.
- $\bar{x}$ is the x distance from the point of origin to the centroidal y axis, while $\bar{y}$ is the y distance from the point of origin to the x centroidal axis (Figure 6-5).

The next section will discuss the location of centroids in typical geometric shapes such that we can apply this principle to further studies.

6.3   Finding Centroids

Since discussing the fundamentals of finding centroids in the previous section, it is now time to discuss the concepts *used* in finding centroids. These concepts find their basis in those used previously in finding the resultants of parallel force systems (Section 2.8). Section 2.8 discussed the fact that the resultant of the parallel force system had to be equivalent to the actual system, relative to all types of movement. Such a resultant had to be located at a specific point, so that the result-

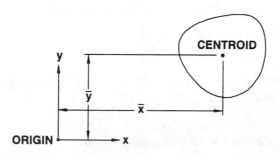

Figure 6-5   x and y Distances Defining the Location of the Centroid

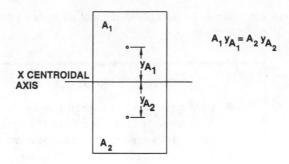

Figure 6-6 Moment of area A_1 Equals Moment of Area A_2 along the Centroidal Axis

ant and actual system would be equal. Centroids are very similar to this idea, since the centroid is located at the specific place in which the moments caused by pieces of area on both sides of the point are equal. This can be demonstrated with the simple rectangle (Figure 6-6). It is already known that the centroid of a rectangle lies in the center of the rectangle. This is so because the moment to the center of the area of each side of the centroidal axes is the same. Each centroidal axis serves as the "balancing line" of the moments of the areas on each side. Therefore the intersection of these axes is the centroid.

In mathematical terms this location of the centroid can be viewed as the location where the moment created by the total area (resultant) is equal to the sum of the moments created by each incremental piece. This can be expressed as follows:

$$A\bar{x} = \Sigma\, ax \qquad\qquad \text{(Eq. 6-1)}$$

$$A\bar{y} = \Sigma\, ay \qquad\qquad \text{(Eq. 6-2)}$$

where

$\bar{x}$ = the x distance from the point of origin to the y centroidal axis

$\bar{y}$ = the y distance from the point of origin to the x centroidal axis

a = incremental pieces of area

A = total of all areas

These formulas will be used extensively in finding centroids for complex or composite shapes.

6.4   Finding Centroids of Composite Areas

A composite area is an area or shape made up of more than one simple area. An example of a composite shape is that of a T-beam (as shown in Figure 6-7). Such

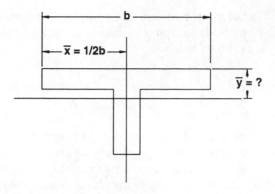

Figure 6-7   Typical Composite Shape where Location of Centroid is needed

a shape is not at all uncommon and the centroid must be located in order to investigate its behavior when subjected to bending.

The process of finding centroids of composite shapes is important, since it will allow a designer to obtain information about nonstandard shapes (shapes other than the simple shapes listed in Table 6-1). The process for finding the centroid of a composite shape is relatively straightforward and can easily be performed by using the step-by-step process outlined next. It should be noted that this process works very nicely when using a tabular format; therefore the author strongly recommends that all students use such a table. The process is as follows:

1. Break the composite shape into common geometric areas—rectangles, triangles, etc. Label each individual area as 1, 2, 3, etc.
2. Choose a point of origin if one is not already given. This will be the point from which $\bar{x}$ and $\bar{y}$ distances are given.
3. Construct a table similar to the one shown.

Shape	Area	x	ax	y	ay
	A =		ax =		ay =

4. For each individual area, fill in the table. List the area of the individual shape and the x and y distances from the point of origin *to the centroid of that individual shape.* Also multiply the individual area by these x and y distances and place those values in the Σax and Σay columns.

5. When all individual areas are complete, sum the *A*, *ax*, and *ay* columns. The centriodal $\bar{x}$ and $\bar{y}$ distances are calculated by Equations 6-3 and 6-4, as shown below.

$$\bar{x} = \Sigma \, ax/A \qquad \text{(Eq. 6-3)}$$

$$\bar{y} = \Sigma \, ay/A \qquad \text{(Eq. 6-4)}$$

It should be stressed that the point of origin must be decided upon at the beginning of the process and must remain the same throughout the entire process. Failure to do this will result in erroneous results. The following examples will illustrate this process of finding centroids for composite shapes.

EXAMPLE 6.1

Find the centroid of the shape shown below. The point of origin is given as the origin of the axes shown.

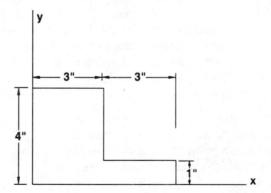

Start the process by breaking up the shape into simple areas. This may be done as shown.

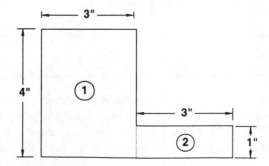

Fill in the table for the individual areas listing area, x and y distances for the point of origin to the centroid of the individual shape, and Σax and Σay. The x and y distances from the point of origin to the centroid of the shapes is sometimes the most common place for a mistake to be made, therefore these distances for each area are as follows:

Area 1-Large rectangle

The x distance from the stated point of origin to the y centroidal axis of area 1 is:

$$x = b/2 = 3 \text{ in.}/2 = 1.5 \text{ in.}$$

The y distance from the stated point of origin to the x centroidal axis of area 1 is:

$$y = h/2 = 4 \text{ in.}/2 = 2 \text{ in.}$$

Area 2-Small rectangle

The x distance from the stated point of origin to the y centroidal axis of area 2 is:

$$x = 3 \text{ in.} + b/2 = 3 \text{ in.} + (3 \text{ in.}/2) = 4.5 \text{ in.}$$

Notice the distance is still taken from the stated point of origin. *This point of origin does not move during the process.*

The y distance from the stated point of origin to the x centroidal axis of area 2 is:

$$y = h/2 = 1 \text{ in.}/2 = 0.5 \text{ in.}$$

The table is then filled out accordingly as shown below.

Shape	Area	x	ax	y	ay
Lg. rect.	12 in^2	1.5 in	18 in^3	2 in	24 in^3
Sm. rect.	3 in^2	4.5 in	13.5 in^3	0.5 in	1.5 in^3
	A = 15 in^3		Σax = 31.5 in^3		Σay = 25.5 in^3

The centroid can then be located by solving Equation 6-1 and 6-2 as follows:

$$\bar{x} = \Sigma \, ax/A \qquad \text{(Eq. 6-3)}$$

$$\bar{x} = 31.5 \text{ in.}^3/15 \text{ in.}^2 = 2.1 \text{ in.}$$

$$\bar{y} = \Sigma \, ay/A \qquad \text{(Eq. 6-4)}$$

$$\bar{y} = 25.5 \text{ in.}^3/15 \text{ in.}^2 = 1.7 \text{ in.}$$

The centroid of the shape is located at an $\bar{x}$ distance of 2.1 in. and a $\bar{y}$ distance of 1.7 in. from the point of origin. This is shown as follows:

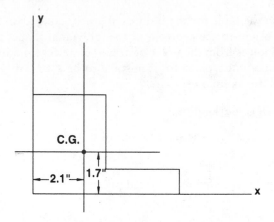

The next example will illustrate a technique which can be used to account for "holes" which may be included in a shape. Many shapes may have holes in them, such as plates used in sheet metal fabrication. Such plates may have circular or rectangular holes punched through them which must be accounted for when calculating the location of the centroid. These holes can be treated as negative areas and therefore should be subtracted from the properties of the full shape. It must be noted that in order to subtract a hole area, one must have area located where the holes are taken out of. For instance, if a plate has a hole in its center, one would calculate the properties for a complete plate and then subtract the properties out of that, to account for the hole. (It would be impossible to subtract a hole from an area in which a hole already existed.) The next example illustrates this technique.

EXAMPLE 6.2

Find the centroid of the shape shown. The point of origin is given as the origin of the axes shown.

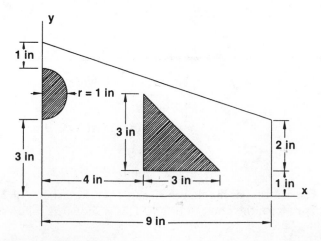

Start the process by breaking up the shape into simple areas. Notice the triangular and semi-circular holes are broken into their own areas, however they will be subtracted from the full area in which they are contained. This may be done as shown.

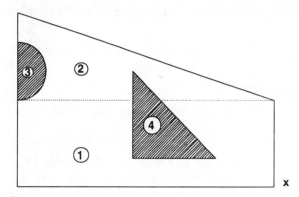

Fill in the table for the individual areas; listing area, x and y distances from the point of origin to the centroid of the individual shape, and ax and ay. The x and y distances are from the point of origin to the centroid of the individual shapes.

Area 1-Large full rectangle

The x distance from the stated point of origin to the y centroidal axis of area 1 is:

$$x = b/2 = 9 \text{ in.}/2 = 4.5 \text{ in.}$$

The y distance from the stated point of origin to the x centroidal axis of area 1 is:

$$y = h/2 = 3 \text{ in.}/2 = 1.5 \text{ in.}$$

Area 2-Large full triangle

The x distance from the stated point of origin to the y centroidal axis of area 2 is:

$$x = b/3 = (9 \text{ in.})/3 = 3 \text{ in.}$$

Notice the distance is still taken from the stated point of origin.
The y distance from the stated point of origin to the x centroidal axis of area 2 is:

$$y = 3 \text{ in.} + h/3 = 3 \text{ in.} + (3 \text{ in.})/3 = 4 \text{ in.}$$

Notice this point of origin does not move during the process.

Area 3-Semi-circular hole

The x distance from the stated point of origin to the y centroidal axis of area 3 is:

$$x = .424 \, r = .424 \, (1 \text{ in.}) = .424 \text{ in.}$$

Notice this distance is referred to as the $\overline{y}$ distance when taken from Table 6-1. This is because the semicircle has been rotated 90°. Still notice that the point of origin stays constant.

The y distance from the stated point of origin to the x centroidal axis of area 3 is:

$$y = 3 \text{ in.} + r = 3 \text{ in.} + 1 \text{ in.} = 4 \text{ in.}$$

Area 4-Triangular hole

The x distance from the stated point of origin to the y centroidal axis of area 4 is:

$$x = 4 \text{ in.} + (3 \text{ in.})/3 = 5 \text{ in.}$$

Again notice that the point of origin stays constant. The y distance from the stated point of origin to the x centroidal axis of area 4 is:

$$y = 1 \text{ in.} + (3 \text{ in.})/3 = 2 \text{ in.}$$

The table is then filled out as shown, with the area of the holes being input as negative values.

Shape	Area	x	ax	y	ay
Lg. rect.	27 in^2.	4.5 in.	121.5 in^3	1.5 in.	40.5 in^3
Lg. tri.	13.5 in^2	3 in.	40.5 in.^{3}	4 in.	54 in.^{3}
Smi-cir hole	–1.57 in.^{2}	0.42 in.	–0.66 in.^{3}	4 in.	–6.28 in.^{3}
Tri-hole	– 4.5 in.^{2}	5 in.	–22.5 in.^{3}	2	–9in.3
	A = 34.43		$\Sigma ax = 138.84$ in.^{3}		$\Sigma ay = 79.22$ in.^{3}

The centroid can then be located by solving Equations 6-3 and 6-4 as follows:

$$\overline{x} = \Sigma \, ax/A \qquad\qquad\qquad\qquad \text{(Eq. 6-3)}$$

$$\overline{x} = 138.84 \text{ in.}^3/34.43 \text{ in.}^2 = 4.03 \text{ in.}$$

$$\overline{y} = \Sigma \, ay/A \qquad\qquad\qquad\qquad \text{(Eq. 6-4)}$$

$$\overline{y} = 79.22 \text{ in.}^3/34.43 \text{ in.}^2 = 2.30 \text{ in.}$$

The centroid of the shape is located at an $\bar{x}$ distance of 4.03 in. and a $\bar{y}$ distance of 2.3 in. from the point of origin. This is shown as follows:

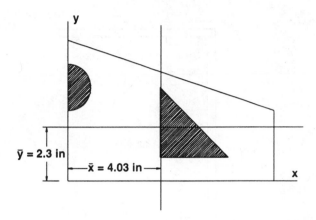

Another possibility is that the location of a centroid is needed for a shape that is partially made from an existing manufactured shape. A good example of this are structural members that may be composed of angles, channels, or wide flange sections. How does one incorporate these "areas" into the method that has been illustrated in this chapter so far? The answer is simply to treat these manufactured shapes as individual areas, since their properties are known and can be found in Appendix A. It should be noted that many steel wide flange shapes (W sections) are symmetrical; therefore their centroid lies in the center of the shape. This is true for all symmetrical shapes. However, shapes such as angles and channels are not symmetrical and therefore the centroids of these sections have to be located. The location of the centroid in standard steel shapes that are not symmetrical has already been done and can be found by looking at the property tables. Notice that the tables for steel angles contain columns listed as $\bar{x}$ and $\bar{y}$. These values are the x and y distances from the back edge of the angle to the location of the centroid for that individual angle. Channels are labeled in a similar fashion. This $\bar{x}$ and $\bar{y}$ value which locates the centroid of that particular shape can be used in the process for finding the centroid of the complete shape. The following example will demonstrate such a use.

EXAMPLE 6.3

Find the centroid of the shape shown below. Use the lower left portion of the channel as the point of origin.

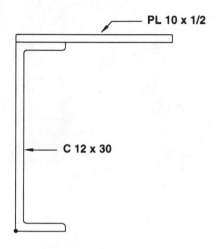

Start the process by breaking up the shape into simple areas. These areas will be the rectangular plate and the channel.

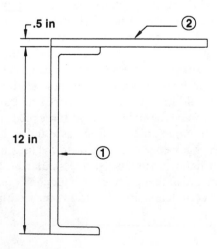

Fill in the table for the individual areas listing area, x and y distances from the point of origin to the centroid of the individual shape, and ax and ay. The x and y distances are from the point of origin to the centroid of the individual shapes.

Channel

The centroid of the channel can be found by using the tables located in Appendix A. Since the channel is symmetrical about its centroidal x axis, we know that this axis has to lie in the center of the section or a numerical value equal to one-half the channel depth. The centroidal y-axis is located off the back of the channel a distance $(\bar{x})$ equal to .674 in., this distance found in the column labeled $\bar{x}$ for the channel. Therefore for the channel shape:

$$x = .674 \text{ in.}$$

$$y = 1/2 \text{ depth} = 1/2(12 \text{ in.}) = 6 \text{ in.}$$

Rectangular Plate

The x distance and the y distance to the centroid of the plate are calculated as previously done with respect to the stated point of origin. These are as follows:

$$x = b/2 = 10 \text{ in.}/2 = 5 \text{ in.}$$

$$y = 12 \text{ in.} + h/2 = 12 \text{ in.} + (0.5 \text{ in.})/2 = 12.25 \text{ in.}$$

The table is then filled out accordingly, as shown below.

Shape	Area	x	ax	y	ay
Channel.	8.82 in.²	0.674 in.	5.94 in.³	6 in.	52.92 in.³
Plate	5 in.²	5 in.	25 in.³	12.25 in.	61.25 in.³
	$A = 13.82$ in.²		$\Sigma ax = 30.94$ in.³		$\Sigma ay = 114.17$ in.³

The centroid can then be located by solving Equation 6-3 and 6-4 as follows:

$$\bar{x} = \Sigma \, ax/A \qquad\qquad \text{(Eq. 6-3)}$$

$$\bar{x} = 30.94 \text{ in.}^3/13.82 \text{ in.}^2 = 2.24 \text{ in.}$$

$$\bar{y} = \Sigma \, ay/A \qquad\qquad \text{(Eq. 6-4)}$$

$$\bar{y} = 114.17 \text{ in.}^3/13.82 \text{ in.}^2 = 8.26 \text{ in.}$$

The centroid of the shape is located at an $\bar{x}$ distance of 2.24 in. and a $\bar{y}$ distance of 8.26 in. from the point of origin. This is shown as follows:

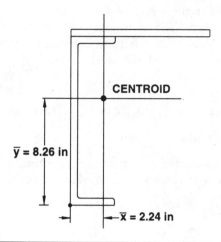

Integral calculus may also be used to find the centroid of a plane area. This method can easily be used if an expression can be written which describes an incremental piece of area and the limits over which it is to be integrated. Because this method is not widely used, it will be discussed briefly in Appendix E.

6.5   Summary

The center of gravity of a three dimensional shape is the location where the resultant of the gravitational force acts. In a two dimensional shape, this location is referred to as the centroid of that shape. The locations of the center of gravity and the centroid are important because it allows us to place the resultant weight in a location that makes this resultant force act exactly the same as the original body. The location of the centroid becomes important because it will be used extensively when bending of an object occurs; since the centroidal axis is also the axis of bending in simple, homogeneous shapes.

EXERCISES

1. Find the centroid of the shapes with respect to the point of origin being shown.

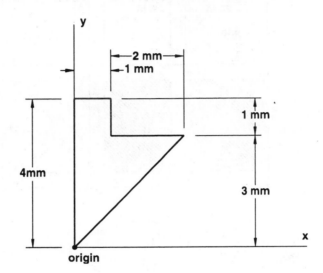

2. Find the centroid of the shapes with respect to the point of origin being shown.

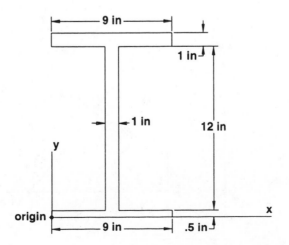

3. Find the centroid of the shapes with respect to the point of origin being shown.

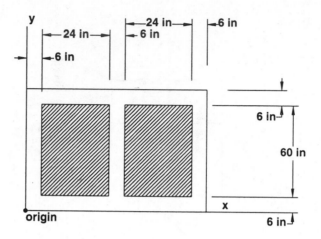

4. Find the centroid of the shapes with respect to the point of origin being shown.

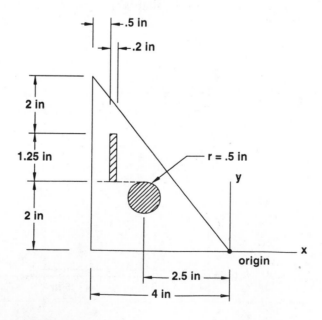

5. Find the centroid of the shapes with respect to the point of origin being shown.

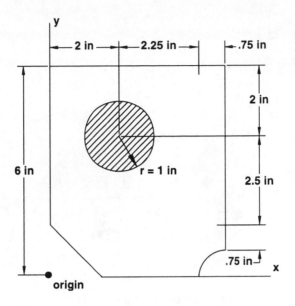

6. Find the centroid of the shapes with respect to the point of origin being shown.

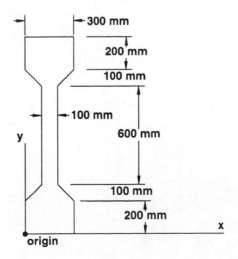

7. Find the centroid of the shapes with respect to the point of origin being shown.

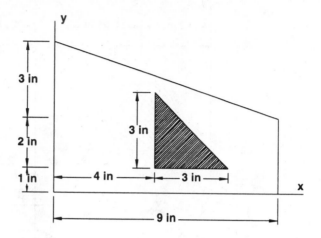

8. Find the centroid of the shapes with respect to the point of origin being shown.

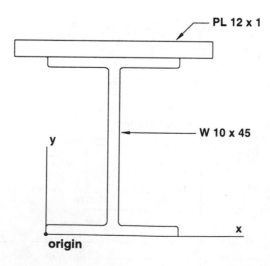

9. Find the centroid of the shapes with respect to the point of origin being shown.

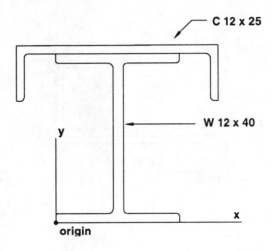

10. Find the centroid of the shapes with respect to the point of origin being shown.

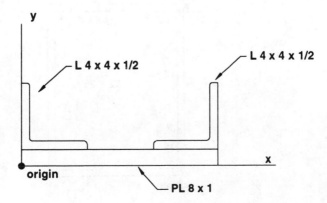

11. Find the centroid of the shapes with respect to the point of origin being shown.

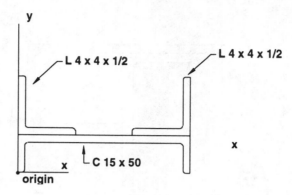

12. Find the centroid of the shapes with respect to the point of origin being shown.

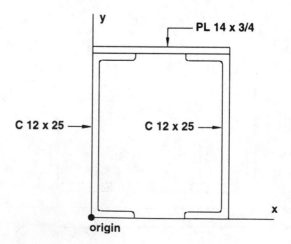

13. Calculate the centroid of the same shapes if the point of origin is changed to the new location as shown.

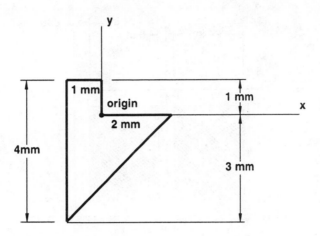

14. Calculate the centroid of the same shapes if the point of origin is changed to the new location as shown.

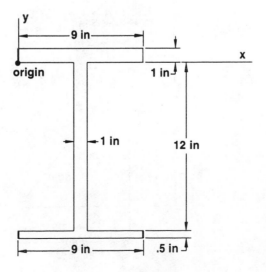

15. Calculate the centroid of the same shapes if the point of origin is changed to the new location as shown.

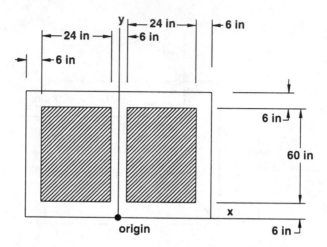

16. Calculate the centroid of the same shapes if the point of origin is changed to the new location as shown.

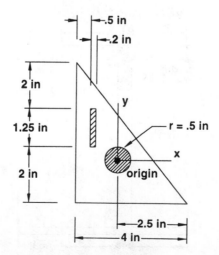

17. Calculate the centroid of the same shapes if the point of origin is changed to the new location as shown.

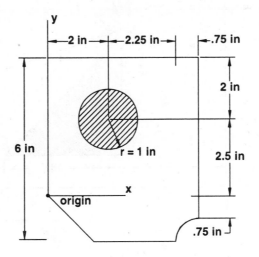

18. Calculate the centroid of the same shapes if the point of origin is changed to the new location as shown.

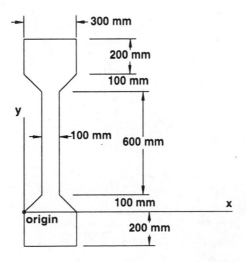

19. Calculate the centroid of the same shapes if the point of origin is changed to the new location as shown.

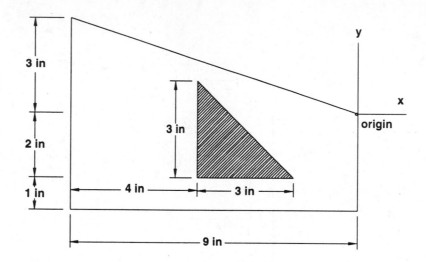

20. Calculate the centroid of the same shapes if the point of origin is changed to the new location as shown.

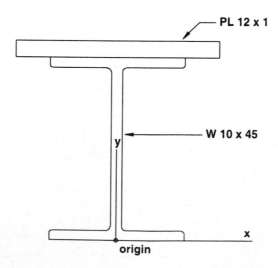

21. Calculate the centroid of the same shapes if the point of origin is changed to the new location as shown.

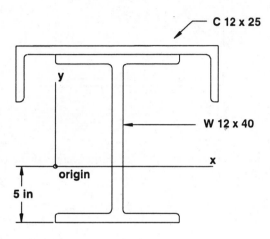

22. Calculate the centroid of the same shapes if the point of origin is changed to the new location as shown.

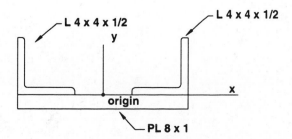

23. Calculate the centroid of the same shapes if the point of origin is changed to the new location as shown.

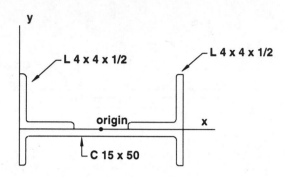

24. Calculate the centroid of the same shapes if the point of origin is changed to the new location as shown.

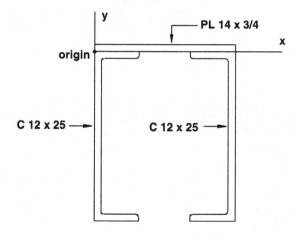

7

MOMENT OF INERTIA OF AREA

7.1   Introduction

Moment of inertia is a mathematical property of the area that describes the second moment of area about an axis. Moment of inertia can be thought of as the areal property that describes a shape's resistance to rotating, or bending about, a given axis. It is the shape property that controls resistance to bending. The larger a shape's moment of inertia, the more resistance to bending the shape will have. Moment of inertia will increase as more area is located away from the axis of bending. This is why a 2 × 4 is very stiff, relative to when bent with the long side vertical; but very flexible when bent with the short side vertical (Figure 7-1). Knowledge of a shape's moment of inertia is important so that the strongest axis in a shape is utilized for bending. The bending axis of a homogeneous member is also the centroidal axis; therefore the ability to locate the centroid of a shape is closely associated with moment of inertia.

Moment of inertia is expressed in units of length to the fourth power. While this value is physically abstract and not very meaningful when thought of individually, it is of vital importance when bending behavior is discussed. For the time being, the student may think of moment of inertia as the "area of resistance" when a shape is subjected to bending. In this chapter we will explore the methods used to calculate the value of moment of inertia for a given shape. The importance of this ability will become evident as the student encounters bending behavior and flexural stresses that are found in practically all structural members.

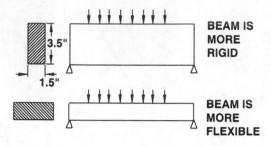

Figure 7-1   Moment of Inertia Relating to Beam Stiffness

7.2   Moment of Inertia of Simple Areas

In the previous section it was discussed how moment of inertia is utilized as the "area of resistance" when a shape is subjected to bending. Without exploring bending behavior in a full manner (as will be done in Chapter 12), how can this property best be described?

When a shape is subjected to bending, a rotation occurs about some axis in the shape referred to as the **neutral axis** or axis of bending. (Many times the neutral axis is the centroidal axis). The area located further away from this neutral axis is more "active" in carrying or resisting the bending forces (this concept will be fully discussed in Chapter 12). In essence, the area away from the neutral axis is more crucial in carrying the bending forces because it has to resist a greater magnitude of force. In a mathematical context, the force carried by a piece of area is dependent on the size of the area and its distance away from the neutral axis. This could be expressed as follows and illustrated in Figure 7-2.

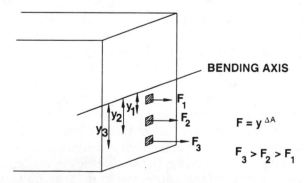

Figure 7-2   Bending Forces Developed by a Beam Increases the Farther from the Bending Axis

$$F = y \, \Delta A$$

where

F = force carried by the incremental area

y = distance away from the neutral axis or axis of bending

ΔA = incremental area being considered

Since moment of inertia is a property describing resistance to bending or rotation, the moment of inertia has to refer to the moment resistance. Therefore, it can be viewed as the sum of all these weighted forces multiplied by the individual moment arms from the neutral axis (Figure 7-3). Moment of inertia is sometimes referred to as the second moment of area and can be expressed in mathematical terms as follows:

$$I_x = \Sigma \, y^2 \, \Delta A \qquad \text{(Eq. 7-1)}$$

where

I_x = moment of inertia about the x axis

y = distance from the neutral axis or axis of bending

ΔA = incremental area being considered

Notice that moment of inertia is always referenced about a particular axis and that moment of inertia increases as the area located further away from this axis increases.

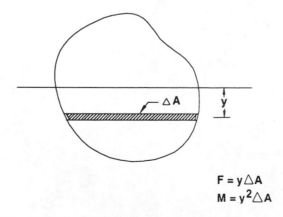

$$F = y \triangle A$$
$$M = y^2 \triangle A$$

Figure 7-3 Moment of Inertia of Incremental Area is in the Second Moment of that Area

Exact values for the moment of inertia are found by integrating equations for the area, which is beyond the scope of this chapter but is explored in Appendix E. However, many typical geometric shapes have standard formulas for moment of inertias and these are listed in Table 7-1. Please notice that these formulas are given for the shape's centroidal axis. If values are desired for an axis other than the centroidal axis, another method must be used. This will be the focus of the next section.

The property tables for rolled steel shapes also contain information regarding moment of inertias for particular rolled members. Appendix A contains many of the more common shapes and the properties associated with these shapes in both U.S. and SI units. Please note that moment of inertias are listed about the x-axis and y-axis. The y-axis for steel shapes is typically referred to as the "weak axis" since the properties of area (such as moment of inertia) are smaller about this axis. If a shape is built up from either a variety of simple shapes or from a variety of standard members, its moment of inertia will have to be calculated on a individual basis. Section 7.4 will explore the method for accomplishing this procedure.

Table 7-1   Moment of Inertia for Simple Shapes about the Centroidal Axis

Shape	Area	Moment of Inertia
	bh	$1/12(b)(h)^3$
	$1/2(b)(h)$	$1/36(b)(h)^3$
	$(\pi/4)d^2$	$(\pi/4)r^4$
	$(\pi/8)d^2$	$.0349(\pi)r^4$

7.3   Parallel Axis Theorem

When dealing with complex shapes made up from more than one of the simple areas that were shown in Table 7-1 or when moment of inertia for a simple area is desired about an axis other than the centroidal axis, the **parallel axis theorem** must be applied. This theorem states that the moment of inertia for an area about some axis that is not the centroidal axis can be defined as its centroidal moment of inertia plus the area multiplied by the square of the distance between the axis desired and the centroidal axis. In a mathematical sense this theorem can be written as follows:

$$I_x = I_{xc} + Ad^2 \qquad\qquad \text{(Eq. 7-2)}$$

or

$$I_y = I_{yc} + Ad^2$$

where

I_x, I_y = moment of inertia desired about some parallel x or y axis, respectively

I_{xc}, I_{yc} = moment of inertia about the shape's centroidal x or y axis, respectively

A = area of the shape

d = the perpendicular distance between the axis of intertest and the centroidal axis

This concept is illustrated in Figure 7-4. The following example demonstrates the use of this important principle.

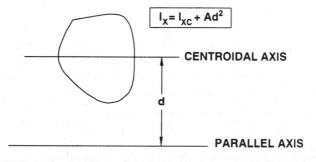

Figure 7-4   Parrallel Axis Theorem

EXAMPLE 7.1

For the shape shown below, calculate the moment of inertia about the x-axis.

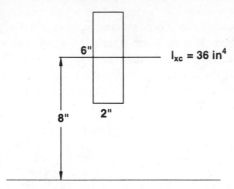

Moment of inertia for the rectangle shown has to be calculated from the parallel axis theorem discussed above. This is because the only standard moment of inertia formula for a rectangle which we now know is for the moment of inertia through the rectangle's centroidal axis. The axis which we want moment of inertia for is about a parallel axis. Begin by calculating the moment of inertia for the rectangle's centroidal axis from Table 7-1 on page 210:

$$I_{xc} = bh^3 / 12$$

$$I_{xc} = (2 \text{ in.})(6 \text{ in.})^3 / 12 = 36 \text{ in.}^4$$

Next apply the parallel axis theorem as follows:

$$I_x = I_{xc} + Ad^2 \qquad\qquad \text{(Eq. 7-2)}$$

where

$$A = 12 \text{ in.}^2$$

$$d = 8 \text{ in.}$$

$$I_x = 36 \text{ in.}^4 + (12 \text{ in.}^2)(8 \text{ in.})^2 = 804 \text{ in.}^4$$

Notice that the distance, d, is between the axis that we want moment of inertia taken about and the centroidal axis of the individual shape. As this distance increases, the moment of inertia or bending resistance of the shape increases very quickly. Also notice that the base dimension, b, in the moment of inertia formula for the shape is the side parallel to the axis in question (in this case it was the x-axis).

From the previous example, the Ad^2 term reflects the increasing importance of area as it is shifted away from the axis under consideration (typically the neutral axis or axis of bending). As area moves away from the bending axis it becomes more effective at resisting bending forces. This concept explains why beams with larger depths can hold larger loads. It is not simply because larger beams have more area, but more importantly because the area is distributed further away from the bending axis, thereby making the section's moment of inertia larger and more effective. The following example will illustrate the concept.

EXAMPLE 7.2

Two 2×4's (actual dimensions 1 ½ in. × 3 ½ in.) can be placed together in one of two manners as shown. Calculate the moment of inertias for each particular case.

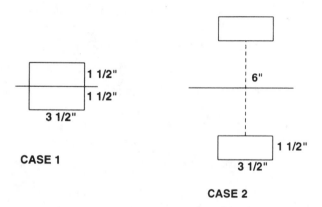

CASE 1

CASE 2

In the first case, the 2×4's are arranged in a manner such that they form one large rectangle measuring 3 in.×3 1/2 in. Calculating the moment of inertia about the x-axis for this shape can be accomplished simply through the use of the standard moment of inertia formula from Table 7-1 as follows:

$$I_{xc} = bh^3 / 12$$

$$I_{xc} = (3.5 \text{ in.})(3 \text{ in.})^3 / 12 = 7.88 \text{ in.}^4$$

In the other case, the 2×4's form areas (much like flanges of a beam) separated by a 6 in. distance. The axis of bending would be 3 in. away from the inner side of each area, therefore the distance from the bending axis to the centroid of each area is 3.75 in.

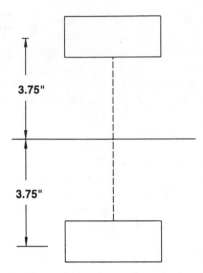

The moment of inertia about this axis can be calculated using the parallel axis theorem. First calculate the moment of inertia for each 2×4 about its own centroidal axis, notice the base length is the 3.5 in. side.

$$I_{xc} = bh^3 / 12 = (3.5 \text{ in.})(1.5 \text{ in.})^3 / 12 = 0.98 \text{ in.}^4$$

Next apply the parallel axis theorem as follows:

$$I_x = I_{xc} + Ad^2 \qquad \qquad \text{(Eq. 7-2)}$$

In this case:

$$I_x = 2 \text{ areas}(I_{xc} + Ad^2)$$

where

$$A = (3.5 \text{ in.})(1.5 \text{ in.}) = 5.25 \text{ in.}^2$$

$$d = 3.75 \text{ in.}$$

$$I_x = 2[(0.98 \text{ in.}^4 + (5.25 \text{ in.}^2)(3.75 \text{ in.})^2] = 149.6 \text{ in.}^4$$

The moment of inertia in the second case is approximately nineteen times larger than the value found in the first case, although the area is exactly the same. This shows the value of moment of inertia is not solely based on area but also on the distribution of that area.

With this knowledge of the parallel axis theorem we can now begin to calculate the moments of inertia for many different, nonstandard shapes. This will be discussed in the following section.

7.4   Moment of Inertias of Composite Areas

Composite shapes are those shapes made up of more than one of the simple areas that were listed in Table 7-1. Most of the shapes used in building and industrial applications are composite because they are made up of a mixture of common shapes. The I-beam shape (Figure 7-5) is a very typical shape used in building construction for beams, columns, and other structural members.

The process of finding the moment of inertia of composite shapes is important since it will allow a designer to obtain information about nonstandard shapes (shapes other than the simple shapes listed in Table 7-1). The process for finding the moment of inertia of a composite shape is relatively straightforward and can easily be performed by using the step-by-step process outlined below. As was the case in the previous chapter with centroids, the process for finding moments of inertia works very nicely when using a tabular format. Therefore the author strongly recommends that all students use such a table. The process can be outlined as follows:

1. Break the composite shape into common geometric areas—rectangles, triangles, etc. Label each individual area as 1, 2, 3, etc.
2. Locate the axis for which the moment of inertia is desired if it is not located already. (Sometimes the centroidal axis for the shape may have to first be located before the process of calculating moment of inertia can begin). Review the process for finding centroids, if necessary.

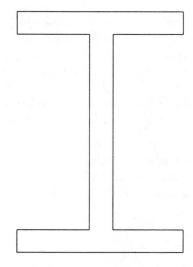

Figure 7-5 Typical Composite Shape Needing Moment of Inertia Calculation

3. Construct tables similar to the ones shown below.

Shape #	$I_{x\,centrd}$	A	d	Ad^2	$I_{x\,total}$

Shape #	$I_{y\,centrd}$	A	d	Ad^2	$I_{y\,total}$

4. For each individual area, fill in the table. Calculate the centroidal moment of inertia of that individual shape and determine the distance, d, between the centroidal axis of the individual shape and the axis for which moment of inertia is desired. The total moment of inertia for that shape will be $I + Ad^2$.

5. When all individual areas are complete, total the I_{total} column. This is the total moment of inertia for the composite shape.

Again, it must be stressed that the axis which moment of inertia is taken about must be located at the beginning of the process and must remain the same throughout the entire process. Failure to do this will result in erroneous results. The following examples will illustrate this process of finding centroids for composite shapes.

EXAMPLE 7.1

For the T-shape shown below, calculate the moment of inertia about the centroidal x and y axes, which are already located.

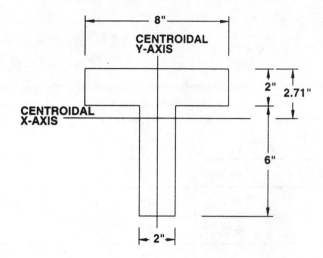

The location of the axes for the moment of inertia needed is already known, so begin by breaking down the composite shape into two simple rectangles.

Once this is accomplished we will begin to calculate the moment of inertia about the centroidal x axis. Sometimes it is convenient to locate the centroidal x axis on each individual shape such that the d distance can be readily observed.

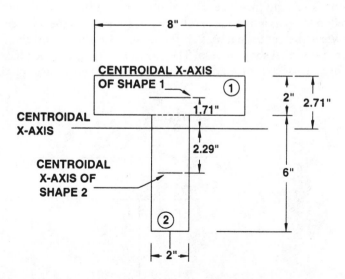

For shape #1:

$$I_{xc} = 1/12(8 \text{ in.})(2 \text{ in.})^3 = 5.33 \text{ in.}^4$$

$$d \text{ distance} = 2.71 \text{ in.} - 1 \text{ in.} = 1.71 \text{ in.}$$

For shape #2:

$$I_{xc} = 1/12(2 \text{ in.})(6 \text{ in.})^3 = 36 \text{ in.}^4$$

$$d \text{ distance} = 3 \text{ in.} - .71 \text{ in.} = 2.29 \text{ in.}$$

Once these values have been determined, the moment of inertia table can be filled out as follows. Realize that the parallel axis theorem is used for each shape and that the total moment of inertia for the entire shape is simply a sum of its individual pieces.

Shape #	Ix centrd	A	d	Ad²	Ix total
1	5.33 in.⁴	16 in.²	1.71 in.	46.79 in.⁴	52.12 in.⁴
2	36 in.⁴	12 in.²	2.29 in.	62.93 in.⁴	98.93 in.⁴
				TOTAL = 151.05 in.⁴	

Therefore, the moment of inertia about the centroidal x axis is 151.05 in.⁴.

The procedure for finding the moment of inertia about the composite shape's y axis is exactly the same. However, it should be stressed that we are only interested in the y axis and the parallel y axis for each shape. Notice that the d distance between the centroidal axis of the composite shape and the centroidal axes for each individual shape is zero. They lie in exactly the same place, therefore there is no need to account for additional moment of inertia through the Ad^2 term.

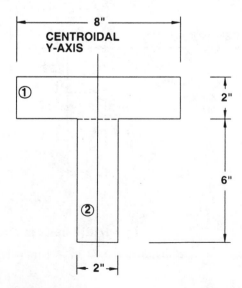

For shape #1:

$$I_{yc} = 1/12(2 \text{ in.})(8 \text{ in.})^3 = 85.33 \text{ in.}^4$$

d distance = 0

For shape #2:

$$I_{yc} = 1/12(6 \text{ in.})(2 \text{ in.})^3 = 4 \text{ in.}^4$$

d distance = 0

Once these values have been determined, the moment of inertia table can be filled out as follows. Realize that the parallel axis theorem is used for each shape and that the total moment of inertia for the entire shape is simply a sum of its individual pieces.

Shape #	Iy centrd	A	d	Ad^2	Iy total
1	85.33 in.^{4}	16 in.^{2}	0	0	85.33 in.^{4}
2	4 in.^{4}	12 in.^{2}	0	0	4 in.^{4}
				TOTAL	= 89.33 in.^{4}

The moment of inertia about the shape's centroidal y axis is 89.33 in.^{4}.

When calculating moments of inertia for composite shapes, a useful tool is the ability to account for "holes" in the shape. The method used is very similar to that which was illustrated with centroids in the previous chapter, namely, to treat the moment of inertia for the holes as a negative moment of inertia. The most important point to remember when using this technique is to only subtract the moment of inertia of the hole from an area where moment of inertia was already included.

The following example will demonstrate this method.

EXAMPLE 7.4

For the shape shown below, calculate the moment of inertia about the centroidal x and y axes which are already located.

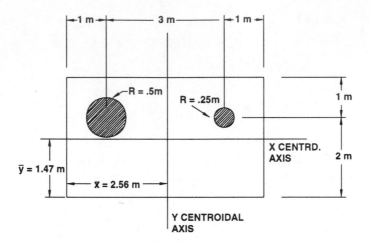

The location of the axes for the moment of inertia needed are known, so begin by breaking down the composite shape into one large rectangle with two smaller holes to be taken out of it.

Once this is accomplished we will begin to calculate the moment of inertia about the centroidal x axis. Sometimes it is convenient to locate the centroidal x axis on each individual shape such that the d distance can be readily observed.

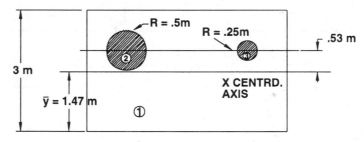

For shape #1 (rectangle):

$$I_{xc} = 1/12(b)(h)^3 \text{ (from Table 7-1)}$$

$$I_{xc} = 1/12(5 \text{ m})(3 \text{ m})^3 = 11.25 \text{ m}^4$$

d distance $= 1.5 \text{ m} - 1.47 \text{ m} = .03 \text{ m}$

For shape #2 (left-circle):

$$I_{xc} = (\pi/4)(r)^4 \text{ (from Table 7-1)}$$

$$I_{xc} = (\pi/4)(r)^4 = (\pi/4)(.5 \text{ m})^4 = -.05 \text{ m}^4 \text{ (remember negative)}$$

d distance $= 2 \text{ m} - 1.47 \text{ m} = .53 \text{ m}$

For shape #3 (right-circle):

$$I_{xc} = (\Pi/4)(r)^4 \quad \text{(from Table 7-1)}$$

$$I_{xc} = (\Pi/4)(r)^4 = (\Pi/4)(.25 \text{ m})^4 = -.003 \text{ m}^4 \text{ (consider this to be negligible)}$$

d distance $= 2 \text{ m} - 1.47 \text{ m} = .53 \text{ m}$

Once these values have been determined, the moment of inertia table can be filled out as follows. Realize that the parallel axis theorem is used for each shape and that the total moment of inertia for the entire shape is simply a sum of its individual pieces. Notice that both of the circular holes are registered as negative moment of inertias.

Shape #	$I_{x \text{ centrd}}$	A	d	Ad^2	$I_{x \text{ total}}$
1	11.25 m⁴	15 m²	.03 m	.01 m⁴	11.26 m⁴
2	−.05 m⁴	−.78 m²	.53 m	−.22 m⁴	−.27 m⁴
3	—	−.2 m²	.53 m	−.06 m⁴	−.06 m⁴
				TOTAL =	10.93 m⁴

Therefore, the moment of inertia about the centroidal x axis is 10.93 m⁴.

The procedure for finding the moment of inertia about the composite shape's y axis is exactly the same. However, it should be stressed that we are only interested in the y axis and the parallel y axis for each shape. Notice that the d distances between the centroidal y axis of the composite shape and the centroidal axes for each individual shape are all different. This is because each one of the individual shapes are different distances from the axis from which we are taking moment of inertia about. Also notice that the rectangle's centroidal moment of inertia changes and that the holes are included as negative values.

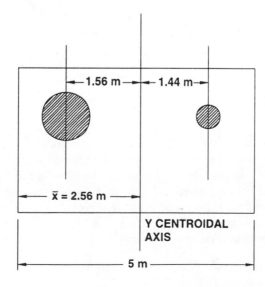

For shape #1 (rectangle):

$$I_{yc} = 1/12(b)(h)^3 \text{ (from Table 7-1)}$$

$$I_{yc} = 1/12(3 \text{ m})(5 \text{ m})^3 = 31.25 \text{ m}^4$$

d distance = 2.56 m – 2.5 m = .06 m

For shape #2 (left-circle):

$$I_{yc} = (\pi/4)(r)^4 \text{ (from Table 7-1)}$$

$$I_{yc} = (\pi/4)(.5 \text{ m})^4 = -.05 \text{ m}^4 \text{ (remember negative)}$$

d distance = 2.56 m – 1 m = 1.56 m

For shape #3 (right-circle):

$$I_{yc} = (\pi/4)(r)^4 \text{ (from Table 7-1)}$$

$$I_{yc} = (\pi/4)(.25 \text{ m})^4 = -.003 \text{ m}^4 \text{ (consider this to be negligible)}$$

d distance = 4 m – 2.56 m = 1.44 m

Once these values have been determined, the moment of inertia table can be filled out as follows. Realize that the parallel axis theorem is used for each shape and that the total moment of inertia for the entire shape is simply a sum of its individual pieces.

Shape #	$I_{y \text{ centrd}}$	A	d	Ad^2	$I_{y \text{ total}}$
1	31.25 m^4	15 m^2	.06 m	.05 m^4	31.30 m^4
2	–.05 m^4	–.78 m^2	1.56 m	–1.90 m^4	–1.95 m^4
3	—	–.2 m^2	1.44 m	–.42 m^4	–.42 m^4
				Total =	28.93 m^4

The moment of inertia about the shape's centroidal y axis is 28.93 m^4.

To complete this coverage on the methods utilized to calculate moment of inertias for various composite shapes, one must finally know how to incorporate standardized shapes into this technique. By standardized shapes, we are primarily referring to manufactured or fabricated shapes that are more complicated than simple geometric shapes but, that also typically have many of their properties (such as moment of inertia) known. If a standardized shape has its moment of inertia already known, the shape can "count" as one area when "breaking down" the composite shape. This greatly simplifies the process of finding moment of inertia when dealing with such materials as steel since many fabricated sections have properties that are known. The student should already be aware that many steel section properties are found in Appendix A of this text. The following example illustrates the utilization of this process when standardized shapes are used.

EXAMPLE 7.5

For the built-up shape shown below, calculate the moment of inertia about the centroidal x and y axes that are already located.

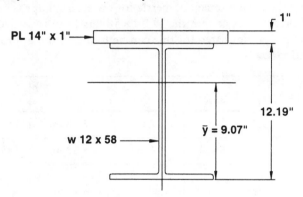

The location of the axes for the moment of inertia needed is already known, so begin by breaking down the composite shape into two shapes —the W 12 x 58 and the rectangular plate. Remember, if the centroidal axes were not already located it would be necessary to find them prior to beginning the process to find moment of inertia. Review of the technique to find the centroids of areas can be found in Chapter 6.

Calculate the moment of inertia about the centroidal x axis. Sometimes it is convenient to locate the centroidal x axis on each individual shape such that the d distance can be readily observed.

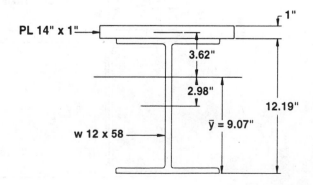

Area #1 – W 12×58:

$$I_{xc} = 475 \text{ in.}^4 \text{ (from the section table in Appendix A)}$$

d distance $= 9.07$ in. $- (12.19$ in. $/ 2) = 2.98$ in.

Area #2 – rectangular plate:

$$I_{xc} = 1/12(b)(h)^3 = 1/12(14 \text{ in.})(1 \text{ in.})^3 = 1.16 \text{ in.}^4$$

d distance $= 12.69$ in. $- 9.07$ in. $= 3.62$ in.

Once these values have been determined the moment of inertia table can be filled out as follows. Realize that the parallel axis theorem is used for each shape and that the total moment of inertia for the entire shape is simply a sum of its individual pieces. Notice that the W 12×58 has been recorded as one separate area and the properties listed for this can be found in the appropriate section table in Appendix A.

Shape #	$I_{x \text{ centrd}}$	A	d	Ad^2	$I_{x \text{ total}}$
1	475 in.^{4}	17 in.^{2}	2.98 in.	150.97 in.^{4}	625.97 in.^{4}
2	1.16 in.^{4}	14 in.^{2}	3.62 in.	183.46 in.^{4}	184.62 in.^{4}
					Total = 810.59 in.^{4}

The moment of inertia about the shape's x centroidal axis is 810.59 in.^{4}.

The procedure for finding the moment of inertia about the composite shape's y axis is exactly the same. Notice that the d distance between the centroidal axis of the composite shape and the centroidal axes for each individual shape is zero. They lie in exactly the same place, therefore there is no need to account for additional moment of inertia through the Ad^2 term.

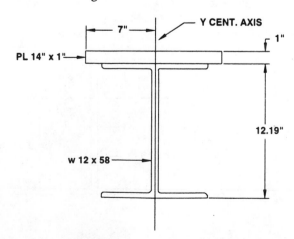

Area #1 W 12 × 58:

$$I_{yc} = 107 \text{ in.}^4 \text{ (from the section tables)}$$

d distance = 0 (centroidal axis of the W 12 section and of the entire shape coincide)

Area #2 rectangular plate:

$$I_{yc} = 1/12(b)(h)^3 = 1/12(1 \text{ in.})(14 \text{ in.})^3 = 228.7 \text{ in.}^4$$

d distance = 0

Once these values have been determined, the moment of inertia table can be filled out as follows. Realize that the Ad^2 term is zero, due to the fact that the centroidal y axes of both the individual pieces and the total composite shape lie in the same place.

Shape #	$I_{y\ centrd}$	A	d	Ad^2	$I_{y\ total}$
1	107 in.^{4}	17 in.^{2}	0	0	107 in.^{4}
2	228.7 in.^{4}	14 in.^{2}	0	0	228.7 in.^{4}
				Total = 335.7 in.4	

The moment of inertia about the shape's y centroidal axis is 335.7 in.^{4}. Notice that the moment of inertia about the shape's x axis is appreciably larger. This is why such a shape is normally bent about the x-axis.

7.5   The Polar Moment Of Inertia

The previous sections have discussed a shape's resistance to bending about an axis being dependent on a property referred to as moment of inertia. In many mechanical applications circular members are more frequently subjected to a twisting or torsional force. The shape property which reflects a shape's resistance to these torsional forces is referred to as the **polar moment of inertia**, J. This property is analogous to the moment of inertia except that twisting takes place about a member's longitudinal axis. Therefore, distance of an incremental piece of area from this point is referred to a radial distance, r. The polar moment of inertia can be mathematical defined as follows:

$$J = \Sigma r^2 \Delta A \qquad \text{(Eq. 7-3)}$$

where

J = polar moment of inertia

r = radial distance from the longitudinal axis to incremental area

ΔA = incremental unit of area

This concept is similar to the standard moment of inertia and can be illustrated as shown in the figure below (Figure 7-6).

By the Pythagorean theorem, the radial distance, r, could be represented by rectangular components as follows:

$$r^2 = x^2 + y^2$$

Therefore, Eq. 7-3 can be rewritten as follows:

$$J = \Sigma\, r^2\, \Delta A \qquad\qquad\text{(Eq. 7-3)}$$

or

$$J = (x^2 + y^2)\, \Delta A$$

From Equation 7-1 it is evident that $I_x = y^2 \Delta A$ and $I_y = x^2 \Delta A$, therefore, the polar moment of inertia is expressed as follows:

$$J = I_x + I_y \qquad\qquad\text{(Eq. 7-4)}$$

Since the polar moment of inertia can be represented in this fashion it will become obvious that circular shapes are ideal for torsional resistance since their moment of inertia about the x axis would be the same as the value about the y axis. The following example will illustrate this calculation of polar moment of inertia.

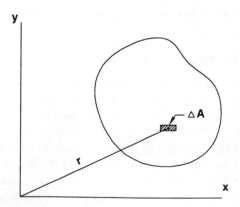

Figure 7-6   Polar Moment of Inertia

EXAMPLE 7.6

Calculate the polar moment of inertia for the two shapes shown below about the center of each. Both shapes have an area approximately equal to 8 in.^{2}.

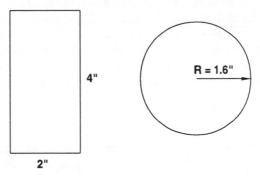

For the circular shape:

$$I_x = I_y = (\pi/4)(r)4 = (\pi/4)(1.6 \text{ in.})^4 = 5.15 \text{ in.}^4$$

$$J = I_x + I_y = 5.15 \text{ in.}^4 + 5.15 \text{ in.}^4 = 10.3 \text{ in.}^4$$

For the rectangular shape:

$$I_x = 1/12(b)(h)^3 = 1/12(2 \text{ in.})(4 \text{ in.})^3 = 10.66 \text{ in.}^4$$

$$I_y = 1/12(b)(h)^3 = 1/12(4 \text{ in.})(2 \text{ in.})^3 = 2.66 \text{ in.}^4$$

$$J = I_x + I_y = 10.66 \text{ in.}^4 + 2.66 \text{ in.}^4 = 13.33 \text{ in.}^4$$

Although both values are relatively close, the circular shape will gain a distinct advantage (relative to the amount of area) as tube sections are used.

7.6   Radius of Gyration

The radius of gyration, r, is a property of a shape's area, which is extensively used in column design. In a physical sense, the radius of gyration is a rather abstract value and is best defined as the relationship between a shape's moment of inertia (about a given axis) and its area. The relationship is as follows:

$$r = \sqrt{I/A} \qquad\qquad \text{(Eq. 7-5)}$$

Because the radius of gyration is a function of a shape's moment of inertia, its value is relative to the particular axis of reference on the shape. Therefore, it is typical that the radius of gyration will have different values about the x and y axis. The radius of gyration for standard steel sections can be found in the tables located in Appendix A.

7.7   Summary

Moment of inertia is a property of an area which reflects the resistance of that area to bending forces. This property is given with respect to a particular axis, which is usually the bending axis. An increase in the moment of inertia for a shape about a given axis will represent an increased resistance to bending forces.

The polar moment of inertia for a given shape represents that shape's resistance to torsion or twisting forces. This property is typically taken about the shape's center, since twisting will occur about this point. This property is very similar in nature to the bending moment of inertia and is used in many mechanical applications.

EXERCISES

1. Calculate the moment of inertias about the centroidal x and the centroidal y axes for the shapes shown below.

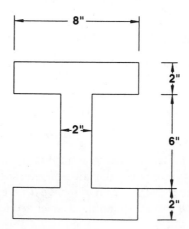

2. Calculate the moment of inertias about the centroidal x and the centroidal y axes for the shapes shown below.

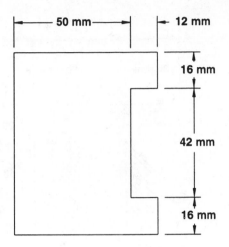

3. Calculate the moment of inertias about the centroidal x and the centroidal y axes for the shapes shown below.

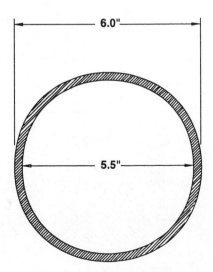

4. Calculate the moment of inertias about the centroidal x and the centroidal y axes for the shapes shown below.

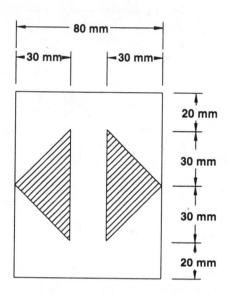

5. Calculate the moment of inertias about the centroidal x and the centroidal y axes for the shapes shown below.

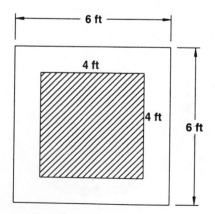

6. Calculate the moment of inertias about the centroidal x and the centroidal y axes for the shapes shown below.

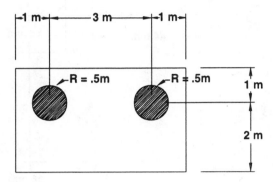

7. Calculate the moment of inertias about the centroidal x and the centroidal y axes for the shapes shown below.

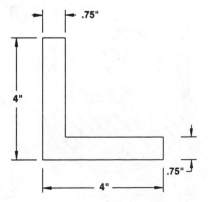

8. Calculate the moment of inertias about the centroidal x and the centroidal y axes for the shapes shown below.

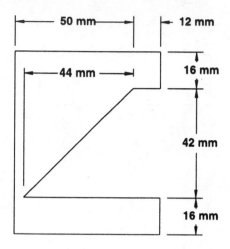

9. Calculate the moment of inertias about the centroidal x and the centroidal y axes for the shapes shown below.

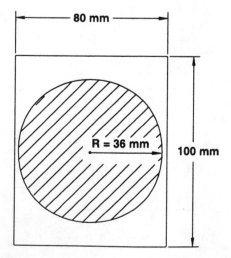

10. Calculate the moment of inertias about the centroidal x and the centroidal y axes for the shapes shown below.

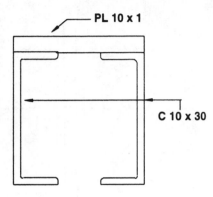

PL 10 x 1

C 10 x 30

11. Calculate the moment of inertias about the centroidal x and the centroidal y axes for the shapes shown below.

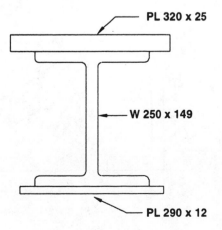

PL 320 x 25

W 250 x 149

PL 290 x 12

12. Calculate the moment of inertias about the centroidal x and the centroidal y axes for the shapes shown below.

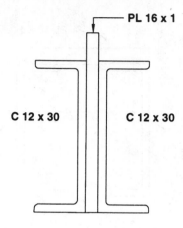

13. Calculate the moment of inertias about the centroidal x and the centroidal y axes for the shapes shown below.

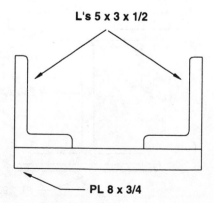

14. Calculate the moment of inertias about the centroidal x and the centroidal y axes for the shapes shown below.

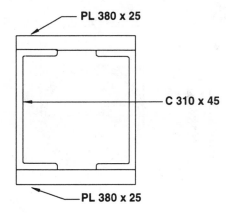

PL 380 x 25

C 310 x 45

PL 380 x 25

15. Calculate the moment of inertias about the centroidal x and the centroidal y axes for the shapes shown below.

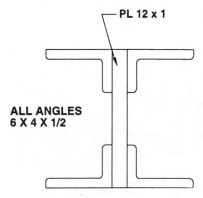

PL 12 x 1

ALL ANGLES
6 X 4 X 1/2

16. Calculate and compare the polar moment of inertia for each of the three shapes shown below. Is the area of each shape proportional to the calculated value of J? Draw a conclusion as to what has the largest impact on the value of polar moment of inertia; area or distribution of area?

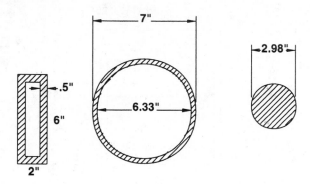

8

STRESS AND STRAIN

8.1 Introduction

The discussion in the previous chapters of this text discussed the fundamental concepts of statics. These concepts focused primarily on the affect of forces and the ensuing reactions on rigid bodies. The basic view of the body or the structure in statics is external. The forces applied to a body are analyzed and the forces necessary for equilibrium are determined. In this chapter we begin to explore the fundamental concepts of strength of materials.

Strength of materials is the study of the internal reaction of the body or the structure to the externally applied forces. This internal reaction of the body will include changes in the internal structure of the body. Such concepts of stress, strain, and deformation will be central to the laws governing the internal changes that a body will undergo. What were assumed to be rigid bodies in statics actually undergo such changes because they are elastic. This being the case, when bodies are acted on by external forces they may possibly stretch, shorten, expand or bend. The study of strength of materials will help explain such behaviors.

This study of a body's internal changes which occurs under the application of forces is the cornerstone for the design of members made from any material. To design a member properly, it is necessary to know more than just the amount of force applied to a member—it is also necessary to predict the behavior of the member under the application of the force. How much will the beam deflect? What temperature will cause the expansion joint to close? The study of strength of materials will give us the fundamental knowledge to find the answers to these and other questions.

8.2   Types of Direct Stress

The most basic concept of strength of materials is that of **stress**. Stress is the internal resistance of a body or a member to an applied force. In Figure 8-1 we have a uniform bar subjected to a tension force, P. To keep from breaking apart, the bar develops an internal resistance to the externally applied force. In this case, the internal resistance is referred to as tensile stress and can be viewed as the spreading out of the total force over the resisting area. The formula for calculating such stress is as follows:

$$\sigma = P / A \qquad\qquad\qquad (Eq.\ 8\text{-}1)$$

where

σ = stress, psi or Pa

P = externally applied force, lbs or N

A = resisting area, in.^{2} or m^2

There are many different types of stresses that can be developed in bodies. The types of stresses (sometimes referred to as direct stresses) that are governed by Equation 8-1 are the following:

- tensile stresses
- compressive stresses
- shear stresses

Tensile and compressive stresses are sometimes referred to as **axial stresses**, since these stresses act in the same direction as the applied load (which typically is along the longitudinal axis of the member). Conversely, shear stresses act in a manner parallel to the applied load. Other internal resistances, such as bending

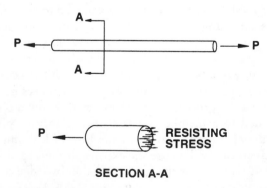

SECTION A-A

Figure 8-1   Stress Developed in a Member

stress and torsional stress, are somewhat more complex and will be the focus of separate chapters.

It is very important that the difference between stress and force be completely understood. The applied load on a member can be very large and the corresponding stress that the member develops can be very small. How can this be? If a tension member has 1,000,000 lbs of load applied to it, most would consider the load to be quite large. However, if the resisting area was 1,000,000 in.^{2} the corresponding stress would be 1 psi. This does not seem all that large. It is imperative to understand that stress is a function of the force applied *and* the resisting area of the member. A second point to dwell on is that stress is actually an internal reaction. A member cannot develop stress without first being subject to an applied load, since stress is the member's internal resistance.

The following examples will demonstrate the calculation of tensile and compressive stresses.

EXAMPLE 8.1

The bar shown below is subjected to a tensile force of 4000 lbs. Calculate the stress in the bar at section *A-A* and *B-B*.

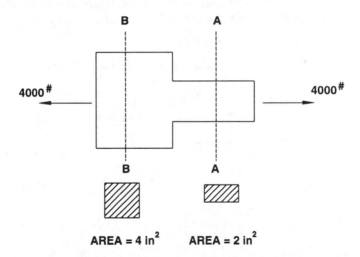

Externally the bar is in equilibrium, since 4000 lbs acts to the right and to the left. Consider a free-body diagram of the bar sectioned at *A-A* and then again at *B-B*.

As can be seen in the figure below, the 4000 lb force has to be resisted at area *A-A* by 2 in.^{2} of the bar's crosssection. From Equation 8-1, the tensile stress at this location can be calculated as:

$$\sigma = P / A \qquad \text{(Eq. 8-1)}$$

$$\sigma = 4000 \text{ lbs} / 2 \text{ in.}^2 = 2000 \text{ psi}$$

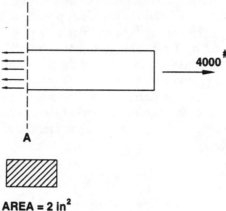

AREA = 2 in^2

Likewise, at section *B-B* the 4000 lb force has to be resisted at area *B-B* by 4 in.^{2} of the bar's cross-section, as shown the figure below. From Equation 8-1 the tensile stress at this location can be calculated as:

$$\sigma = P / A \quad \text{(Eq. 8-1)}$$

$$\sigma = 4000 \text{ lbs} / 4 \text{ in.}^2 = 1000 \text{ psi}$$

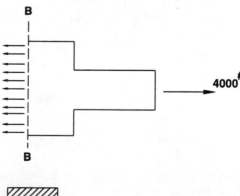

AREA = 4 in^2

Notice the stress developed at each section is actually the exact amount needed for equilibrium to be maintained.

EXAMPLE 8.2

A column measuring 250 mm × 250 mm is attached to a rigid concrete footing as shown. The footing is 1.5 m × 1.5 m. Calculate the stress in the column and the stress on the soil beneath the footing.

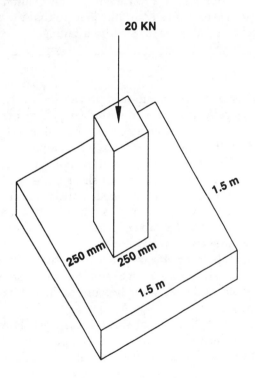

The column is acted upon by an applied compressive force of 20 KN, which causes a compressive stress in the structure shown. On the column, this stress can be calculated from Equation 8-1 as follows:

$$\sigma = P / A \qquad \text{(Eq. 8-1)}$$

with

$$A = .25 \text{ m} \times .25 \text{ m} = .0625 \text{ m}^2$$

$$\sigma = 20 \text{ KN} / .0625 \text{ m}^2 = 320 \text{ KN} / \text{m}^2 \text{ or } 320 \text{ KPa}$$

The soil is acted on by the same 20 KN force (neglecting selfweight of the members) over the 1.5 m × 1.5 m area. So the pressure on the soil would be calculated from Equation 8-1 as follows:

$$\sigma = P / A \qquad \text{(Eq. 8-1)}$$

with

$$A = 1.5 \text{ m} \times 1.5 \text{ m} = 2.25 \text{ m}^2$$

$$\sigma = 20 \text{ KN} / 2.25 \text{ m}^2 = 8.89 \text{ KN} / \text{m}^2 \text{ or } 8.89 \text{ KPa}$$

Notice the decrease in pressure from the column to the soil. The concept behind footings is to "spread out" the load so that the weaker material (soil) will have the ability to carry the load without failing.

Another type of stress mentioned previously was shear stress. The shearing of a member, typically, is associated with the ripping or tearing apart of the body in question. Shear stress is calculated from Equation 8-1 just like tensile and compressive stresses. However, shear stress is developed parallel to the direction of the applied load. The internal resistance of the body develops such that the body does not rip apart. A common example of shear stresses developing in response to a shear force is that of a common bolted connection (Figure 8-2). A bolt is used to connect two members together and to transmit the force from one member to the other member. The transmission of force from one member to the other can occur through shear stresses being developed over the area of the bolt.

Many times bolts or pins will have more than one area per bolt, resisting the applied shear stresses. A common example of this is shown in Figure 8-3. This situation is known as **double shear** because each bolt has two planes resisting the applied shear force.

Many other applications of shear stress exist. However, the student should realize that the shear stress is always acting parallel to the applied force. The following examples will illustrate the calculation of this type of stress.

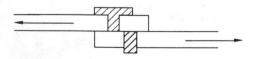

Figure 8-2   Shear Forces "Ripping" a Bolt in a Two-Plate Connection

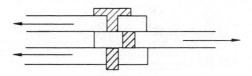

Figure 8-3   Shear Forces Acting on a Bolt in a Three-Plate Connection

EXAMPLE 8.3

The plates shown below are subjected to an applied load of 15 kips. The connection is made using four, 1/2 in. diameter bolts. Calculate the shear stress which must be developed in the bolts to keep the connection intact.

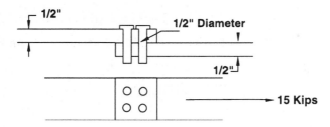

The shear force on the bolts is 15 kips and each bolt has one area of resistance. Therefore, the total area of the bolts resisting shear is:

$$A = (\pi /4)(0.5 \text{ in.})^2 \times 4 \text{ bolts} = .785 \text{ in.}^2$$

Therefore the shear stress on the bolts is calculated from Equation 8-1 as:

$$\sigma = P / A \qquad\qquad\qquad (\text{Eq. 8-1})$$

$$\sigma = 15 \text{ kips} / .785 \text{ in.}^2$$

$$\sigma = 19.1 \text{ ksi}$$

Realize that other stresses exist in this problem, such as the tension on the plates and possibly a bearing (compressive) stress of the bolts onto the plates.

EXAMPLE 8.4

The butt joint shown below is subjected to a tension force of 50,000 lbs. The bolts are 3/4 in. diameter. Calculate the shear stress developed by the bolts and the tension stress developed by the inside and outside plates.

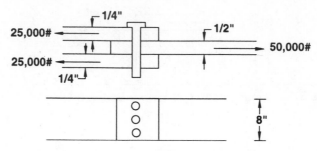

The first item of note is that the bolts are in a condition of double shear, which means that every bolt has two areas resisting the shear force. It must also be recognized that the 50,000 lb force is resisted equally by the top plate and the bottom plate. The total shear force on the bolts is 50,000 lbs and each bolt has two areas of resistance (each area resists 25,000 lbs). Therefore, the total area of the bolts resisting shear is:

$$A = (\pi/4)(.75 \text{ in.})^2 \times 3 \text{ bolts} \times 2 \text{ areas per bolt} = 2.65 \text{ in.}^2$$

Therefore, the shear stress on the bolts is calculated from Equation 8-1 as:

$$\sigma = P/A \qquad\qquad \text{(Eq. 8-1)}$$

$$\sigma = 50,000 \text{ lbs} / 2.65 \text{ in.}^2$$

$$\sigma = 18,863 \text{ psi}$$

The tension stress of the outside plates is again calculated from Equation 8-1 as follows:

$$A = 8 \text{ in.} \times 1/4 \text{ in.} = 2 \text{ in.}^2$$

$$P = 25,000 \text{ lbs (half of the 50,000 load goes to each outside plate)}$$

$$\sigma = P/A$$

$$\sigma = 25,000 \text{ lbs} / 2 \text{ in.}^2$$

$$\sigma = 12,500 \text{ psi}$$

The tension stress of the inside plate is again calculated from Equation 8-1 as follows:

$$A = 8 \text{ in.} \times 1/2 \text{ in.} = 4 \text{ in.}^2$$

$$P = 50,000 \text{ lbs}$$

$$\sigma = P/A \qquad\qquad \text{(Eq. 8-1)}$$

$$\sigma = 50,000 \text{ lbs} / 4 \text{ in.}^2$$

$$\sigma = 12,500 \text{ psi}$$

The stress on the inside and outside plates is equal, due to the fact that twice as much load is carried by twice as much area (for the inside plate).

EXAMPLE 8.5

A circular machine punch is 2 inches in diameter and punches a hole through a 1/4 in. thick plate. Determine the force needed to punch through the plate if the maximum shear stress developed by the plate is 10,000 psi.

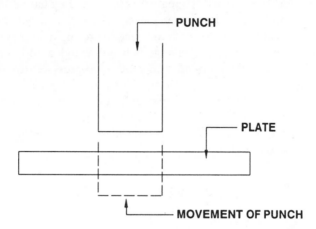

To create the hole, the force applied to the punch must make the shear stresses in the plate exceed 10,000 psi. The shear stress in the plate is developed over the surface area of the hole left after the punch goes through the plate. This area goes around the perimeter of the punch and extends over the depth of the plate. The area is:

$$A = (\pi)(d)(\text{plate thickness})$$

$$A = (\pi)(2 \text{ in.})(0.25 \text{ in.}) = 1.57 \text{ in.}^2$$

Therefore, the force needed can be found by rearranging Equation 8-1 as follows:

$$\sigma = P / A \qquad \qquad \text{(Eq. 8-1)}$$

or

$$P = \sigma \times A$$

$$P = 10,000 \text{ psi} \times 1.57 \text{ in.}^2 = 15,708 \text{ lbs}$$

Any force exceeding 15,708 lbs will be enough to make the punch work adequately.

8.3   Introduction to Strain

In the previous sections it was discussed how a member or body develops an internal resistance to applied forces. This internal resistance was referred to as stress. This development of stress will always be accompanied by another basic behavior—deformation.

Deformation (δ) is the change of length that a body will undergo as it is pulled on, pushed on, or subjected to bending. Deformation is caused anytime a force is applied to a body. The force causes the atomic planes[1] of the body to begin to slip past each other and this "slippage" results in the change in length, known as deformation. Deformation occurs even under the application of small loads, although it will be so small it is undetectable to the naked eye.

Deformation is the total change in length which a body undergoes. However, in many instances it is more advantageous to express deformation on a unit basis. For example, if one is concerned over how much deformation will cause fracture of a material, it would be more meaningful to express deformation of a body per unit in., unit mm, etc. Such an expression of deformation per unit length of the body is referred to as **strain**. The definition of strain is deformation per unit length and can be written as follows:

$$\varepsilon = \delta / L \qquad\qquad\qquad \text{(Eq. 8-2)}$$

where

ε = strain

δ = deformation or change in length, in., mm, etc.

L = original length

Strain is expressed in units of length per length, such as in./in. or mm/mm. Strain can best be explained as the deformation that occurs over every increment of length on the body, whereas deformation is the sum of all these incremental changes.

When axial loads (tension or compression) are applied, the strain is referred to as an axial strain. Axial strains will be the focus of the discussion in this chapter. The deformation caused by shear stress is referred to as shear strain and this will be discussed further in Chapter 10. The following example will illustrate this concept of axial strain.

EXAMPLE 8.6

A column that is 10 ft. in height has compressive loads applied and shortens by 1/10 of an inch. Calculate the axial strain in the column.

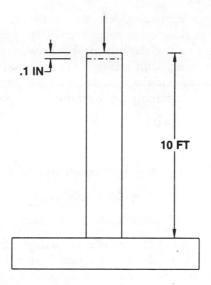

Strain, by definition, is deformation per unit length or as follows per Equation 8-2

$$\varepsilon = \delta / L \qquad \text{(Eq. 8-2)}$$

In this case

L = 10 ft or 120 in.

δ = 0.1 in.

Therefore, strain can be calculated as:

ε = 0.1 in. / 120 in. = .000833 in. / in.

This means that every inch of the column deforms (shortens) by .000833 in. (When 120 of the deformations per inch are summed up the total deformation of the column is 0.1 in.)

Note: Although somewhat less typical, the strain could also be calculated as:

ε = 0.1 in. / 10 ft. = .01 in. / ft.

8.4   Modulus of Elasticity

Previously the concepts of stress and strain were discussed in a separate manner. However, the relationship between stress and strain is one of the most distinguishable and important properties of a material. Although the ideas of stress and strain were not known at the time, Robert Hooke, in the seventeenth century, recognized that there was a specific and constant relationship between an applied load and the corresponding deformation in a material. This idea of a constant relationship has since come to be known as Hooke's law. In the early 1800's, Thomas Young defined this relationship between the stress placed on a body and the subsequent strain. This was found to be a constant material property which is now known as the **modulus of elasticity, E** (sometimes also referred to as Young's modulus). This relationship can be stated as follows:

$$E = \sigma / \varepsilon \qquad\qquad\qquad \text{(Eq. 8-3)}$$

where

E = modulus of elasticity, psi, ksi, KPa, etc.

σ = stress, psi, ksi, KPa, etc.

ε = strain, in. / in., mm / mm

The importance of the modulus of elasticity, E, is that it allows engineers the ability to predict the behavior of a given material under various values of stress. For most materials, the value of modulus of elasticity is a known and constant value. With this being the case, engineers and designers have the ability to determine what amount of strain (or deformation) will occur under a particular amount of stress (or load). This is an extraordinary tool, which lies at the heart of all types of design. Could you imagine trying to design with some material that had no information concerning its relationship between load and deformation? It would be impossible (or foolish) to do so.

Average values for modulus of elasticity and a list of common materials can be found in Table 8-1. A more complete table is listed in Appendix B. A typical value of E for steel is 29×10^6 psi, while for aluminum a typical value is 10×10^6 psi. What do these values really mean? In a comparison between these two values, one can see that the modulus for steel is 2.9 times larger than that for aluminum. Referring back to the definition of modulus, this means that to achieve 1 "unit" of strain in each material it would take a stress 2.9 times larger for the steel as com-

Table 8-1   Typical Material Values for Modulus of Elasticity, E

Material	PSI	MPa
Steel	29×10^6	200×10^3
Aluminum	10×10^6	70×10^3
Brass	15×10^6	100×10^3
Copper	17×10^6	105×10^3
*Timber	1×10^6	7×10^3
*Concrete	3×10^6	21×10^3

*Approximate Values
Note: 1 psi = 6.895 KPa

pared to the aluminum. Reality aside, if it took 1 psi of stress to make an aluminum member strain 1 in. / in., it would take 2.9 psi of stress to make an equivalent steel member to strain likewise. Modulus of elasticity defines the stiffness or rigidity that a particular material possesses. It is not necessarily a measure of strength, although many materials with high E values are also considered "strong".

The following examples will illustrate the use of this relationship between stress and strain.

EXAMPLE 8.7

A standard 6 in. × 12 in. cylinder (6 in. diameter and 12 in. height) of concrete has a modulus of elasticity of 3×10^6 psi. Calculate the axial load necessary that would cause a .001 in deformation (shortening) of the cylinder.

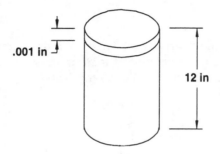

The following are known about the cylinder:

$$A = (\pi/4)(6 \text{ in.})^2 = 28.3 \text{ in.}^2$$

$$L = 12 \text{ in.}$$

$$\delta = .001 \text{ in.}$$

$$E = 3 \times 10^6 \text{ psi} \quad \text{(from Table 8-1)}$$

Since the deformation and original length are given, the strain associated with the required condition can be calculated from Equation 8-2 as:

$$\varepsilon = \delta / L \qquad\qquad \text{(Eq. 8-2)}$$

$$\varepsilon = .001 \text{ in.} / 12 \text{ in.} = .000083 \text{ in.} / \text{in.}$$

Rearranging Equation 8-3, we could solve for the stress required to cause this strain in the concrete cylinder as:

$$E = \sigma / \varepsilon \qquad\qquad \text{(Eq. 8-3)}$$

or

$$\sigma = E(\varepsilon)$$

$$\sigma = 3 \times 10^6 \text{ psi } (.000083 \text{ in. / in.}) = 250 \text{ psi}$$

From rearranging the stress equation, Equation 8-1, we can find the load, P, as follows:

$$\sigma = P / A \qquad\qquad\qquad\qquad\qquad \text{(Eq. 8-1)}$$

$$P = \sigma \times A$$

Therefore, the load to cause a .001 in. deformation is:

$$P = 250 \text{ psi}(28.3 \text{ in.}^2) = 7075 \text{ lbs}$$

A similar design-type problem is to find the amount of deformation that will occur under a given amount of load or stress. By manipulating the modulus of elasticity equation we can find the following:

$$\varepsilon = \delta / L$$

$$\sigma = P / A$$

$$E = (P / A) / (\delta / L)$$

Rearranging this, deformation can be expressed as follows:

$$\delta = (P / A) / (E / L)$$

or

$$\delta = PL / AE \qquad\qquad\qquad\qquad\qquad \text{(Eq. 8-4)}$$

This deformation equation is a widely used formula in many design oriented applications. The following example will illustrate such a use.

EXAMPLE 8.8

A balcony weighs 100 KN and is supported by 4 circular steel cables with an equal amount of load being carried by each. Determine the minimum diameter of the cables, such that the deformation does not exceed 10 mm. The cables are 5 m long and E_{steel} = 207,000 MPa.

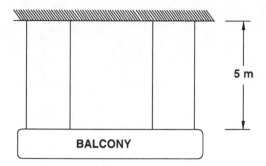

Since the problem asks to determine minimum diameter, we must realize that this is a component of area. We are given the following:

$$L = 5 \text{ m}$$

$$\delta = 10 \text{ mm} = .01 \text{ m}$$

$$P = 25 \text{ KN} = .025 \text{ MN per cable}$$

$$E = 207000 \text{ MPa} = 207000 \text{ MN / m}^2$$

Equation 8-4 is written to determine deformation, however, by rearranging we can solve for area as follows:

$$\delta = PL / AE \qquad\qquad\qquad\qquad\text{(Eq. 8-4)}$$

$$A = PL / \delta E$$

$$A = (.025 \text{ MN})(5 \text{ m}) / (.01 \text{ m})(207000 \text{ MN / m}^2)$$

$$A = 60.4 \times 10^{-6} \text{ m}^2 \text{ or } 60.4 \text{ mm}^2$$

Solving for the diameter we find:

$$60.4 \text{ mm}^2 = \pi/4 (d)^2$$

$$d = 8.77 \text{ mm}$$

Therefore, the cables should have a minimum diameter of 8.77 mm.

EXAMPLE 8.9

Calculate the elongation of a 1/2 inch diameter steel cable which supports an elevator weighing 4000 pounds. The length of the cable is 300 feet.

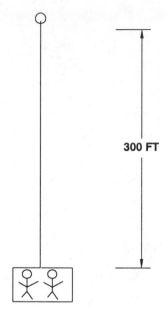

Elongation (deformation) of the cable is a part of axial strain, which is a function of stress on the cable and the modulus of elasticity, E.

The following are given in this problem:

$P = 4000$ lbs.

$A = (\pi/4)(d)^2 = .196$ in.^{2}

$L = 300$ ft.

$E = 29 \times 10^6$ psi

Utilizing the known information from Equation 8-4 we can find the elongation as follows:

$$\delta = PL \: / \: AE \qquad\qquad \text{(Eq. 8-4)}$$

$$\delta = PL \: / \: AE = (4000 \text{ lb})(300 \text{ ft.}) \: / \: (.196 \text{ in.}^2)(29 \times 10^6 \text{ psi})$$

$$= .21 \text{ ft. or } 2.53 \text{ in.}$$

8.5   The Tension Test, Stress-Strain Diagrams

Previously, the importance of the stress-strain relationship in a material has been discussed. The relationship between the stress and strain in a given material is truly the cornerstone for design, since this is really an attempt to select the "right" members for adequate behavior. If the right members are selected, the behavior of the system will be satisfactory to the owner.

It was assumed previously that the relationship between stress and strain was constant and this constant was referred to as the modulus of elasticity, *E*. However, the constant relationship between stress and strain does not occur in all materials and does not remain constant indefinitely. Because of this fact, all materials need to be tested to determine the full range of their stress-strain behavior. The most common method of testing a material's stress-strain relationship is the static tensile test.

In the static tensile test, or tension test, a specimen is placed into a specialized testing apparatus (Figure 8-4). The test specimen may be of circular, square, or rectangular cross-section; although with metal specimens, the circular cross-section is most common. Marks or indentations will be made on the specimen at a known distance. The distance between the marks is referred to as the gage length. A device called an extensometer is then clipped onto the specimen at these marks. The extensometer will measure the deformation that occurs as the specimen is pulled. The specimen is pulled at a slow rate, with the load and deformation being recorded at intervals or continually on a chart recorder. The measurements of load can easily be converted to stress by dividing the recorded load by the original cross-sectional area of the specimen. The measurements of deformation can also be converted into strain by knowing the original length (which many times is the gauge length). A stress-strain diagram for a low-carbon steel (steels having a carbon content less than .25%) would be similar to the one shown in Figure 8-5.

The low-carbon steel is chosen for descriptive purposes because it illustrates the different areas of the stress-strain diagram, which we will want to discuss. Other metals and materials may have very different shapes and values from their own stress-strain diagram. This points out the purpose for knowing this behavior on each type material. The following points on a stress-strain diagram are important and should be understood.

Elastic Range

This is the first portion of the diagram extending from the origin to the point that is referred to as the elastic limit. There are two major items of interest in this range. First, a material which undergoes stresses within this range behaves in an elastic manner or "like a rubber band". If stresses within this range are applied to a material, the material exhibits a change of length. But, if the stresses are removed, the material will return to its original length. Secondly, on many materials the elastic range is practically linear, meaning that there is an almost constant relationship between stress and strain. Previously we have discussed the modulus of elasticity, *E*, as being the relationship between stress and strain. With stress usually recorded on the y-axis and strain on the x-axis, the slope (m) of this straight line can be written as follows:

$$m = \Delta y / \Delta x$$

(Eq. 8-5)

Figure 8-4 Universal Testing Machine and Extensometer used in Tension Test
(Courtesy of Tinius Olsen Testing Machine Co., Inc., Willow Grove, PA.)

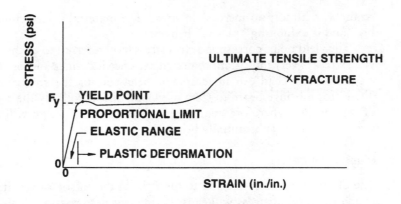

Figure 8-5   Ideal Stress-Strain Curve for Mild-Carbon Steel

where

$$m = \text{slope of the line}$$

$$\Delta y = \text{change in the } y \text{ values}$$

$$\Delta x = \text{change in the x values}$$

Referring back to Equation 8-3, we see that:

$$E = \sigma / \varepsilon \qquad\qquad\qquad \text{(Eq. 8-3)}$$

Therefore, the change in the stress units (y values) divided by the change in the strain units (x values) will yield the modulus of elasticity. The slope of this straight-line portion of the elastic range is the graphical representation of a material's modulus of elasticity, E.

Proportional Limit

The proportional limit is the point on the stress-strain diagram that is truly the end of the constant relationship between stress and strain. This is the last point used to determine the usual value of a material's modulus of elasticity, E. After this point, the straight-line portion of the graph ends and the diagram begins to curve. This point is usually slightly below a material's elastic limit.

Yield Point

The yield point is the point on the graph where the strain begins to drastically increase with little or no increase in stress. This point marks the location where the graph begins to turn almost horizontal, which signifies a huge increase in strain with almost no increase in stress. The material is stretching "uncontrollably" and is exhibiting "plastic" behavior.

The yield point corresponds to a stress level referred to as the yield stress. The yield stress is very important to designers, since it defines a level of stress that corresponds to a "bad" or unacceptable behavior for the structure. (Would it be "bad" for a bridge beam to start stretching uncontrollably the moment your car drove onto it?) Therefore, many methods of member design will involve limiting stresses to values substantially below the yield stress.

Plastic Range

The plastic range is the region to the right of the elastic range. It is the region in which the level of stress will cause permanent deformation. This means that the specimen will no longer return to its original length. Although a structural member having significant amounts of permanent deformation would not be extremely useful, materials with large plastic ranges are advantageous because they are considered to be ductile. A ductile material is one that deforms or stretches a significant amount before actually breaking. This behavior is favorable in a structural member because it would allow the structure additional time before ultimately collapsing.

Ultimate Tensile Strength

The ultimate tensile strength is the highest point on the curve corresponding to the maximum stress level that the material can carry. Soon after this stress level is reached, many materials exhibit fracture or breaking. Therefore, the ultimate tensile strength is considered to be the maximum strength level for a particular material.

The aforementioned items are considered to be the most important parts of the stress-strain diagram, although it should be noted that such diagrams will vary substantially based on the material being tested. For instance, brittle materials do not exhibit well defined yield points or large plastic deformations, as compared to their more ductile counterparts (Figure 8-6). Yield points in these cases are determined typically by one of three standardized methods per the American Society of Testing Materials (ASTM). These methods are referred to as the 0.1% offset method, the 0.2% offset method, and the 0.5% extension under load method. When using the offset methods, the yield point could be found by offsetting a line parallel to the straight line portion of the stress-strain diagram at a strain value of 0.001 or 0.002. The point where the offset line intersects the stress-strain diagram is then determined to be the yield point. Similarly, the extension under load method will find the yield point by intersecting a vertical line with the stress-strain diagram at a strain value of 0.005. This point is then considered to be the yield point (Figure 8-7).

The following example will illustrate the construction and location of values in a typical tension test.

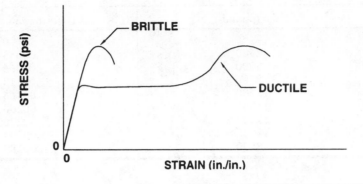

Figure 8-6   Difference in Stree-Strain Curves for a Brittle vs. Ductile Material

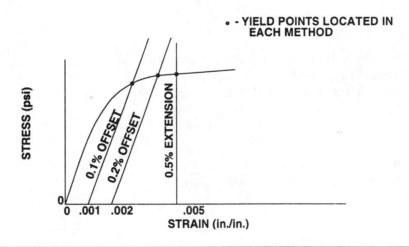

Figure 8-7   Yield Points Located by the Three ASTM Methods

EXAMPLE 8.10

The following data is tabulated from a tension test of a material. Calculate corresponding stress and strain values if the area of the specimen is 0.2 in.^{2} and the original gage length is 2 inches. Plot the values, calculate the modulus of elasticity, and determine the yield point visually.

Load (pounds)	Deformation (inches)
0	0
2500	.00033
5000	.00066
8500	.00113
12500	.00166
15000	.00200
17500	.00220
18000	.00244
18500	.00280

The load versus deformation data will need to be converted into stress-strain data. Load is converted into stress by applying Equation 8-1 and using the original area of the specimen as 0.2 in.^{2}. Therefore this is as follows:

$$\sigma = P / A \qquad \text{(Eq. 8-1)}$$

$$\sigma = P / 0.2 \text{ in.}^2$$

The deformation is converted into strain by recognizing that the original length being measured on the specimen (gage length) is 2 inches. Therefore applying Equation 8-2 using the original length as 2 inches, strain is as follows:

$$\varepsilon = \delta / L \qquad \text{(Eq. 8-2)}$$

$$\varepsilon = \delta / 2 \text{ in.}$$

The stress-strain data would be as follows:

Stress (psi)	Strain (inches/inch)
0	0
12500	.000166
25000	.000333
42500	.000565
62500	.000833
75000	.000100
87500	.00110
90000	.00122
92500	.00140

Plotting this data (with stress on the ordinate axis and strain on the abscissa), we would see the following:

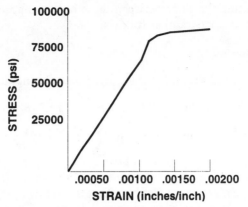

The modulus of elasticity, *E*, was defined as the slope of the initial straight-line portion of the stress-strain diagram. This was defined previously as:

$$m = \Delta y / \Delta x \qquad \text{(Eq. 8-5)}$$

or

$$E = \sigma / \varepsilon \qquad \text{(Eq. 8-3)}$$

Choosing any two points along the *straight line* region is necessary to determine the slope; for our purpose we will choose the points as shown below:

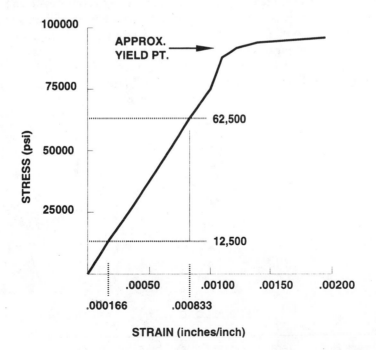

Calculating the slope between these points is accomplished by the deviation of y (stress) and x (strain) as follows:

$$m = \Delta y\ /\ \Delta x \qquad\qquad \text{(Eq. 8-5)}$$

$$m = (62500\text{psi} - 12500\text{psi})\ /\ (.000833\ \text{in./in.} - .000166\ \text{in./in.})$$

$$m = 50000\ \text{psi}\ /\ .00066\ \text{in./in.} = 75 \times 10^6\ \text{psi}$$

Therefore, the modulus of elasticity, E, is 75×10^6 psi.

The yield point would be approximately where the curve begins to turn "horizontal", indicating a large amount to strain. From the figure, this hypothetical material appears to reach yield at a stress level of 87,500 psi. If the diagram was less abrupt in this region (more rounded or curved), another method of determination, such as the offset methods, might be employed.

8.6   Poisson's Ratio

Previously we discussed the concept of axial strain that occurred when a member was loaded (and therefore stressed) axially. The strain that was produced manifested itself by the deformation of the member. This deformation was either an elongation (due to tension) or a shortening (due to compression). Since the volume of a member remains virtually constant, the member will also strain in a lateral direction in conjunction with the primary strain. This means that as a member is pulled on, it elongates in the direction of pull, but also "thins down" in the lateral direction as well. Conversely, when a member is compressed, it will shorten in the direction of load but will also "bulge out" laterally (Figure 8-8). This effect can

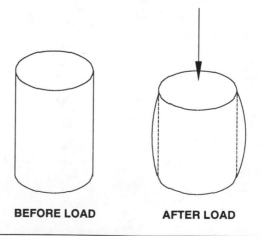

BEFORE LOAD   **AFTER LOAD**

Figure 8-8   Lateral Straining Caused by Axial Stress

Table 8-2

Material	μ
Steel	0.25
Copper	0.35
Aluminum	0.33
Cast Iron	0.26
Concrete	0.15-0.25
Brass	0.35

be very important when trying to define the exact stress or the true behavior in a material.

The magnitude of strain in the lateral direction will be smaller than the primary axial strain and is typically expressed as a decimal value. This decimal value is referred to as **Poisson's ratio (μ)** and can be defined as follows:

$$\mu = \text{lateral strain} \, / \, \text{primary (axial) strain}$$

or

$$\mu = \varepsilon_{\text{lateral}} \, / \, \varepsilon_{\text{primary}} \qquad \text{(Eq. 8-6)}$$

This ratio of lateral strain to primary strain is a constant material property. A list of the typical values for Poisson's ratios are listed in Table 8-2.

The following example will illustrate the use of Poisson's ratio.

EXAMPLE 8.11

A steel bar that is 25 mm in diameter is originally 500 mm in length. The bar is pulled on with a tensile force of 20 KN. Determine the new diameter of the bar if $\mu = 0.25$ and $E = 200 \times 10^6$ KPa.

The problem is to determine the new diameter, which will be smaller than the original diameter of 25 mm (the bar will decrease in diameter as it is stretched). The decrease in diameter is a deformation (δ) in the lateral direction of the bar. Therefore, to find the strain in the lateral direction ($\varepsilon_{\text{lateral}}$), we must first calculate the primary strain ($\varepsilon_{\text{primary}}$). This can be accomplished through rearranging Equation 8-4 as follows:

$$\delta = PL \, / \, AE \qquad \text{(Eq. 8-4)}$$

Rearranging:

$$\varepsilon = \delta / L = (P / A) / E$$

$$\varepsilon_{primary} = (20 \text{ KN} / .00049 \text{ m}^2) / 200 \times 10^6 \text{ KPa}$$

$$\varepsilon_{primary} = .000204$$

Since the lateral strain ($\varepsilon_{lateral}$) is a function of Poisson's ratio, we can rearrange Equation 8-6 as follows:

$$\mu = \varepsilon_{lateral} / \varepsilon_{primary} \qquad \text{(Eq. 8-6)}$$

or

$$\varepsilon_{lateral} = \mu (\varepsilon_{primary})$$

$$\varepsilon_{lateral} = 0.25 (.000204) = .000051$$

Strain is deformation divided by original length, therefore the deformation in the lateral direction can be calculated as follows:

$$\varepsilon_{lateral} = \delta / L$$

$$\delta = \varepsilon_{lateral} (L)$$

$$\delta = .000051(25 \text{ mm}) = .00127 \text{ mm}$$

This means that the 25 mm diameter shrinks by .00127 mm, or that the new diameter is:

$$d = 25 \text{ mm} - .00127 \text{ mm} = 24.998 \text{ mm}$$

8.7   Summary

Strength of materials is a subject that considers the internal changes that a body undergoes as it is acting on by external forces. The most fundamental internal change is the concept of stress. Stress (σ) is the internal reaction of a body that keeps the body in equilibrium and therefore, from falling apart. There are many types of stress; such as tensile, compressive, shear, and bending. As a body develops stresses to react against external loads, it also undergoes strain. Strain (ε) is the deformation (stretching, etc.) per unit length along the body. The relationship in a material between stress and strain is typically constant over some range of stress. The constant relationship is referred to as the modulus of elasticity, E. This material constant allows designers to predict a behavior in a material as it is subjected to various amounts of stress.

EXERCISES

1. A 1 in. × 1 in. square bar is pulled on with 5000 lbs of tensile force. Calculate the tensile stress in the bar.
2. A circular bar is compressed axially under a load of 100 KN and is 75 mm in diameter. Calculate the stress in the bar.
3. The connection shown below is pulled on with 80 kips of force. Calculate the shear stresses that will need to be developed in the ¾ in. diameter bolts.

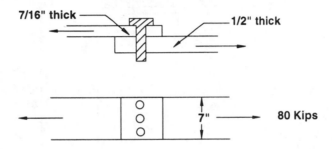

4. In the previous problem, calculate the tensile stress that exists in both plates.
5. A circular tube with an outside diameter of 50 mm and an inside diameter of 30 mm is compressed under a load of 100 KN. Calculate the compression stress in the tube.
6. Calculate the maximum load that a standard concrete cylinder (6 in. × 12 in.) would support if the concrete could withstand a stress up to 3800 psi.
7. Calculate the deformation of the circular tube in exercise 8.5 if the tube is made of aluminum and has an original length of 600 mm. $E = 70 \times 10^3$ MPa
8. Calculate the stress necessary to make a 1/2 in. diameter steel cable stretch a total of 0.25 in. if the cable is originally 100 ft. long.
9. It has been determined that a specimen of concrete with a cross-sectional area of 1.5 in.^{2} will break when reaching a strain of .003 in. / in. If the modulus of elasticity is 3.5×10^6 psi for the specimen, calculate the load needed to cause failure.

10. A column which has a cross-sectional area of 400 mm^2 and an original length of 2 m shortens by 15 mm when loaded with 30 KN. Calculate the material's modulus of elasticity.

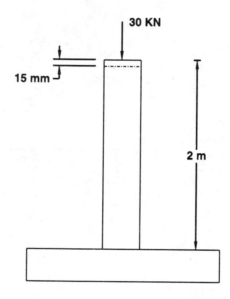

11. A circular cable that is 50 ft. long is made from steel and has to hold a total of 40000 lbs. The cable can elongate no more than 0.34 inches. Determine the diameter of the cable to meet this requirement.

12. A 4 × 4 timber post ($A = 12.25$ in.^{2}) can only be subjected to a maximum stress of 600 psi and deform no more than 0.12 inches. Determine the maximum load that the post can carry if its original length is 8 ft. $E = 1.5 \times 10^6$ psi

13. The bracket shown below is supported by pins at each support and both members are two-force members. Calculate the stress in both members if the cross-sectional area in both is 3 in.^{2}. Also determine the change of length in each member if $E = 30 \times 10^6$ psi.

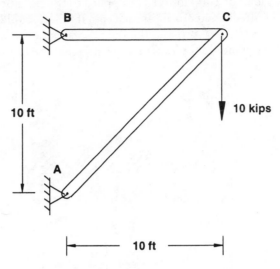

14. A simple truss is loaded as shown below. Determine the cross-sectional area required in each member if the maximum tension stress allowed is 1000 psi and the maximum compression stress allowed is 3000 psi.

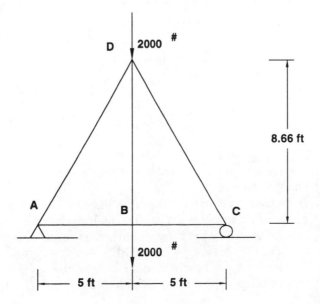

15. A circular punch measuring 35 mm in diameter is to punch through a 15 mm thick plate. If the maximum shear stress developed in the plate is 13800 KPa, determine the force needed to drive the punch through the plate.

16. A load of 1000 lbs is to be applied to the rigid beam that is supported by a 1 in. diameter cable at each end. If the left cable is made from aluminum ($E = 10 \times 10^6$ psi) and the right cable made from steel ($E = 30 \times 10^6$ psi), where should the load be applied so that the beam will remain horizontal?

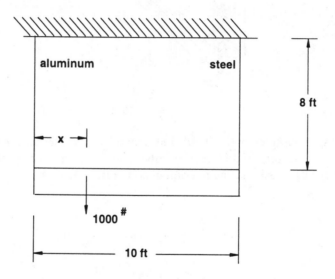

17. Repeat the previous problem if the length of the aluminum rod was changed to 6 ft.

18. A circular steel bar is 50 mm in diameter and 5 m long. A force P is applied that stretches the bar 75 mm. If the bar was made from aluminum instead of steel, what force would make the bar stretch the same amount? $E_s = 210,000$ MPa, $E_a = 70,000$ MPa

19. In exercise 8.18, how would the answer change if the bar was switched from steel to copper? $E_c = 105,000$ MPa

20. A tension test is performed on a 1/2 in. diameter specimen with the results listed below. If the gauge length was 2 in., plot the stress-strain diagram, calculate the modulus of elasticity, and determine the yield stress using the 0.2% offset method.

Load (lbs)	Deformation (inches)
0	0
1500	.0003
3000	.0006
4500	.0009
6000	.0013
7500	.0017
9000	.0022
10500	.0027
11000	.0035
12000	.0045
13000	.0059
13600	.0076

21. A tension test is performed on a 1/2 in. diameter specimen with the results listed below. If the gauge length was 2 in., plot the stress-strain diagram, calculate the modulus of elasticity, and determine the yield stress using the 0.1% offset method.

Load (lbs)	Deformation (inches)
0	0
1000	.00033
3000	.00068
5000	.00104

(continued on next pg.)

(continued from previous page)

Load (lbs)	Deformation (inches)
7000	.00143
9000	.00181
11000	.00223
13000	.00271
14000	.00324

22. A circular steel bar that is 800 mm long is compressed under a load of 25 KN. If the bar has an original diameter of 40 mm, calculate the increase in diameter due to the load. $E = 200 \times 10^6\ KPa$, $\mu = 0.25$.

23. A 2 in. × 2 in. square aluminum member is pulled on with a 10 kip tensile force. If the member is originally 18 in. in length, calculate the new side dimensions after being pulled. $E = 10 \times 10^6$ psi, $\mu = 0.33$

24. A 16 in. long circular bar is pulled on and deforms by a total of .07 inches. The bar's original 1 in. diameter measures 0.9993 in. after being pulled. Determine the Poisson's ratio for this bar.

REFERENCE

1. Van Vlack, Lawrence H., *Elements of Material Science and Engineering*, Addison-Wesley Publishing, 4th edition, 1980, p.193.

CHAPTER

9

FURTHER APPLICATIONS OF STRESS AND STRAIN

9.1   Introduction

In the previous chapter the concepts of stress and strain were introduced. The relationships between stress and strain were discussed on a fundamental level and the modulus of elasticity was found to be the material constant relating these two concepts.

This chapter further explores stress and strain as these concepts relate to thermal changes and also to indeterminate problems. The knowledge of stress-strain relationships used in conjunction with the laws of equilibrium is very useful in solving statically indeterminate problems.

9.2   Thermal Deformation and Stresses

As the temperature rises and falls, materials used in our structures and machines will change in size. Most materials will expand when the temperature rises and contract when the temperature falls. This potential effect of expansion and contraction must be accounted for in the design of practically all types of building and structures. This can be accomplished through the judicious use of expansion or contraction joints, which can be seen in many different applications (Figure 9-1).

The expansion and contraction of materials due to changes in temperature is considered thermal deformation. Thermal deformation, if considered as a unit deformation over the member's original length (L), may be expressed as a thermal

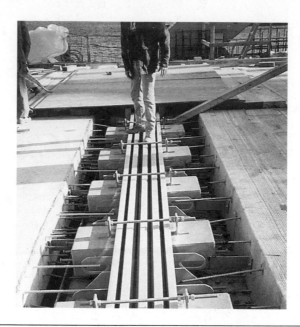

Figure 9-1 Bridge Expansion Joint (Courtesy D.S. Brown Company)

strain. This may be generalized by the strain equation of Equation 8-2 and repeated as follows:

$$\varepsilon = \delta\,/\,L \qquad\qquad \text{(Eq. 8-2)}$$

Thermal strain is not developed due to an axial stress, but rather a temperature change referred to as ΔT. The greater the temperature change that occurs, the greater the strain will be in a given material. However, temperature change alone is not the only influence on thermal strain. Just as important is the type of material involved. Every material is inclined to expand and contract at its own individual rate. This rate of expansion or contraction is a material constant referred to as the **coefficient of thermal expansion, α.** Typical values of this coefficient for various materials are listed in Table 9-1 and also in Appendix B. The higher the coefficient reflects a greater sensitivity to temperature change, or that the material expands or contracts at a higher rate.

Since thermal strain is dependent on the magnitude of temperature change (ΔT) and the material's coefficient of thermal expansion (α), it can be expressed as follows:

$$\varepsilon_{\text{thermal}} = \alpha\,\Delta T \qquad\qquad \text{(Eq. 9-1)}$$

Table 9-1 Typical Coefficients of Thermal Expansion for Standard Materials

Material	Coefficient of Expansion	
	in/in/°F	mm/mm/°C
Steel	6.5×10^{-6}	11.7×10^{-6}
Concrete	5.5×10^{-6}	9.9×10^{-6}
Cast Iron	6.4×10^{-6}	11.5×10^{-6}
Copper	9.0×10^{-6}	16.2×10^{-6}
Aluminum	12.7×10^{-6}	23×10^{-6}

where

$$\varepsilon_{thermal} = \text{thermal strain (in./in., mm/mm, etc.)}$$

$$\alpha = \text{coefficient of thermal expansion}$$

$$\Delta T = \text{change in temperature}$$

Rearranging Equation 8-2 to solve directly for deformation can be shown as follows:

$$\varepsilon = \delta / L \qquad\qquad\qquad (Eq. 8\text{-}2)$$

$$\delta = \varepsilon L$$

Sustituting Equation 9-1 into the previous equation, we find:

$$\delta_{thermal} = \alpha \Delta T L \qquad\qquad\qquad (Eq. 9\text{-}2)$$

where

$$\delta_{thermal} = \text{deformation caused by temperature change (thermal deformation)}$$

The use of the thermal strain and deformation formulas (Equation 9-1 and 9-2) is very useful in many different applications. The following example will illustrate these aforementioned concepts.

EXAMPLE 9.1

A surveyor's steel tape is exactly 100.00 ft. long when the temperature is 68°F. If the temperature is 95°F on a hot summer's day, what is the actual length of the tape?

This problem is a very common application of the concept of thermal deformation. Using the information as stated, we can list the known information as follows:

Length of tape = 100.00 ft.

Steel's α = .0000065 in./in./ °F   (from Table 9-1)

Change of temperature, ΔT = 27°F (rise)

Using Equation 9-2, we can find the thermal deformation of the steel tape as:

$$\delta_{thermal} = \alpha\,\Delta T\,L \qquad\qquad\qquad\qquad (Eq.9\text{-}2)$$

$$\delta_{thermal} = (.0000065 \text{ in./in./ °F})(27°F)(100.00 \text{ ft.}) = .00176 \text{ ft.}$$

Therefore, the steel tape on a 95° day would actually be 100.00176 ft. long. This would be very important to a surveyor measuring a significant distance because it could introduce an error of several inches.

EXAMPLE 9.2

If a slab of concrete is 60 ft. long, calculate the temperature rise needed to have a deformation of 0.25 in. Assume α_c = .0000055 in./in./ °F.

A temperature change will lead to thermal deformation and therefore strain. Equation 9-2 will calculate thermal deformation as follows:

$$\delta_{thermal} = \alpha\,\Delta T\,L \qquad\qquad\qquad\qquad (Eq.9\text{-}2)$$

Rearranging to solve for change in temperature:

$$\Delta T = \delta\,/\,\alpha\,L$$

Knowing the amount of deformation required in the problem, we can solve for the temperature as follows:

$$\Delta T = (0.25 \text{ in.})\,/\,(.0000055 \text{ in./in./ °F})(\,60 \text{ ft.} \times 12 \text{ in. /ft.})$$

$$\Delta T = 63.1°F$$

Therefore, an increase in temperature of approximately 63° will lead to the concrete slab deforming .25 in.

Thermal deformation and thermal strain will occur anytime there is a temperature change. This is a normal and predictable occurrence that should not cause a designer much of a problem unless the members in the building or structure are prevented from moving freely. If structures are restrained against thermal deformation, a large increase of stress will occur within the members. This can be illustrated by considering the bar that is attached to the two walls as shown in Figure 9-2. Consider the walls to be immovable and able to resist any magnitude of force imposed on them. If the temperature in this situation rises by 20°F, the bar will

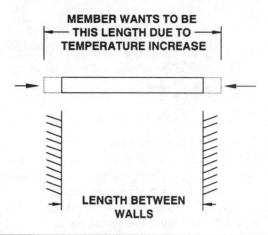

Figure 9-2   Development of Stress Due to Restraint of Thermal Movement

try to thermally deform (expand) based on the aforementioned principles. Since the bar wants to expand but cannot, there are significant stresses developed within the bar. This may better be illustrated by assuming that the bar could expand to its new length. What would have to be done to get the bar back in between the immovable walls? The bar would have to be "squeezed" by a compressive force large enough to make it fit between the walls.

The above scenario reveals that by not allowing a member to move freely under a temperature change, a stress can be developed in an axial fashion on the member. The development of additional stresses due to temperature is not usually accounted for in design and if these stresses occur it can have disastrous consequences for a structure. That is why the designing of expansion or contraction joints is part of any structural design for members subjected to significant temperature changes.

When a member is prevented from straining due to thermal changes, the amount of thermal strain which is prevented has to be the same amount caused due to axial stresses. As such, the following formula for the stress developed in a member restrained from thermal deformation can be derived as follows:

$$\sigma = E(\varepsilon)$$ (Eq. 8-3)

From earlier in this chapter in Equation 9-1 it was stated:

$$\varepsilon_{thermal} = \alpha \, \Delta T$$ (Eq. 9-1)

Substituting this into Equation 8-3, the following formula is obtained:

$$\sigma_{thermal} = \alpha \, \Delta T \, E$$ (Eq. 9-3)

where

$\sigma_{thermal}$ = thermal axial stress due to restrained deformation

α = coefficient of thermal expansion

ΔT = change in temperature

E = material's modulus of elasticity

It should be stated that this formula applies *only to members made from one material*. If members should be made from more than one material (composite members), the amount of thermal strain in each material would not equal the amount of axial strain in each. Therefore, using Equation 9-3 would be incorrect since this formula is based on equivalent values of thermal and axial strain. The composite member problem is indeterminate and its solution will be addressed in upcoming sections.

The following example will illustrate the calculation of axial stress due to restraining thermal deformation. Notice that the members are consistently made from one material.

EXAMPLE 9.3

The steel bar shown below is attached to walls that are immovable. The existing temperature is 50°F. What stress will be created in the steel bar if the temperature increases to 100°F?

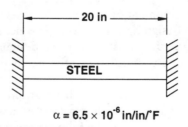

$\alpha = 6.5 \times 10^{-6}$ in/in/°F

The bar will want to increase in length because the temperature rises by 50°F, but will not be able to because the walls will not move. Therefore, the walls will exert a compressive force on the bar and the bar will resist this force by creating an internal stress. Since the member is made from one material (steel) in this problem, the amount of potential thermal strain is offset by an equal amount of axial strain due to the force applied by the unyielding walls. Therefore, the stress can be calculated by Equation 9-3 as follows:

$$\sigma_{thermal} = \alpha \, \Delta T \, E \hspace{4cm} \text{(Eq. 9-3)}$$

with

$\alpha = .0000065$ in./in./°F   (for steel from Table 9-1)

$\Delta T = 50°$

$E = 29000$ ksi   (for steel from Table 8-1)

$\sigma_{thermal} = (.0000065 \text{ in./in./°F})(50°F)(29 \times 10^3 \text{ ksi}) = 9.43 \text{ ksi}$

Therefore, the compressive stress in the bar is 9.43 ksi. An alternate approach to this problem could be to assume that the bar was allowed to deform a certain amount due to the temperature increase. After this, one could calculate the axial stress needed to reduce this increase in length back to the bar's original length.

The bar would deform due to a temperature increase by the following:

$$\delta_{thermal} = \alpha \, \Delta T \, L \hspace{4cm} \text{(Eq. 9-2)}$$

$\delta_{thermal} = (.0000065 \text{ in./in./°F})(50°F)(20 \text{ in.}) = 0.0065 \text{ in. (increase in length)}$

The axial deformation needed to offset this increase in length due to temperature change was given in Chapter 8 by the following:

$$\delta = PL \, / \, AE \hspace{4cm} \text{(Eq. 8-4)}$$

Rearranging Equation 8-4 to isolate for stress (P/A):

$$P \, / \, A = \delta \, E \, / \, L$$

Solving:

$$P \, / \, A = [(.0065 \text{ in.}) \, (29000 \text{ ksi})] \, / \, 20 \text{ in.} = 9.43 \text{ ksi}$$

As one can see, the answers are indeed the same.

9.3   Indeterminate Problems: Members in Series

Many times a structure or member is made up of more than one material and therefore is referred to being a **composite member**. In such a problem it may be important to find the stresses (or forces) on each piece of the member in order to perform an adequate design. However, this problem is not solvable using only the laws of statics and equilibrium because the forces on each member may, in fact, be different. This makes the problems indeterminate. Because of this inability to solve such a problem using statics, it becomes necessary to use our knowledge of strength of materials to provide a relationship that will help in finding the solution. This section will examine the use of strength of materials in solving indeterminate problems with respect to members arranged *in series*.

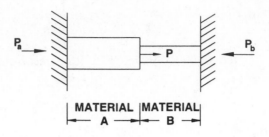

Figure 9-3 Member Made of Two Materials in Series

A member with individual pieces arranged in series is shown in Figure 9-3. A member in series has its pieces arranged in an end-to-end fashion. A common indeterminate problem involving members in series is that of a member placed between two immovable walls with each piece acted on by different forces. Consider the case of the member shown in Figure 9-3.

In this case, the member has an applied force, P, acting on it, which would cause reactions at the walls. We will refer to the reactions at the walls as P_A and P_B. Trying to solve the system using the laws of equilibrium, one would undoubtedly start with summing forces in the x direction as follows:

$$\Sigma F_x = 0$$

$$P_A - P_B + P = 0$$

However, with this equation alone, one cannot solve for the unknown wall reactions, P_A and P_B. Therefore the problem is indeterminate. How might we use the principles of strength of materials (stress, strain, deflection etc.) to help in developing a relationship between P_A and P_B?

Since the walls are immovable, it can be stated that any change (elongation) in length of piece A must be offset by a corresponding decrease in length by piece B. This relationship could be stated formally as follows:

$$\delta_A = \delta_B$$

Since this deformation is caused by an axial force, we can rewrite the previous relationship in terms of axially caused deformation as follows:

$$\delta_A = \delta_B$$

$$(PL / AE)_A = (PL / AE)_B \qquad \text{(Eq. 9-4)}$$

Rearranging Equation 9-4 to isolate the term P_A, we would have the following:

$$P_A = (PL \,/\, AE)_B \,(AE \,/\, L)_A$$

or

$$P_A = P_B \,(L \,/\, AE)_B \,(AE \,/\, L)_A \qquad \text{(Eq. 9-5)}$$

This now defines a relationship that must exist between the two materials, A and B. Knowing some information about the two pieces (A, E, L), we can solve this relationship and use it to substitute into and solve the original equilibrium equation. This utilization of a strength of materials concept can be useful in solving many types of indeterminate problems. The following example will demonstrate this method.

EXAMPLE 9.4

The member shown below is made from two materials, steel and aluminum. If a 20 KN force is applied as shown, calculate the tensile force created in the aluminum section and the compressive force generated in the steel section. Also calculate the stresses in each section.

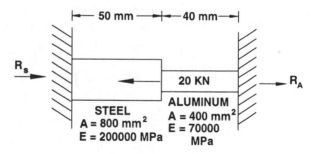

The equilibrium equation is based on the forces being in the x direction and can be expressed as follows:

$$\Sigma F_x = 0$$

$$R_S + R_A - 20 \text{ KN} = 0$$

Being unable to directly solve this equation for the unknown forces R_A and R_B, a relationship must be found through the use of material properties. This can be done by realizing:

$$\delta_S = \delta_A$$

From Equation 9-5 we are able to find a relationship between R_A and R_S:

$$P_A = P_B \, (L \, / \, AE)_B(AE \, / \, L)_A \tag{Eq. 9-5}$$

or in terms of R_A and R_S:

$$R_S = R_A \, (L \, / \, AE)_A(AE \, / \, L)_S$$

Substituting in the known values we find:

$$R_S = R_A \, [.040 \text{ m} / (.0004 \text{ m}^2)(70,000 \text{ MPa})][(.0008 \text{ m}^2)(200,000 \text{ MPa}) / (.050 \text{ m})]$$

$$R_S = 4.57 \, R_A$$

Substituting this into the original statics equation we find:

$$\Sigma F_x = 0$$

$$R_S + R_A - 20 \text{ KN} = 0$$

$$4.57 \, R_A + R_A = 20 \text{ KN}$$

$$5.57 \, R_A = 20 \text{ KN}$$

$$R_A = 3.59 \text{ KN}$$

$$R_S = 4.57 \, R_A$$

$$R_S = 4.57 \, (3.59 \text{ KN}) = 16.41 \text{ KN}$$

Therefore, the compressive force in the steel is 16.41 KN and the tension force in the aluminum is 3.59 KN. The stresses in each can be found as follows:

$$\sigma = P \, / \, A$$

$$\sigma_A = 3.59 \text{ KN} / .0004 \text{ m}^2 = 8,975 \text{ KPa}$$

$$\sigma_S = 16.41 \text{ KN} / .0008 \text{ m}^2 = 20,513 \text{ KPa}$$

This type of "members in series" problem can also be coupled with a thermal stress problem. These are handled in a similar manner, that is by writing an indeterminate equation of static equilibrium. The problem typically assumes that supports are immovable and therefore the following strength of material concept can again be used:

$$\delta_A = \delta_B$$

The difference in this problem is that the δ_A (and likewise δ_B) is made up of two separate and contrasting effects. The thermal deformation effect will generally act opposite of the axial deformation effect and leave each of the members with a "net" deformation. The net deformation of each material is dependent on the thermal deformation and also the deformation associated with axial stress.

These "net" deformations for member A and B will have to be equal to each other and can be viewed as follows:

$$\delta_{A\ net} = \delta_{B\ net}$$

or

$$\delta_{A\ net} - \delta_{B\ net} = 0 \text{ (assumung the walls are immovable)}$$

Consolidating the thermal deformation and axial components of each material and assigning expansion a positive algebraic sign while contraction is negative, the following can be stated:

$$[(\alpha\,\Delta T\,L)_A + (\alpha\,\Delta T\,L)_B] - [(PL\,/\,AE)_A + (PL\,/\,AE)_B\,] = 0 \quad \text{(Eq. 9-6)}$$

This will allow a relationship to be determined beween the force in member A and the force in member B. This relationship can then be used to solve the indeterminate statics equation. It should be noted that the positive and negative signs used in Equation 9-6 are arbitrary and can be changed based on the given problem. (The text will assume that an effect causing expansion or elongation is positive). The following example will illustrate the aforementioned solution process.

EXAMPLE 9.5

The member shown below is made from two materials, steel and copper. If the existing temperature drops 80° and the member is rigidly attached to immovable walls, calculate the tensile force created in the two materials. Also calculate the stresses in each section.

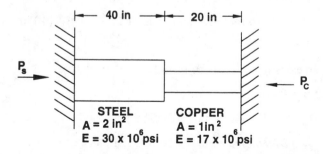

The equilibrium equation is based on the forces being in the x direction and can be expressed as follows:

$$\Sigma F_x = 0$$

$$+P_S - P_C = 0$$

or

$$P_s = P_C$$

This means that the force through the member is constant. Being unable to directly solve this equation for the unknown forces, P_S and P_C, a relationship must be found through the use of material properties. This can be done by realizing:

$$\delta_{A \, net} - \delta_{B \, net} = 0 \text{ (assuming the walls are immovable)}$$

Assuming the temperature drop causes contraction of the two materials, the thermal deformation will be assigned a negative sign and the axial deformations (since it has to keep the members elongated) are assigned postive values. This can be expressed as follows:

$$-[(\alpha \, \Delta T \, L)_A + (\alpha \, \Delta T \, L)_B] + [(PL \, / \, AE)_A + (PL \, / \, AE)_B \,] = 0 \quad \text{(Eq. 9-6)}$$

Rearranging this equation, we find:

$$[(PL \, / \, AE)_C + (PL \, / \, AE)_S] = [(\alpha \, \Delta T \, L)_C + (\alpha \, \Delta T \, L)_S \,]$$

Entering the known data for α, ΔT, and L we find:

$$(PL \, / \, AE)_C + (PL \, / \, AE)_S = [6.5 \times 10^{-6}(80°)(20 \text{ in.})] + [9.3 \times 10^{-6}(80°)(40 \text{ in.})]$$

$$[(PL \, / \, AE)_C + (PL \, / \, AE)_S] = .04016 \text{ in.}$$

Since we know that $P_C = P_S$, we can isolate P as follows:

$$P[(L \, / \, AE)_C + (L \, / \, AE)_S] = .04016 \text{ in.}$$

Solving for P:

$$P = .04016 \text{ in.} / [(L \, / \, AE)_C + (L \, / \, AE)_S]$$

$$P = .04016 \text{ in.} / [(20 \text{ in.} / (1 \text{ in.}^2)(17 \times 10^6)) + (40 \text{ in.} / (2 \text{ in.}^2)(30 \times 10^6)]$$

$$P = 21,789 \text{ lbs}$$

The stresses in each can be found as follows:

$$\sigma = P \, / \, A$$

$$\sigma_C = 21,789 \text{ lbs} / 1 \text{ in.}^2 = 21,789 \text{ psi}$$

$$\sigma_S = 21,789 \text{ lbs} / 2 \text{ in.}^2 = 10,895 \text{ psi}$$

9.4   Indeterminate Problems: Members in Parallel

In the previous section we discussed the solution of forces in indeterminate composite members arranged in series. These problems were unable to be solved using

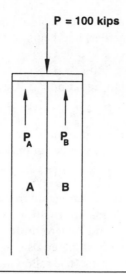

Figure 9-4   Member Made of Two Materials in Parallel

only the laws of statics and equilibrium because the forces on each member may in fact be different. Because of this inability to solve such a problem using statics it becomes necessary to use our knowledge of strength of materials to provide a relationship that will help in finding the solution. This section will examine a similar indeterminate problem with the members arranged *in parallel*.

A member with individual pieces arranged in parallel is shown in Figure 9-4. As one can see, the pieces are arranged in a parallel fashion. A common indeterminate problem would involve members in parallel, such as a compression member made up of two materials. An example of this type of problem would be a reinforced concrete column, acted on by a force that is resisted in some manner by the different materials. Consider the case of the member shown in Figure 9-4.

In this case, the applied force would definitely cause reactions in each material making up the member, which we will refer to as P_A and P_B. Trying to solve using the laws of equilibrium, one would undoubtedly start with summing forces in the y direction as follows:

$$\Sigma F_y = 0$$

$$P_A + P_B - 100 \text{ kips} = 0$$

However, with this equation alone, one cannot solve for the unknown forces, P_A and P_B. Therefore, the problem is indeterminate. How might we use the principles of strength of materials (stress, strain, deflection, etc.) to help in developing a relationship between P_A and P_B?

In this type of problem with the materials arranged in parallel, it is practically mandatory that the deformation of each material be equal. Without this occurring, the composite member would lose its structural intregrity (the member would not work together) and begin to fail. If the following is assumed:

$$\delta_A = \delta_B$$

In a manner similar to Equation 9-5, we can deduce the following:

$$P_A = P_B \, (L \, / \, AE)_B \, (AE \, / \, L)_A \qquad \text{(Eq. 9-7)}$$

If the lengths of the two materials are of equal lengths, which many times is the case, the above equation can be simplified to:

$$P_A = P_B \, (AE)_A \, / \, (AE)_B$$

This now defines a relationship that must exist between the two materials, A and B. By knowing some information about the two materials (A, E, L), we can solve this relationship and use it to substitute into the original statics equation. This utilization of a strength of materials concept can be useful in solving many types of indeterminate problems. The following example will demonstrate this method.

EXAMPLE 9.6

The reinforced concrete column shown below has four steel reinforcing bars, each with 1 in.^{2} of area. If 80,000 lbs of force is distributed over the column in a uniform manner, determine the load resisted by the steel bars and the concrete. $E_{steel} = 30 \times 10^6$ psi, $E_{con} = 3 \times 10^6$ psi

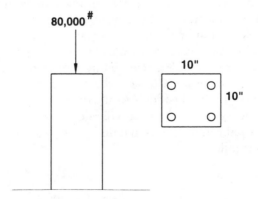

This problem invoves a member made of two materials arranged side by side or in parallel. An attempt to solve for the force resisted by the steel bars (P_s) and the concrete (P_c), using the laws of equilibrium, would yield the following:

$$\Sigma F_y = 0$$

$$P_s + P_c - 80,000 \text{ lb} = 0$$

or

$$P_{s} + P_{c} = 80,000 \text{ lb}$$

The previous equation is indeterminate since there are two unknowns. However, by using some strength of materials concepts we are able to define a relationship between the steel and the concrete, if the following is assumed:

$$\delta_{A} = \delta_{B}$$

From this assumption, Equation 9-7 can be derived and the forces can be determined as follows:

$$P_{A} = P_{B} (L / AE)_{B} (AE / L)_{A} \qquad \text{(Eq. 9-7)}$$

$$P_{s} = P_{c}(A_{s} / A_{c})(E_{s} / E_{c})$$

$$P_{s} = P_{c}(4 \text{ in.}^{2} / 96 \text{ in.}^{2})(30 \times 10^{6} \text{ psi} / 3 \times 10^{6} \text{ psi})$$

$$P_{s} = .416 \, P_{c}$$

Substituting this back into the equilibrium equation that was written earlier, the following is found:

$$P_{s} + P_{c} = 80,000 \text{ lb}$$

$$.416 P_{c} + P_{c} = 80,000 \text{ lb}$$

$$1.416 P_{c} = 80,000$$

$$P_{c} = 56,740 \text{ lb}$$

Therefore P$_{s}$ can be found as follows:

$$P_{s} = 80,000 \text{ lb} - 56,740 \text{ lb} = 23,260 \text{ lb}$$

This type of parallel problem can also be coupled with the temperature considerations that were discussed earlier. The basic solution process remains consistent with the process that has already been discussed. An equilibrium equation is written and will be found to be indeterminate. Using an assumption of equal deformations between materials, a relationship between the materials is found. With temperature-induced deformations also involved, the assumption of equal deformations is as follows:

$$\delta_{A \text{ net}} = \delta_{B \text{ net}}$$

This net deformation will now consist of a thermal deformation component and an axial deformation component. These components will generally act in opposite directions and this text will assume that a positive algebraic sign will be assigned to an increase or potential increase in deformation. (It should be noted that this convention is completely arbitrary).

$$(\alpha \, \Delta T \, L)_A + (PL \, / \, AE)_A = (\alpha \, \Delta T \, L)_B - (PL \, / \, AE)_B \qquad \text{(Eq. 9-8)}$$

The following example will illustrate the solution of such a problem.

EXAMPLE 9.7

The member shown below consists of a wooden exterior shell that is bonded rigidly to a steel bar in its center. The member is attached to an immovable wall at one end. If the temperature increases by 60°C, what force will be developed in the steel and wooden pieces. Also calculate the stress in each piece.

$\alpha_s = 11.7 \times 10^{-6}$ mm/mm/ °C, $\alpha_w = 5.4 \times 10^{-6}$ mm/mm/°C

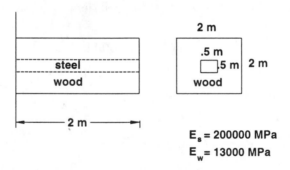

$E_s = 200000$ MPa

$E_w = 13000$ MPa

This problem involves a member made of two materials arranged side by side, or in parallel, subjected to a temperature rise. Both materials will attempt to expand, but at different rates, based on their individual coefficients of thermal expansion, α. Because the steel has a higher coefficient it will expand faster than the wood, thereby inducing a force into the wood. In essense, because the steel is expanding faster it will "pull" on the wood. In contrast to this, the wood will expand at a slower rate, thereby inducing a force into the steel. The wood will cause a "drag" on the steel. These forces are the only forces acting in this system. An attempt to solve for the force resisted by the steel bar (P_s) and the wood (P_w), using the laws of equilibrium, would yield the following:

$$\Sigma F_x = 0$$

$$-P_s + P_w = 0$$

or

$$P_s = P_w$$

It should be noted that the force in the steel was considered negative since it is the force resisting from movement to the right.

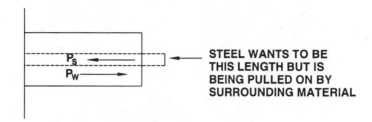

As one can see, the above equation is indeterminate since there are two unknowns. However, by using some strength of materials concepts we are able to define a relationship between the steel and the concrete if the following is assumed:

$$\delta_{s\ net} = \delta_{w\ net}$$

From this assumption, the forces can be determined from Equation 9-8 as follows:

$$(\alpha \, \Delta T \, L)_A + (PL \, / \, AE)_A = (\alpha \, \Delta T \, L)_B - (PL \, / \, AE)_B \qquad \text{(Eq. 9-8)}$$

$$(\alpha \, \Delta T \, L)_s - (PL \, / \, AE)_s = (\alpha \, \Delta T \, L)_w + (PL \, / \, AE)_w$$

Rearranging with $P_s = P_w = P$

$$(\alpha \, \Delta T \, L)_s - (\alpha \, \Delta T \, L)_w = P[(L \, / \, AE)_s + (L/AE)_w]$$

$$.000756 \text{ m} = P(.000081 \text{ m} / \text{MN})$$

$$P = 9.33 \text{ MN}$$

The stresses in each are then:

$$\alpha_s = 9.33 \text{ MN} / .25 \text{ m}^2 = 37.3 \text{ MPa}$$

$$\alpha_w = 9.33 \text{ MN} / 3.75 \text{ m}^2 = 2.5 \text{ MPa}$$

9.5   Stresses on Oblique Planes

Throughout the last two chapters, it was assumed that the stress developed by an axially loaded member was always perpendicular to the direction of applied force.

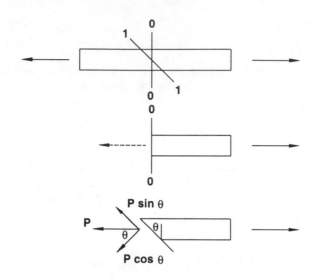

Figure 9-5 Development of Normal and Shear Forces on an Oblique Plane in an Axial Loaded Member

This perpendicular plane was chosen because it represented the maximum tension or compressive stresses that would be developed by the member. However, there are many planes throughout the member that develop stress in response to an applied force. Consider the tension member shown in Figure 9-5. When such a tensile force is applied, typically we have always discussed the stress developed perpendicular to the direction of force. In reality there are many planes that could be considered to develop stresses other than the perpendicular plane. If a plane at an angle of θ was considered (plane 1-1), it would be seen that the tensile force was inclined to the plane and could be broken down into its rectangular components relative to the plane. The component which is perpendicular to plane 1-1 is referred to as the **normal force** and the component parallel to plane 1-1 is referred to as the **shear force**.

Considering the direct stress formula as applied to plane 1-1, we would find the normal stress (σ_{normal}) needed to resist the normal force as:

$$\sigma_{normal} = P_{normal} / A_{normal}$$

where

$$P_{normal} = P \cos \theta$$

$$A_{normal} = A / \cos \theta$$

Therefore the normal stress on any plane θ can be expressed as follows:

$$\sigma_{normal} = P \cos \theta / (A / \cos \theta) = P \cos^2 \theta / A \qquad \text{(Eq. 9-9)}$$

Along this oblique plane there is also a parallel force which has to be resisted by a shearing stress, σ_{shear}. Based on the direct stress formula, the shearing stress, σ_{shear}, on any oblique plane can be calculated as follows:

$$\sigma_{shear} = P_{shear} / A_{normal}$$

where

$$P_{shear} = P \sin \theta$$

$$A_{normal} = A / \cos \theta$$

Such that the shear stress on any plane at angle θ can be expressed as follows:

$$\sigma_{shear} = P \sin \theta / (A / \cos \theta) = P \sin \theta \cos \theta / A \qquad \text{(Eq. 9-10)}$$

As one can see from Equations 9-9 and 9-10; for axially loaded members, the resisting normal stress reaches a maximum on a plane where shear stresses are zero. This corresponds to the perpendicular plane, which was always considered previous to this section. Conversely, the shear stress would reach a maximum at a value of 45° and attain a value equal to 1/2 P/A. This is important in materials that are sensitive or weak with regard to shearing stresses, since this may indicate the plane on which failure will occur. More discussion on normal and shear stresses will be undertaken in Chapter 14 and will focus on the combination of these two and other stress effects.

The following example will illustrate the the calculation of stresses on oblique planes.

EXAMPLE 9.8

A cylindrical member with a cross-sectional area of 10 in.^{2} is subjected to a 50 kip compressive force. Calculate the normal and shear stress on planes of 0°, 30°, and 45°.

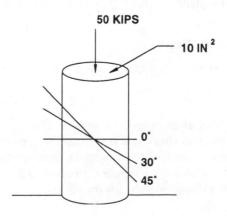

50 KIPS

10 IN ^{2}

0°

30°

45°

From Equation 9-9, the normal stress on any plane can be calculated as follows:

$$P \cos^2 \theta / A \qquad \text{(Eq. 9-9)}$$

Entering the values for the angles of each plane, we find:

For 0°

$$P \cos^2 \theta / A = 50 \text{ kips (cos 0°)}^2 / 10 \text{ in.}^2 = 5 \text{ ksi}$$

For 30°

$$P \cos^2 \theta / A = 50 \text{ kips (cos 30°)}^2 / 10 \text{ in.}^2 = 3.75 \text{ ksi}$$

For 45°

$$P \cos^2 \theta / A = 50 \text{ kips (cos 45°)}^2 / 10 \text{ in.}^2 = 2.5 \text{ ksi}$$

From Equation 9-10, the shear stress on any plane can be calculated as follows:

$$P \sin \theta \cos \theta / A \qquad \text{(Eq. 9-10)}$$

Entering the values for the angles of each plane, we find:

For 0°

$$P \sin \theta \cos \theta / A = 50 \text{ kips(sin 0°)(cos 0°)} / 10 \text{in.}^2 = 0 \text{ ksi}$$

For 30°

$$P \sin \theta \cos \theta / A = 50 \text{ kips(sin 30°)(cos 30°)} / 10 \text{in.}^2 = 2.16 \text{ ksi}$$

For 45°

$$P \sin \theta \cos \theta / A = 50 \text{ kips(sin 45°)(cos 45°)} / 10 \text{in.}^2 = 2.5 \text{ ksi}$$

Again it can be seen that maximum normal stresses are developed perpendicular to the applied force when axial forces are applied and shear stresses develop a maximum at a plane located 45° from the maximum normal plane.

9.6 Stress Concentrations

The direct stress formula (Equation 8-1) is as follows:

$$\sigma = P / A \qquad \text{(Eq. 8-1)}$$

This formula has been used exclusively thus far in the determination of tensile, compressive, and shear stresses. However, a thorough review of this formula shows it to be an average stress formula, that is, it contemplates the total load being spread out uniformly over the entire resisting area. Is this concept of stresses being spread out uniformly always the case?

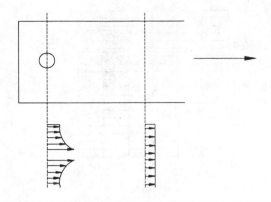

Figure 9-6   Stress Concentration Around a Hole

The simple answer is "no". The direct stress formula works well where the resisting area is continuous and uninterrupted. However, when the area is interrupted by holes, cracks, welds, and other obstructions, the stress distribution is very nonuniform, with the highest stress being located adjacent to the obstructions (Figure 9-6).

These obstructions are sometimes referred to as **stress-raisers**. It might be helpful to consider stresses much in the same way that water flows in a river. As a river channel is uniform and undisturbed, the water flows in a smooth and tranquil manner. However, should the river be interrupted by bridge piers, rocks, or other impediments, the flow becomes visibly rougher and more turbulent. The impact of discontinuities on stress within a member can be viewed in the same manner. Some experimental studies have shown the stress levels adjacent to obstructions to be as much as 2 to 5 times higher than the levels determined from the direct stress equation.[1] Although an in-depth study of stress concentrations is beyond the scope of this text, the student should be aware that discontinuities in a member will "attract" and accumulate stresses in a very rapid manner. As such, discontinuities such as holes, sharp corners, and cracks, must be studied carefully to determine the actual stress distribution and its impact on the member.

The maximum stress in the vicinity of a discontinuity is typically determined through experiemntal study or mathematical theory. Stress concentration factors, K_t, have been developed and are found in many references[2] for various types of discontinuites and have found to be independent of material properties[3]. Typical stress concentration factors for are shown in Figure 9-7. The maximum stress in the vicinity of a discontinuity is then calculated as follows:

$$\sigma_{max} = (K_t)\sigma_{nom} \qquad \text{(Eq. 9-11)}$$

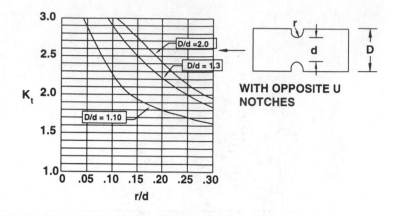

Figure 9-7 Example of Stress Concentration Factor Chart

where

$$\sigma_{max} = \text{maximum stress}$$

$$K_t = \text{stress concentration factor}$$

$$\sigma_{nom} = \text{average stress calculated from standard equations}$$
$$\text{such as Equation 8-1}$$

Stress concentrations can lead to failure in materials because the stresses may exceed the yield strength or ultimate tensile strength of the material under consideration. Failure could initiate by fracture in a brittle material or by excessive deformation in ductile materials. Therefore, the impact of discontinuites in members should be thoroughly evaluated.

9.7 Summary

Temperature changes will always induce thermal strains on a material. Usually, as the temperature rises, materials will tend to expand and when the temperature falls, materials will tend to contract. If a material is allowed to strain freely when the temperature fluctuates, there will be no stresses developing due to this change. However, if a material is restrained or prevented from the strain associated with a temperature change, significant stresses can build up within a member. It is important to be able to account for these stresses in the design or analysis of a member.

Many problems that may be encountered are statically indeterminate, which means that the forces that act on the member cannot be found through only the use of the laws of equilibrium. A common example of this would be a composite member, which is a member made from more than one material. In such cases, it will be advantageous to find a relationship between the two materials that comprise the member. A common relationship is found by exploring the amount of deformation and strain that must occur in each material. The usual assumption is that these deformations must be approximately equal, and from this a suitable relationship can be found.

EXERCISES

1. Two pieces of steel guardrail, each measuring 130.00 ft., are to be connected using bolts in the field. (A standard hole has a 7/8" in. diameter). A slotted bolt hole will be used to account for thermal change. If the temperature change should account for a 50°F increase and a 50° decrease, how long should the slot be? $\alpha_s = 6.5 \times 10^{-6}$ in. /in./°F

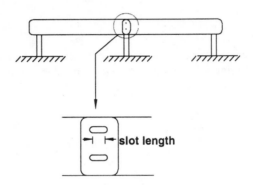

2. A driveway slab measures 150.0 ft. in length and butts flush into the curb of the street (no expansion joint). What size expansion joint would be necessary, adjacent to the garage slab, such that problems will not occur when the temperature rises 60°F ? $\alpha_c = 5.5 \times 10^{-6}$ in. /in./°F

3. A bar made from an unknown material expands by 10 mm when the temperature rises by 30°C. If the bar is originally 2 m in length, calculate the coefficient of thermal expansion.

4. Concrete pavement slabs measuring 20 m in length are placed end-to-end in the construction of a road. What size should the joints between slabs be, if a temperature rises of 25°C is expected? $\alpha_c = 10.8 \times 10^{-6}$ mm /mm/°C

5. In exercise #9.4, calculate the stress developed in the pavement slab if the temperature rises by 40°C instead of the expected 25°C. $E_c = 30,000$ MPa. Consider the slabs making contact after 25° C.

6. A steel bar measuring 36 in. is held rigidly between two immovable walls. If the temperature falls by 30°F, calculate the stress developed in the bar. $E_s = 30 \times 10^6$ psi

7. The two steel bars shown below have a gap of 0.01 in. between them. If the temperature is presently at 80°F, calculate the temperature needed to have the bars make contact. Then calculate the temperature at which the

stress in the bars will reach 1000 psi. $E_s = 30 \times 10^6$ psi, $\alpha_s = 6.5 \times 10^{-6}$ in./in./°F

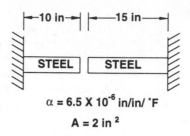

$\alpha = 6.5 \times 10^{-6}$ in/in/ °F

A = 2 in²

8. The copper bar shown below is attached to an immovable wall as shown. Calculate the temperature rise needed such that the stress in the bar reaches 100 MPa. $E_c = 105{,}000$ MPa, $\alpha_c = 25 \times 10^{-6}$ mm/mm/ °C

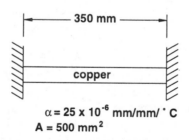

$\alpha = 25 \times 10^{-6}$ mm/mm/ ° C

A = 500 mm²

9. If the wall on the right side of the bar in exercise #8 was deflected laterally by 5 mm, at what temperature would the bar reach 100 KPa?

10. A 0.5 in diameter steel cable measuring 75.0 ft. has a 1 kip load at its end which rests on the ground. If initially the cable has no stress in it (the ground is supporting the full 1 kip) determine the temperature decrease needed to make the load rise a total of 0.25 inch off the ground. $E_s = 30 \times 10^6$ psi, $\alpha_s = 6.5 \times 10^{-6}$ in./in./°F

11. How would the answer in exercise #10 change if initially the cable supported the full 1 kip load (the load was just barely making contact with the ground)?

12. The beam shown below weighs 50 MN and is supported by two aluminum cables and one steel cable. All cables are 25 mm in diameter. Calcu-

late the force in each cable if the beam is to remain horizontal. $E_s = 200,000$ MPa. $E_a = 70,000$ MPa

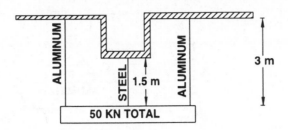

13. A 4 in. × 4 in. steel tube (wall thickness 0.25 in.) is filled with concrete to make a column. Calculate the force in the steel and the concrete if 65,000 lbs is loaded on the column. $E_s = 30 \times 10^6$ psi, $E_c = 2.8 \times 10^6$ psi

14. If the column in exercise #13 had allowable stress levels of 2 ksi on the concrete and 20 ksi for the steel, what would be the maximum load allowed on the column?

15. The compression member shown below is made of a piece of wood with two steel plates attached on the outside. If this column is subjected to 100 kips, determine the force in the wood and the steel. $E_s = 30 \times 10^6$ psi $E_w = 1.5 \times 10^6$ psi

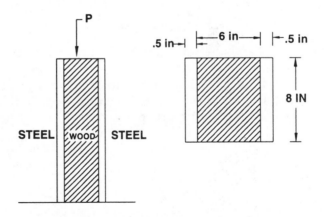

16. A copper tube with a 3 in. O.D. and 2.5 in. I.D. fits over a 2.5 in. steel bar. Both members are the same length and firmly bonded to each other. Calculate the stress in each material if the temperature increases by 100° F. $E_s = 30 \times 10^6$ psi, $E_c = 17 \times 10^6$ psi, $\alpha_s = 6.5 \times 10^{-6}$ in./in./ °F, $\alpha_c = 11.3 \times 10^{-6}$ in./in./° F

17. The member shown is made of aluminum and steel pieces rigidly attached between immovable walls. If a 40 KN force is applied as shown, deter

mine the force in the steel and the aluminum. $E_s = 200,000$ MPa, $E_a = 70,000$ MPa

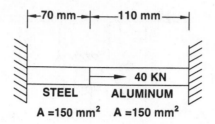

18. A member is made from steel and cast iron, as shown below, and fits in between immovable walls. If the temperature decreases 120 °F, calculate the stress in each bar. $E_s = 30 \times 10^6$ psi, $E_{ci} = 10 \times 10^6$ psi, $\alpha_s = 6.5 \times 10^{-6}$ in./in./ °F, $\alpha_{ci} = 6 \times 10^{-6}$ in./in./ °F

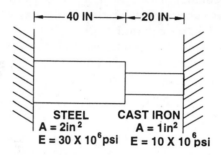

19. If the righthand wall in exercise #18 can deflect freely 0.10 in., what would the stress be?

20. A 20 ft. long column is made of aluminum and steel, as shown below. Calculate the load needed to cause a deflection of 0.05 in. $E_s = 30 \times 10^6$ psi, $E_a = 10 \times 10^6$ psi

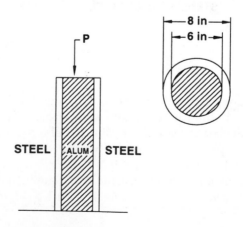

21. If the column in exercise #9.20 had a compressive load on it of 100 kips what would the total deflection of the column be?

22. A cable is 100.00 ft. long, with the top 50.00 ft. made from steel and the bottom 50.00 ft. made from aluminum. The cable has a 1 in. diameter and supports 2000 lbs. Calculate the temperature drop necessary to raise the cable 0.25 in. $E_s = 30 \times 10^6$ psi, $E_a = 10 \times 10^6$ psi, $\alpha_s = 6.5 \times 10^{-6}$ in./in./ °F, $\alpha_a = 12.7 \times 10^{-6}$ in./in./°F. Consider the cable to support the full load before the temperature drops.

23. A 2 in. long copper block ($A = 2$ in.^{2}) rests upon a 4 in. long steel block ($A = 3$ in.^{2}). If 5 kips is placed on top of the blocks, calculate the new length of each block and the total deflection. $E_s = 30 \times 10^6$ psi, $E_c = 17 \times 10^6$ psi

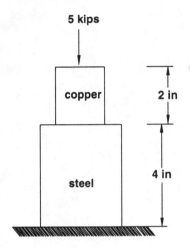

24. A tension member with a cross-sectional area of 3 in.^{2} resists a load of 10 kips. Calculate the normal and shear stress on planes making a 0°, 20°, 45°, and 60° angle with the transverse section.

25. A tension member with a cross-sectional area of 30 mm^2 resists a load of 83 KN. Calculate the normal and shear stress on planes making a 0°, 20°, 45°, and 60° angle with the transverse section.

REFERENCES

1. Olsen, Gerner A., *Elements of Mechanics of Materials*, Englewood Cliffs, NJ, Prentice-Hall, 1974, p.520.

2. Peterson, R.E., *Stress Concentration Factors*, New York, John Wiley and Sons, 1974.

3. *Marks' Standard Handbook for Mechanical Engineers*, McGraw-Hill, 8th Ed., 1978, p.5-6.

CHAPTER
10
TORSION

10.1 Introduction

In the previous two chapters we have discussed the fundamental concepts of direct stress, which was governed by the following equation:

$$\sigma = P / A \qquad \text{(Eq. 8-1)}$$

Direct stress was found to manifest itself typically as tension, compression, or shear stresses that were developed in a body to resist the application of an external force. However, there can also be other stresses that are more complex in their distribution over the resisting area in a body. The most common of these "complex" stresses will develop to resist the application of a torque (twisting moment) or a bending moment. These stresses are referred to as **torsional stresses** and **bending stresses**. The focus of this chapter will be the study of torsional stresses on a circular shaft.

Circular shafts are a common mechanism used to transmit power from one part of a machine to another. A common example of such a mechanism is the drive shaft on an automobile, which transmits power to the axle in order to move the machine. These circular shafts are commonly turned through the application of a torque or "twisting moment". A torque is expressed in units of length and force. Common units which torque may be expressed are inch-pounds, inch-kips, Newton-meters, etc. The torque on a shaft is applied through pulleys or gears that may be located at different places along the length of the shaft (Figure 10-1). The net amount of torque applied at any location along the shaft can be calculated by summing the external torque on each side of the shaft. This will be an important con-

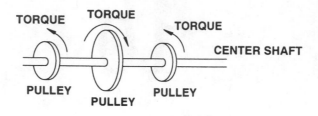

Figure 10-1   Torsion on a Drive Shaft

cept because the student will find that drive shafts will deliver varying amounts of torque to a shaft.

The general concept of torsional stress is similar to that of the direct stresses (tension, compression, etc.), which has been discussed previously. This general concept can be summarized by the following statement. As members are acted on by some external force or moment, they must develop internal stresses to counteract and resist it. In this chapter the external "moment" that is applied to the member is a torque and the circular shaft will resist this by developing an internal torque through the build up of torsional stresses over the cross-sectional area. The following section will explore the development of these torsional stresses.

10.2   Torsion on a Circular Shaft

To explore the development of torsion stresses, we will begin on a fundamental level by considering the stresses developed by a circular member that is fixed at one end and free at the other. A circular shaft that is subjected to torque will develop no tensile, compression, or bending stresses, but instead be placed in a state of pure shear. The students may observe this phenomena by rolling their notebooks into a cylinder and, while holding one end firmly, applying a torque to the other end. What happens to the individual pages of the notebook? You will notice that the individual pages tend to "slip" past each other. Although this is a relatively crude example, the slippage and displacement of the individual sheets relative to each other is a sign of shear stresses being applied. This behavior is analogous to that which happens in a circular shaft, whereby the shaft can be thought of as being made up of individual "rings of area" that are bonded together. When a member is subjected to a torque, these "rings" have a tendency to displace relative to one another, developing a shear stresses across the individual planes of contact (Figure 10-2). Imagine if these shear stresses did not develop. The individual rings would be able to spin uncontrollably with respect to each other and the shaft would not be in equilibrium. Instead, these shearing stresses are developed and the integrity of the shaft is maintained. The nature of these shear stresses will be developed over the remainder of this chapter.

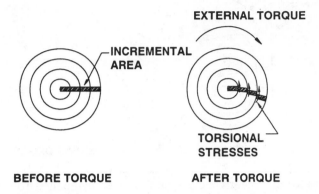

Figure 10-2 Torsional Stresses Developed in "Rings of Area"

It stands to reason that the intensity of the shear stresses produced by torsion should be directly proportional to the magnitude of torque applied to the member. Simply stated, this means if more torque is applied to the member, the intensity of the shear stresses must also increase. However, what other physical properties of the member will dictate the magnitude of shear stresses that are to be developed? To explore the development of shear stress we will consider the circular shaft shown in Figure 10-3. This shaft will be fixed at one end and free at the other, with a certain amount of torque, T, applied to the free end.

Consider the location of two points, A and B, before and after the torque, T, is applied. Before application of the torque, point A and B lie in a horizontal plane, with point A located on the outside surface of the shaft and point B lying midway

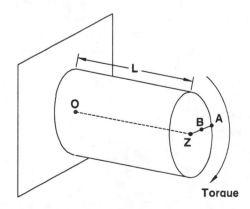

Figure 10-3 Torque on a Circular Shaft

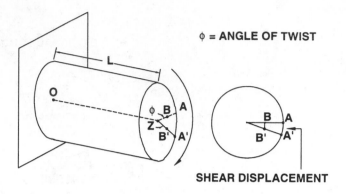

Figure 10-4   Shear Displacement Caused by applied Torque

between the outside of the shaft and the longitudinal axis of the shaft (O-Z). As the torque, T, is applied, points A and B will rotate to their new positions, A' and B' (Figure 10-4). If these displacements (A-A' and B-B') are considered relative to the length of the shaft (L), the shear strain could be calculated as follows:

$$\varepsilon_{s\ A\text{-}A'} = A\text{-}A' \,/\, L$$

$$\varepsilon_{s\ B\text{-}B'} = B\text{-}B' \,/\, L$$

where

$\varepsilon_{s\ A\text{-}A'} =$ shear strain at A-A' (similar for B-B')

$A\text{-}A' =$ displacement of A to A' (similar for B-B')

$L =$ length of the shaft

As one can see from the figure, the calculated shear strain of B-B' is one-half that of A-A'. If the torque, T, has been low enough to keep the stresses under the material's proportional limit, then Hooke's law can be applied. Hooke's law would relate shear strains to shear stresses and therefore, since the shear strain of B-B' is one-half that of A-A', the shear stress at B-B' should be one-half that of A-A'. This proportionality of shear stresses and strains is related through a known material constant referred to as the **shear modulus, G.** The shear modulus, G, has a value approximately 35%–40% of the modulus of elasticity (E) and can be given by the following equation:

$$G = E \,/\, 2(1 + \mu) \hspace{3cm} \text{(Eq. 10-1)}$$

where

$E =$ material's modulus of elasticity

$\mu =$ material's Poisson ratio

Typical material values for the shear modulus are found in Appendix B.

The aforementioned statement regarding the proportionality of shear strains to shear stresses has very important implications that must be grasped in order to understand the behavior of shear stresses caused by torsion. The most important implication is that shear stresses caused by torsion are proportional to the distance away from the longitudinal axis of the shaft. This means that the "outer rings" of the shaft will have the highest torsional strains and therefore will develop the highest shear stresses over the cross-sectional area of the shaft. The other implication is that there must be a property of area that will describe the resistance of a shape to the effects of torsion. These will be looked at in depth in the upcoming section.

10.3   Shear Stresses on a Circular Shaft

As discussed in the previous section, shear stresses developed by torsion on a circular shaft vary over the cross-section of the shaft. These stress are zero at the longitudinal axis of the shaft and maximum at the extreme outside of the shaft (Figure 10-5). When an external torque, T, is applied to the shaft, the shaft must internally develop a counteracting torque such that equilibrium is maintained. These shear stresses that vary over the cross-section of the shaft will be an integral part of the internal resisting torque. The internal resisting torque will develop such that the areas that have higher shear stresses (near the outside surface of the shaft) will develop more of the resisting torque than areas closer to the center of the shaft.

From the discussion in Chapter 8, it was found that stress is equal to a force over some incremental unit of area ($\sigma = P / A$). Conversely, force applied to an incremental piece of area is equal to the stress on that area times that area

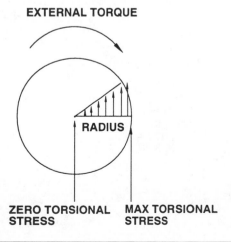

Figure 10-5   Distribution of Shear Stresses Over a Circular Shaft

$(P = \sigma \times A)$. If one would consider a small incremental piece of area somewhere on the shaft (Figure 10-6), it could be seen that the shear stress on that area is proportional to the distance from the longitudinal axis of the shaft. This can be described as follows:

$$\sigma_{area} = \sigma_s \, (r \, / \, c) \qquad\qquad\qquad\text{(Eq. 10-2)}$$

where

σ_{area} = shear stress on the incremental area

σ_s = maximum shear stress on the outside of the shaft

r = distance from the longitudinal axis to the incremental area

c = radius of the shaft (1/2 diameter)

To calculate the resisting shear force developed by this incremental piece of area, we need only to take the shear stress in Equation 10-2 and multiply it by the area as follows:

force = stress × area

$$F = \sigma_{area} \times A$$

or

$$F = \sigma_s \, (r \, / \, c) \times A \qquad\qquad\qquad\text{(Eq. 10-3)}$$

This incremental shear force, F, is only one of many incremental shear forces that would be developed over the circular shaft to resist the externally applied torque. However, forces alone do not resists a torque. Moments have to be generated in order to resist the externally applied torque. The internal resisting moment of the incremental area is calculated as the incremental force, F, multi-

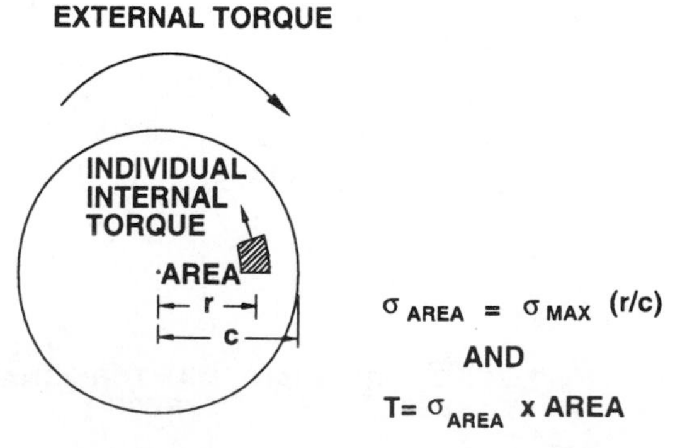

EXTERNAL TORQUE

INDIVIDUAL INTERNAL TORQUE

·AREA

$\sigma_{AREA} = \sigma_{MAX} \, (r/c)$

AND

$T = \sigma_{AREA} \times AREA$

Figure 10-6   Torque Created over Incremental Area

plied by the radial distance, r. This would make the internal resisting moment of the incremental area as follows:

$$T = F \times r$$

or

$$T = (\sigma_s \, (r / c) \times A)(r) = \sigma_s \, (r^2 / c) \times A \qquad \text{(Eq. 10-4)}$$

where

$$T = \text{incremental resistance to torque}$$

Although the above expression is only for one incremental piece of area, the total internal resistance of the shaft can be thought of as the sum of all areas. This would change Equation 10-4 to the following:

$$\Sigma T = \Sigma(\sigma_s / c) \, (Ar^2) \qquad \text{(Eq. 10-5)}$$

where

$$\Sigma T = \text{total internal resistance to torque}$$

From Chapter 7, it was learned that the term Ar^2 is a property of area referred to as the **polar moment of inertia, J**. This property defines a shape's resistance to a twisting moment or torque; the higher this property becomes will reflect in a larger resistance to torque. Substituting this value into Equation 10-5, we find:

$$\Sigma T = \Sigma(\sigma_s / c)(J) \qquad \text{(Eq. 10-6)}$$

Since internal resistance to torque and externally applied torque must be equal in cases of equilibrium, it can be stated that $\Sigma T = T$. Equation 10-6 can also be rearranged to solve for the shear stress on the outside surface as:

$$\sigma_s = T(c) / J \qquad \text{(Eq. 10-7)}$$

where

σ_s = torsional shear stress on the outside of the shaft

T = applied torque

c = radius of the shaft (1/2 diameter)

J = polar moment of inertia

It should also be recognized that since torsional stresses are proportional to their distance from the center of the shaft, the stress at any location can be found by proportions. Equation 10-7 can be used in many applications to either calculate stresses or to design the polar moment of inertia, J, that is required under a certain amount of torque. For such design situations, an allowable value of shear stress that the material can withstand would have to be given in order to provide a safe design.

The following examples will illustrate the use of these torsional stress formulas.

EXAMPLE 10.1

Calculate the shear stress developed on the outside of a solid, 2 in. diameter, circular shaft that is under an applied torque of 25,000 in.-lbs. Also find the shear stress at a distance 0.5 in. away from the center of the shaft.

The polar moment of inertia for a circular cross-section is:

$$J = \pi(d^4)/32$$

For the 2 in. diameter shaft in this problem, the polar moment of inertia is calculated as:

$$J = \pi(2 \text{ in.})^4 / 32 = 1.57 \text{ in.}^4$$

Using Equation 10-7, the torsional shear stress on the outside of the shaft can be calculated as:

$$\sigma_s = T(c) / J \qquad\qquad\qquad \text{(Eq. 10-7)}$$

$$\sigma_s = 25,000 \text{ in.-lb}(1 \text{ in.}) / 1.57 \text{ in.}^4 = 15,915 \text{ psi}$$

The shear stress at a distance of 0.5 in. away from the center would be:

$$\sigma_{.5} = 15,915 \text{ psi } (0.5 \text{ in.} / 1 \text{ in.}) = 7958 \text{ psi}$$

EXAMPLE 10.2

Calculate the shear stress developed on the outside of a hollow, 80 mm diameter, circular shaft (I.D. = 40 mm) that is under an applied torque of 4500 Newton-meters. Also find the shear stress on the inside surface of the shaft.

The polar moment of inertia for a circular cross-section:

$$J = \pi(d_o^4 - d_i^4)/32$$

For the shaft in this problem, the polar moment of inertia is calculated as:

$$J = (\pi(.080 \text{ m})^4 - (.040 \text{ m})^4)/32 = 3.77 \times 10^{-6} \text{ m}^4$$

Using Equation 10-7, the torsional shear stress on the outside of the shaft can be calculated as:

$$\sigma_s = T(c) \, / \, J \qquad\qquad\qquad\text{(Eq. 10-7)}$$

$$\sigma_s = 4500 \text{ N-m}(.04 \text{ m}) \, / \, 3.77 \times 10^{-6} \text{ m}^4 = 4.77 \times 10^7 \text{ Pa}$$

or 47.7 MPa

The shear stress on the inner surface of the shaft would be:

$$\sigma_{in} = 47.7 \text{ MPa } (20 \text{ mm} \, / \, 40 \text{ mm}) = 23.9 \text{ MPa}$$

EXAMPLE 10.3

A circular steel shaft has an allowable shear stress of 15,000 psi. Determine the minimum diameter for the solid shaft if the applied torque it needs to resist is 50,000 in.-lbs.

The property of area that needs to be designed in this example is the polar moment of inertia, J. For a solid circular area this is known to be:

$$J = \pi(d^4) / \, 32$$

Rearranging Equation 10-7 to solve for J, we have:

$$\sigma_s = T(c) \, / \, J$$

or

$$J = T(c) \, / \, \sigma_s$$

Substituting the allowable shear stress and realizing that c is equal to one-half diameter, we have the following:

$$J = T(d/2) \, / \, 15,000 \text{ psi}$$

$$J = 50,000 \text{ in.-lbs}(d/2) \, / \, 15,000 \text{ psi}$$

$$\pi(d^4) \, / \, 32 = 50,000 \text{ in.-lbs}(d/2) \, / \, 15,000 \text{ psi}$$

$$d^3 = 32 \, (50,000) \, / \, \pi(2)(15,000)$$

$$d = 2.57 \text{ in.}$$

Therefore, the minimum diameter is 2.57 in. for the shaft not to exceed 15,000 psi in torsional shear stress.

EXAMPLE 10.4

The 2 in. diameter, solid, circular shaft shown below is driven by pulleys A, B and C. Calculate the torque on the shaft between pulleys A and B and also calculate the maximum shear stress on the shaft between these pulleys.

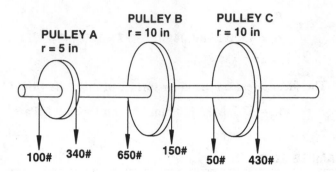

The belt tension shown on each pulley must first be used to calculate the torque at each location. This can be accomplished as follows:

Pulley A

$$T = (340 \text{ lbs} \times 5 \text{ in.}) - (100 \text{ lbs} \times 5 \text{ in.}) = 1200 \text{ in.-lbs (CW)}$$

Pulley B

$$T = (650 \text{ lbs} \times 10 \text{ in.}) - (150 \text{ lbs} \times 10 \text{ in.}) = 5000 \text{ in.-lbs (CCW)}$$

Pulley C

$$T = (430 \text{ lbs} \times 10 \text{ in.}) - (50 \text{ lbs} \times 10 \text{ in.}) = 3800 \text{ in.-lbs (CW)}$$

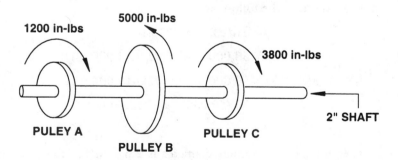

The applied torque at any point along the shaft is simply the sum of the torque up to that point. Between pulley A and pulley B the sum of the torque would simply be 1200 in.-lbs. Therefore, the shear stress on the shaft at this point can be calculated with Equation 10-7 as follows:

$$\sigma_s = T(c) \, / \, J \qquad\qquad \text{(Eq. 10-7)}$$

with

$$T = 1200 \text{ in.-lbs}$$

$$J = \pi d^4 / 32 = \pi (2 \text{ in.})^4 / 32 = 1.57 \text{ in.}^4$$

$$c = d / 2 = 1 \text{ in.}$$

$$\sigma_s = (1200 \text{ in.-lb})(1 \text{ in.}) / 1.57 \text{ in.}^4 = 764.3 \text{ psi}$$

Therefore the maximum shear stress created by the pulleys at this point is 764.3 psi.

10.4 The Torsion Test

The shear modulus, G, of a material relates the proportionality of shear stress to shear strain within a given material. In this manner, it is analogous to a material's modulus of elasticity, E, which relates the proportionality of axial stress to axial strain. In Chapter 9 it was discussed how a material's modulus of elasticity, E, could be determined by subjecting a material to an axial tension test. Similarly, a material's shear modulus, G, can be found through what is known as a torsion test.

A torsion test is performed on a specimen that typically is a hollow circular shaft of some length, L. The specimen is subjected to an increasing amount of torque while the angle of twist (φ) is simultaneously measured. The angle of twist can be illustrated as shown in Figure 10-7.

From Eq. 10-7, the following relationship has been determined:

$$\sigma_s = T(c) / J \qquad\qquad (\text{Eq. 10-7})$$

Since

$$\sigma_s = \varepsilon G$$

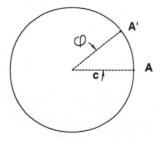

Figure 10-7   Angle of Twist

Equation 10-7 can be rearranged as follows:

$$\varepsilon G = T(c) / J \qquad \text{(Eq. 10-8)}$$

The shearing strain, ε, at small angles can be expressed in terms of the angle of twist (φ) and the length of the shaft (L), as follows:

$$\varepsilon = \varphi\, c / L$$

Substituting this expression in for Equation 10-8, we find the following:

$$(\varphi\, c / L)G = T(c) / J$$

or

$$\varphi = TL / GJ \qquad \text{(Eq. 10-9)}$$

where

φ = angle of twist, radians

T = applied torque

L = length of specimen

G = shearing modulus

J = polar moment of inertia for the shaft

The values plotted in the torsion test are torque (T) versus angle of twist per unit length of the shaft (φ / L) (Figure 10-8). Equation 10-9 could also be expressed as:

$$GJ = T / (\varphi / L)$$

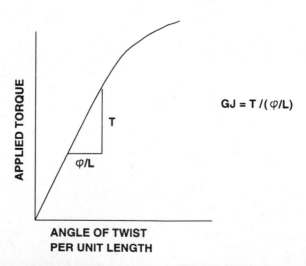

Figure 10-8   Plotted Data from Torsion Test

By plotting the values of torque versus angle of twist per unit length, one can determine a material's value of GJ; since this will simply be the slope of the line (y / x). Knowing the polar moment of inertia of the specimen tested, the value of shear modulus, G, can easily be determined.

This test is useful in the determination of the shear modulus, G, for a given material. This determination of shear modulus is important to accurately predict the behavior of shafts having applied torsion forces.

10.5   Transmitting Power through Shafts

Power is commonly transmitted through a pulley or shaft by an applied torque. As the shaft is rotated, work is performed. The relationship between torque and power will be developed in this section.

The torque applied to a pulley is a product of the force on the pulley, F, and the distance from the center of rotation, c. This can be seen in Figure 10-9 and stated as follows:

$$T = (F)(c)$$

When a belt or cable moves along a pulley due to an applied torque, work is done because the cable has moved a distance, x. This work accomplished by the torque is equal to the product of the force, F, and the distance moved, x. This can be stated as follows:

$$\text{Work} = (F)(x) \qquad \text{(Eq. 10-10)}$$

The distance x is also equal to the arc length that the pulley has rotated, which can be expressed as the product of radial distance to the outside of the shaft, c, and the angle of rotation, φ. Therefore, the work perfomed by the applied torque can be expressed by rewriting Equation 10-10 as shown below:

$$\text{Work} = (F)(x) \qquad \text{(Eq. 10-10)}$$

$$x = (c)(\varphi)$$

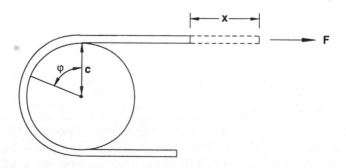

Figure 10-9   Description of Work on a Pulley

Therefore

$$\text{Work} = (F)(c)(\varphi)$$

Since torque has already been expressed as the produce of force, F, and distance, c; work can be stated as follows:

$$\text{Work} = (T)(\varphi) \qquad \text{(Eq. 10-11)}$$

where

T = applied torque on the shaft

φ = angle of rotation, expressed in radians

Power is defined as work per unit time. The unit of power in the U.S. customary system is **horsepower** which is 33,000 ft.-lb per minute. The amount of horsepower produced under an applied torque can be expressed as follows:

$$HP = TN \,/\, 63,000 \qquad \text{(Eq. 10-12)}$$

where

T = applied torque (in.-lb)

N = revolutions per minute

HP = horsepower produced

The unit of power in the SI system is a **watt** which is 1 N-m per second or **kilowatt (kW)** which is 1 KN-m per second. The number of kilowatts produced under an applied torque can be expressed as follows:

$$P = TN \,/\, 9550 \qquad \text{(Eq. 10-13)}$$

where

T = applied torque (N-m)

N = revolutions per minute (rpm)

P = power produced in kilowatts (1000 watts)

The following example will illustrate the concept of power transmission in a pulley.

EXAMPLE 10.5

Calculate the amount of horsepower transmitted through a 1.5 in. diameter, solid steel shaft that is 24 in. long. The allowable shear stress on the shaft is 10,000 psi and the allowable angle of twist is .026 radians. The shaft is turning at 400 rpm. Use $G = 12 \times 10^6$ psi

Begin by calculating the limiting amount of torque which can be applied. The limiting amount of torque will either be governed by the allowable shear stress in the shaft or the allowable angle of twist. The limiting amount of torque by the shear stress is calculated from Equation 10-7 as follows:

$$\sigma_s = T(c) / J \qquad\qquad\qquad \text{(Eq. 10-7)}$$

Rearranging:

$$T = \sigma_s J / c$$

Solving:

$$T = (10,000 \text{ psi})(.497 \text{ in.}^4) / 0.75 \text{ in.} = 6626 \text{ in.-lb}$$

where $J = \pi(d)b^4 / 32$

The limiting amount of torque as governed by the allowable angle of twist can be found from Equation 10-9 as shown:

$$\varphi = TL / GJ \qquad\qquad\qquad \text{(Eq. 10-9)}$$

Rearranging:

$$T = \varphi GJ / L$$

Solving:

$$T = (.026 \text{ radians})(12 \times 10^6 \text{ psi})(.497 \text{ in.}^4) / 24 \text{ in.} = 6461 \text{ in.-lb}$$

Therefore, the limiting amount of torque on the shaft is controlled by the angle of twist (6461 in.-lb).

The horsepower produced can be determined using Equation 10-12;

$$HP = TN / 63,000 \qquad\qquad\qquad \text{(Eq. 10-12)}$$

$$HP = (6461 \text{ in.-lb})(400 \text{ rpm}) / 63,000 = 41 \text{ hp}$$

10.6 Summary

The twisting of a circular shaft occurs by applying a torsional force, or torque. This condition sets up a state of pure shear throughout the shaft and is resisted by the shaft developing its own internal torque, which counteracts the applied torque. The shear stresses that develop throughout the shaft vary in their intensity. The shear stresses are zero at the center of the shaft and are a maximum at the outside surface of the shaft.

The stiffness of a member relative to twisting is dependent on two properties—the shear modulus of the material (G) and the polar moment of inertia (J) of the member. The shear modulus (G) is a material property defining the proportionality between shear stress to shear strain in a given material. It is analogous to the modulus of elasticity for axial stresses and strains. The polar moment of inertia is a mathematical property of an area that describes an area's torsional resistance. A member with more area away from its longitudinal axis will have a higher polar moment of inertia.

EXERCISES

1. Calculate the shear stress developed on the outside of a solid 3.5 in. diameter circular shaft which is under an applied torque of 25,000 inch-pounds. Also find the shear stress at a distance 0.5 in. away from the outside of the shaft.

2. Calculate the shear stress developed on the outside of a hollow 2 in. diameter circular shaft (I.D. = 1.25 in.) which is under an applied torque of 40,000 inch-pounds. Also find the shear stress at a distance 0.25 in. away from the outside of the shaft.

3. Calculate the shear stress developed on the outside of a solid 100 mm diameter circular shaft which is under an applied torque of 2500 Newton-meters. Also find the shear stress at a distance 10 mm away from the outside of the shaft.

4. Calculate the shear stress developed on the outside of a hollow 150 mm diameter circular shaft (I.D. = 100 mm) which is under an applied torque of 10,000 Newton-meters. Also find the shear stress at a distance 10 mm away from the outside of the shaft.

5. The socket wrench shown below is used to tighten bolts with an applied force of 200 pounds. If the socket being used is 0.50 in. diameter (0.40 in. I.D.), determine the maximum shear stress developed in the socket.

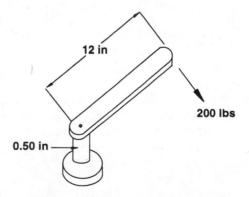

12 in

200 lbs

0.50 in

6. If the socket used in the previous exercise 5 was made of steel with an allowable shear stress of 15,000 psi, determine the maximum force that could be applied to the end of the wrench. The distance is kept at 12 inches.

7. Calculate the diameter needed for a solid circular shaft that has an applied torque of 4500 Newton-meters. The allowable shear stress for the material that the shaft is made out of is 70,000 KPa.

8. Calculate the maximum inside diameter for a hollow circular shaft that has an outside diameter of 3 inches. The applied torque is 45,000 inch-lbs and the allowable shear stress for the material is 12,000 psi.

9. A industrial drive shaft is a solid circular section made out of steel with an allowable shear stress of 60 KPa. If the torque on the shaft is 100 KN-m determine an adequate shaft size for the drive.

10. A 2 in. diameter solid circular shaft is made from aluminum having an allowable shear stress of 5000 psi. Determine the maximum allowable torque which can be applied to the shaft.

11. A 50 mm diameter solid circular shaft is made from aluminum having an allowable shear stress of 30 KPa. Determine the maximum allowable torque that can be applied to the shaft.

12. The torques shown below are applied to a 2.25 in. solid circular shaft through the four pulleys labelled A, B, C, and D. Calculate the maximum torsional stress on the shaft and where does it occur along the length of the shaft.

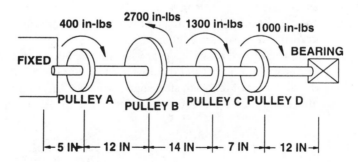

13. If the shaft in exercise 12 is made from steel with a $G = 11,000$ ksi, determine the angle of twist (φ) between sections A-B, B-C, and C-D.

14. The torques shown below are applied to a 110 mm solid circular shaft through the four pulleys labelled A, B, C, and D. Calculate the maximum torsional stress on the shaft and the location where it occurs along the length of the shaft.

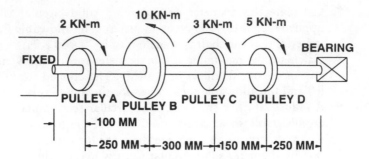

15. If the shaft in exercise 14 is made from steel with a $G = 65,000$ MN, determine the angle of twist (φ) between sections A-B, B-C, and C-D.

16. A torque of 2 KN-m is applied to a solid circular shaft and develops a maximum shearing stress of 30,000 KPa. Determine the diameter of the shaft.

17. A torque of 25 kip-ft. is applied to a solid circular shaft and develops a maximum shearing stress of 12,000 psi. Determine the diameter of the shaft.

18. Calculate the amount of horsepower transmitted through a 1.0 in. diameter solid steel shaft that is 36 in. in length. The allowable shear stress on the shaft is 8000 psi and the allowable angle of twist is .045 radians. The shaft is turning at 600 rpm. Use $G = 12 \times 10^6$ psi.

19. Calculate the power transmitted through a 100 mm diameter solid steel shaft that is 1 m in length. The allowable shear stress on the shaft is 70 MPa and the allowable angle of twist is .040 radians. The shaft is turning at 600 rpm. Use $G = 83,000$ MPa.

20. A motor is used to drive a gear assembly at 100 rpm. If the power needed at output A is 30 hp and the power needed at output B is 60 hp, calculate the diameter of the solid steel shafts from AC and BC. Consider an allowable shear stress of 8000 psi to control the torque in the steel shafts.

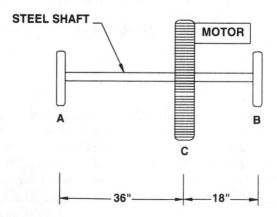

21. If the steel shaft in exercise # 20 was 1 inch in diameter, and the motor was to deliver 40% of its power to output *A* and the other 60% of its power to output *B*, calculate the maximum power which can be delivered to the shaft. The allowable shear stress controls the torque and is still 8000 psi.

22. A motor is used to drive a gear assembly at 300 rpm. If the power needed at output *A* is 40 kW and the power needed at output *B* is 80 kW, calculate the diameter of the solid steel shafts from *AC* and *BC*. Consider the allowable shear stress of 65 MPa to control the torque in the steel shafts.

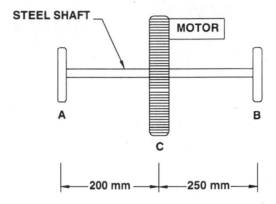

23. If the steel shaft in exercise #22 was 50 mm in diameter, and the motor was to deliver 40% of its power to output *A* and the other 60% of its power to output *B*, calculate the maximum power which can be delivered to the shaft. The allowable shear stress controls the torque and is still 65 MPa.

11

SHEAR FORCES AND BENDING MOMENTS

11.1   Introduction

One of the most common elements used in buildings and other structures is the beam. Beams can be made from almost any type of material and are used throughout the construction world (Figure 11.1). Before the student attempts to learn about the behavior of this structural element, it is imperative that they first learn about the forces (and moments) which act upon the beam. This chapter will explore the two fundamental influences on a beam—**shear force** and **bending moment**. These two influences will be studied to identify their nature and characteristics on typical beams.

11.2   Types of Loads on a Beam

A beam can have many different types of applied loadings. When dealing with the subject of beam design, a load may be classified as either a **dead load** or **live load** based on the permanancy of the load. Dead loads are all permanent loads which are imposed on a structure. Such loads will include the structure's selfweight, piping, conduit, permanent equipment, and permanent furniture. Live loads are simply instantaneous or moveable loads, produced in a structure by the occupancy of people or mobile equipment, such as furniture. Live loads can vary and be somewhat unpredictable with regard to magnitude, location, and duration.

Figure 11-1 Steel Beams in Construction. (Courtesy Bethlehem Steel Corporation)

The subject of strength of materials is more concerned with how particular loads will influence beam behavior. Therefore, the two basic types of loads, as far as the behavior of a beam is concerned, are **concentrated (or point) loads** and **uniformly distributed loads** (Figure 11-2). A concentrated load applies a force to the beam in a very specific location, while a uniform load will apply its force to the beam over an area. A real life example of a concentrated load would be a reaction from a column post which bears directly on a beam. Since the force in the column post applies the column's force to the beam in a relatively small area, it is considered a concentrated load. An example of a uniform load may be the weight of snow on a roof. This snowfall, measuring some depth, would weigh a certain amount which is "spread out" over the roof and therefore applied to the roof beams over a broad area.

There can be several types of distributed loads besides the uniformly distributed load. Such types could include triangular, trapezoidal, and parabolic. The main focus of this chapter will deal with concentrated and uniformly distributed loads, although the other types of distributed loads will also be briefly mentioned.

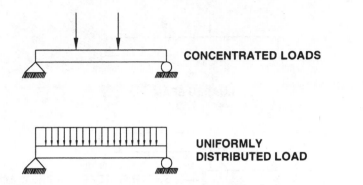

CONCENTRATED LOADS

UNIFORMLY DISTRIBUTED LOAD

Figure 11-2   Representation of Concentrated versus Uniformly Distributed Loads

11.3   Shear Forces Along a Beam

A standard beam is designed to resist loads that are usually applied perpendicular to its longitudinal axis. As these loads are applied, forces develop at the supports such that the beam is in equilibrium. (In Chapter 4 the principles of equilibrium were used to solve the support reactions.) Although the beam is in equilibrium externally, it also has to resist forces that are trying to rip the beam apart internally as well. This type of force is referred to as a shear force.

Shear forces occur across the length of the beam, generally varying in magnitude. A shear force acting in a vertical direction, if large enough to overcome a beam's resistance, would literally rip a beam apart by displacing one side of the beam relative to the other (Figure 11-3). Vertical shear forces are also accompanied by horizontal shear forces, which can lead to several types of failure in a beam (Figure 11-4). One type of shear failure is due to the vertical shear force and the horizontal shear force forming a resultant, which may be characterized by a resulting diagonal type of fracture in some materials. Concrete beams are particularly susceptible to this diagonal shear fracture because of the material's inherent weakness against such forces. Another type of shear failure may occur solely due to the horizontal shear force overcoming a material weakness. Such can be the case with timber beams, where the horizontal component of shear force can cause a splitting of the wood between grains in areas of high shear. Therefore, shear force is a major concern in the design of beams and must be studied such that the designer can predict both its magnitude and locations of occurrence.

Before a designer can proceed with the design of a beam, the magnitudes of shear force across the length of the beam must be known. The *shear force at any location on the beam is simply the sum of all forces either to the right or the left of that location.* This definition is important to remember. To illustrate this con-

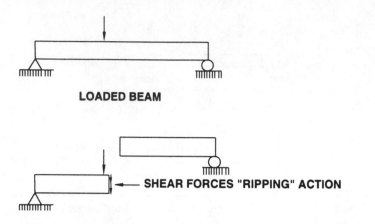

LOADED BEAM

SHEAR FORCES "RIPPING" ACTION

Figure 11-3   Conceptual Shear Failure in a Beam

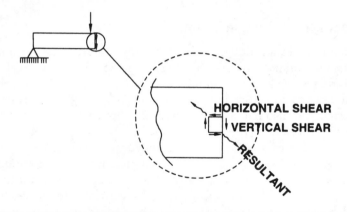

HORIZONTAL SHEAR

VERTICAL SHEAR

RESULTANT

Figure 11-4   Coupling of Vertical and Horizontal Shear

cept, let's consider the beam shown in Figure 11-5. If the shear force was desired at locations *A*, *B*, and *C* we could consider an imaginary cut occurring at each of those points. The shear force at each of those points, by the aforementioned definition, is simply the sum of all forces either to the right or the left of that location. For simply supported beams, it is typical to work from left to right, considering forces acting downward as negative. Therefore the shear forces are as follows:

Point A

$$\Sigma V_A = +14 \text{ kips} - 0 = +14 \text{ kips}$$

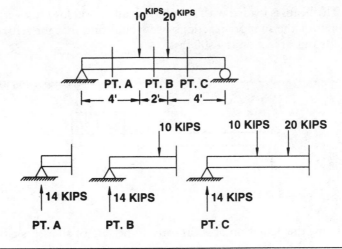

Figure 11-5   Shear at Points Along a Beam

Point B

$$\Sigma V_B = +14 \text{ kips} - 10 \text{ kips} = +4 \text{ kips}$$

Point C

$$\Sigma V_C = +14 \text{ kips} - 10 \text{ kips} - 20 \text{ kips} = -16 \text{ kips}$$

The positive and negative signs are arbitrary, although they follow the author's convention used throughout the text where forces going up are positive and down are negative.

This illustration showed that the shear forces at any location on a beam loaded with concentrated forces can be calculated through the creation of smaller free-bodies of the beam. The following example will utilize this same technique on a beam loaded with a uniform load.

EXAMPLE 11.1

Determine the shear force on the beam shown at 2 ft. intervals.

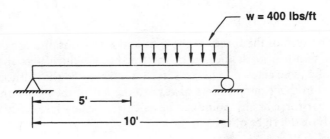

The beam is loaded with a uniform load of 400 lbs / ft. over the right five feet. Solving for the support reactions, we would find that the left reaction is 500 lbs and the right reaction is 1500 lbs.

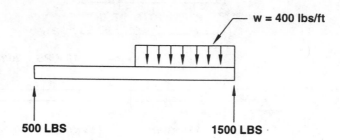

Using the definition of shear force ("the sum of all forces either to the right or the left of the desired locations"), we will proceed to free bodies at the 2 ft., 4 ft., 6 ft., and 8 ft. locations.

The sum of the forces would indicate that the shear forces are as follows:

$$\Sigma V_{2ft.} = +500 \text{ lbs}$$

$$\Sigma V_{4ft.} = +500 \text{ lbs}$$

$$\Sigma V_{6ft.} = +500 \text{ lbs} - 400 \text{ lb/ft.}(1 \text{ ft.}) = +100 \text{ lb}$$

$$\Sigma V_{8ft.} = +500 \text{ lbs} - 400 \text{ lb/ft.}(3 \text{ ft.}) = -700 \text{ lb}$$

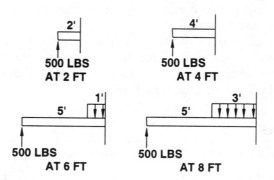

The sum of the forces at the 6 ft. and 8 ft. locations considered the portions of the uniform loads located of the free-body, which was shown. The uniform load of course acted downwards. It should be noted that although the forces considered in this example were always to the left of the point under consideration, the shear force at any point can be calculated using free-body diagrams either to the left or the right of the point.

11.4   Bending Moment Along a Beam

In addition to the shear forces that act on a beam, another important influence must be considered before design can proceed. Bending moments are also caused by the application of loads on the beam. The moment about a point is the product of a force and a distance. Bending moments on a beam occur simultaneously with the aforementioned shear forces, although the type of failure such moments may cause is dramatically different than those caused by shear.

Bending moments that occur along the length of the beam cause the beam to rotate or deflect by some magnitude at each particular location. In essesse, bending moments will try to rotate the beam and, unless internal resistance is achieved, this rotation would becomes excessive and the beam will collapse (Figure 11-6). Because of this behavior it is important to determine the magnitude and location of bending moments as they occur over a beam. Bending moments will vary in magnitude over the length of the beam, although the designer is usually most interested in the maximum value. This determination of the magnitude and location of the bending moment that varies over a beam will be critical for many reasons; including the selection of proper beam size, the proper location of lateral bracing, and even where a beam should be spliced.

The *bending moment at any location on a beam is simply the sum of the moments either to the left or to the right of that point.* The sum of the moments is calculated by taking all the forces (either to the left or right of the point under consideration) and multiplying by the distance from that force to the point where bending moment is being considered. To illustrate this concept, let's consider the beam shown in Figure 11-7. If the bending moment was desired at locations *A*, *B*, and *C,* one could consider an imaginary cut occurring at each of those points. The bending moment at each of those points, by the aforementioned definition, is sim-

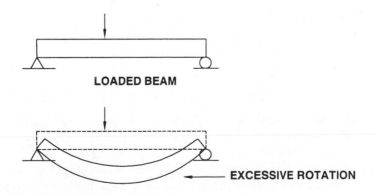

Figure 11-6   Potential Bending Failure in a Beam

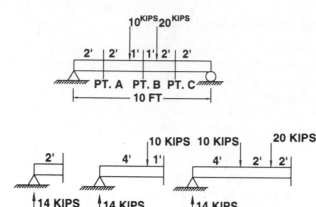

Figure 11-7 Bending Moment at Points along the Beam

ply the sum of all moments either to the right or the left of that location. Therefore, working from the free-bodies shown, the bending moments are as follows:

Point A

$$\Sigma M_A = +14 \text{ kips}(2 \text{ ft.}) = +28 \text{ kip-ft.}$$

Point B

$$\Sigma M_B = +14 \text{ kips}(5 \text{ ft.}) - 10 \text{ kips}(1 \text{ ft.}) = +60 \text{ kip-ft.}$$

Point C

$$\Sigma M_C = +14 \text{ kips}(8 \text{ ft.}) - 10 \text{ kips}(4 \text{ ft.}) - 20 \text{ kips}(2 \text{ ft.}) = +32 \text{ kip-ft.}$$

The positive and negative signs are arbitrary, although they follow the author's convention used throughout the text where moments in the clockwise direction are positive and counterclockwise are negative.

This illustration showed that the bending moments at any location on a beam loaded with concentrated forces can be calculated through the creation of smaller free-bodies of the beam. The following example will utilize this same technique on a beam loaded with a uniform load.

EXAMPLE 11.2

Determine the bending moment on the beam shown at 2 ft. intervals.

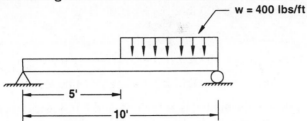

The beam is loaded with a uniform load of 400 lbs / ft. over the right five feet. As we have shown previously, the support reactions are 500 lbs on the left and 1500 lbs on the right.

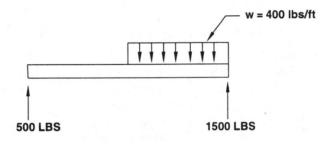

Using the definition of bending moment (" the sum of all moments either to the right or the left of the desired locations") we will proceed to draw free bodies at the 2 ft., 4 ft., 6 ft., and 8 ft. locations.

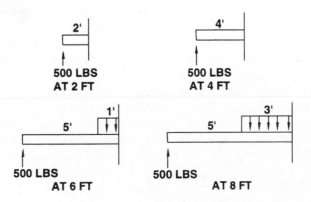

The sum of the moments would indicate that the shear forces are as follows (CW is positive):

$$\Sigma M_{2ft.} = +500 \text{ lb}(2 \text{ ft.}) = +1000 \text{ ft.-lb}$$

$$\Sigma M_{4ft.} = +500 \text{ lb}(4 \text{ ft.}) = +2000 \text{ ft.-lb}$$

$$\Sigma M_{6ft.} = +500 \text{ lb}(6 \text{ ft.}) - 400 \text{ lb/ft. } (1 \text{ ft.})(0.5 \text{ ft.}) = +2800 \text{ ft.-lb}$$

$$\Sigma M_{8ft.} = +500 \text{ lb}(8 \text{ ft.}) - 400 \text{ lb/ft.}(3 \text{ ft.})(1.5 \text{ ft.}) = +2200 \text{ ft.-lb}$$

The sum of the moments at the 6 ft. and 8 ft. locations considers the portions of the uniform loads located off the free-body that was shown. The uniform load of course acts downwards and its resultant lies in the center of the uniformly loaded portion. For the moment at the 8 ft. section, the 3 ft. long section of this uniform load would weigh 1200 lbs (400 lb/ft. × 3 ft.) and act in the center of that 3 ft. section. This would be 1.5 ft. away from the section.

11.5   Shear Diagrams and Bending Moment Diagrams

The previous two sections investigated the magnitude of shear force and bending moment that stems from forces being applied to a beam. Knowledge regarding these two items is necessary so the designer can know the forces and bending moments for which a beam must be designed. However, the method that was used to calculate the shear force and bending moment was tedious and time-consuming. There is, however, another method used to calculate the shear and bending moments that is much easier and at the same time shows the magnitude of shear and moment at all points along a beam. This method presents itself in the graphical technique referred to as shear and moment diagrams.

Shear diagrams are graphical illustrations of the shear force over the length of a beam. The shear diagram is typically drawn directly underneath the free-body diagram of the beam and will visually describe how the shear force varies in magnitude over the length of the beam. Similarly, the moment diagram will graphically illustrate the magnitude of bending moment over the length of a beam. It will usually be drawn beneath the shear diagram and will describe how the moment will vary over the beam length. Shear and moment diagrams are the cornerstone of beam design, since both convey important information to the designer. This information includes the magnitude of the maximum shear and bending moment as well as all other values. The diagrams will also detail the location where such values occur, which is very important for the construction of many types of beams. For instance, if beams are to be spliced together it is usually advisable for the splice to be located away from the location of maximum moment since this would tend to "pull" the flanges apart. The following sections will describe the construction of each diagram in detail.

Shear Diagrams

Shear diagrams are graphical representations of how the shear force varies over the length of the beam. To best illustrate the relationship between the applied loads on the beam and the shear forces these diagrams are usually shown directly beneath the free-body diagram of the beam.

It has already been stated that the shear force at any point along the beam is simply the sum of all force either to the left or right of that point. To begin with, let's consider the effect of a concentrated load on the sum of the forces on any given beam. The application of a concentrated force would algebraically increase (if it acted upwards) or decrease (if it acted downwards) the sum of the forces by the magnitude of the force. This increase or decrease would occur at the location of the concentrated force. Therefore, if we are attempting to graphically illustrate the magnitude of shear forces along a beam, we can state with certainty that *a concentrated load will increase or decrease the shear diagram at the location of the concentrated force by the magnitude of the force. This increase or decrease will occur at the location of the force.* This means that the shear diagram will move vertically under the influence of the concentrated force by the magnitude of the concentrated force.

Consider the effect of a uniformly distributed load on the sum of forces along on any given beam. A uniformly distributed load typically acts downward and at a uniform magnitude per unit length. For instance, a uniformly distributed load of 5 kips/ft. would decrease the net shear by 5 kips for every foot of its length. To graphically illustrate the effect of a uniformly distributed load on the shear force, we can state with certainty that *a uniformly distributed load will change the shear force at a uniform rate equal to the intensity of the load.* This means that the shear diagram will be represented by a diagonal line that has a slope equal to the intensity of the load. A uniformly distributed load having an intensity of 4 kips/ft. will have a slope twice as great as a uniformly distributed load of 2 kips/ft.

The following examples will illustrate the construction of shear forces on a number of different beams.

EXAMPLE 11.3

Construct a shear diagram for the following beam.

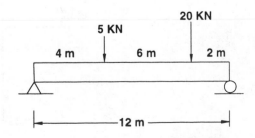

Begin by solving the external support reactions. This is accomplished by using the laws of equilibrium and we would find that the left reaction is 6.67 KN, upwards, while the right reaction is 18.33 KN upwards. The free-body solution would be as follows:

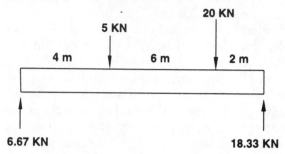

The shear diagram for this beam is a graphical representation of the shear force across the beam. Since the beam is loaded only by concentrated loads (the applied loads and the support reactions), the shear diagram will move up and down (vertically) under the application of each load. Starting on the left side, these variations of shear could be catalogued as follows:

- Left reaction causes a shear force upwards of +6.67 KN
- Shear force does not change until the first 5 KN load, the shear decreases at this point to +1.67 KN as follows:

$$+6.67 \text{ KN} - 5 \text{ KN} = +1.67 \text{ KN}$$

- Shear force does not change again until the 20 KN load, where the shear would decrease by 20 KN to a value of –18.33 KN as follows:

$$+1.67 \text{ KN} - 20 \text{ KN} = -18.33 \text{ KN}$$

- Shear force does not change again until the right reaction where an upwards force of 18.33 KN will make the sum of all force at that point zero:

$$-18.33 \text{ KN} + 18.33 \text{ KN} = 0$$

These steps will be represented in the shear diagram as follows:

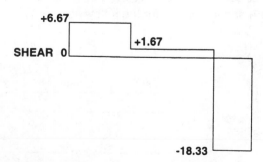

Therefore, the maximum shear along the beam is 18.33 KN and this shear occurs over the rightmost 2 m of the beam. Note that the shear diagram always ends at zero. This must occur for the beam to be in equilibrium.

The previous example illustrated the technique of constructing a shear diagram when all the forces on the beam were concentrated. How does this technique change when uniformly distributed loads are applied? Remember that a distributed load will decrease the shear force by an amount and rate equal to the intensity of load. Although uniformly distributed loads will cause continual variations of shear force (meaning that the magnitude of shear changes from point to point), the overall technique remains consistent with that which has already been discussed. The following examples will demonstrate the construction of shear forces on beams with uniformly distributed loads.

EXAMPLE 11.4

Construct a shear diagram for the beam shown below.

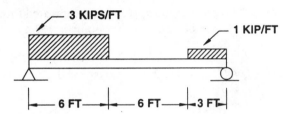

Begin by solving the external support reactions. This is accomplished by using the laws of equilibrium and we would find that the left reaction is 14.7 kips upwards, while the right reaction is 6.3 kips upwards. The free-body solution would be as follows:

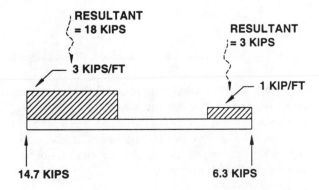

The shear diagram for this beam is a graphical representation of the shear force across the beam. Since the beam is loaded with concentrated loads at the reactions, the shear diagram will move up (vertically) under the application of force at each reaction. The uniformly distributed loads will decrease by a rate equivalent to their intensity. This means the left-hand uniform load will decrease by 3 kips/ft. or by a total of 18 kips over its length. Similarly, the right-hand uniform load will decrease by 1 kip every foot or by a total of 3 kips. Starting on the left side, these variations of shear could be catalogued as follows:

- The left reaction causes an increase of 14.7 kips, shear moves from zero to +14.7 kips
- Going to the right, the shear changes (decreases) at a constant rate of 3 kips/ft., dropping a total of 18 kips over the 6 ft. length of the uniform load. This decreases shear to −3.3 kips as follows:

$$14.7 \text{ kips}-(3 \text{ kips/ft.})(6 \text{ ft.}) = -3.3 \text{ kips}$$

- No change of shear force occurs until the final uniform load, which decreases shear at a constant rate of 1 kip/ft., dropping a total of 3 kips over the 3 ft. length. This decreases shear to −6.3 kips, as follows:

$$-3.3 \text{ kips}-(1 \text{ kip/ft.})(3 \text{ ft.}) = -6.3 \text{ kips}$$

- Finally, the right reaction applies a concentrated force of 6.3 kips upwards, bringing the sum of forces at that point to zero, as shown below:

$$-6.3 \text{ kips} + 6.3 \text{ kips} = 0$$

These steps will be represented in the shear diagram as follows:

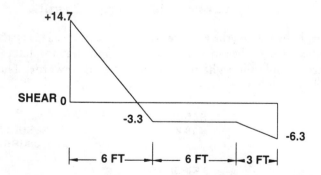

The diagram shows a maximum shear at the left reaction of 14.7 kips. It will later become important to know the location of zero shear. Notice that the shear begins and ends at zero, but it is also zero somewhere in the "middle" of the 3 kip/ft. uniform load. Knowing that the load decreases along a straight line at a "slope" of −3 kip/ft., we can find where the location of zero shear is by using the following straight line relationship:

$$y - mx = 0$$

where

> m = slope of the diagonal line (equal to the intensity of the uniform load)
>
> y = magnitude (y value) of shear , typically at the beginning of the load
>
> x = distance from the beginning of the load to the point of zero shear

Rearranging to solve for x:

$$x = y / m$$

$$x = 14.7 \text{ kips} / (3 \text{ kip/ft.}) = 4.9 \text{ ft.}$$

Therefore, the shear is zero at 4.9 ft. from the beginning of the uniform load.

EXAMPLE 11.5

Construct a shear diagram for the beam shown below.

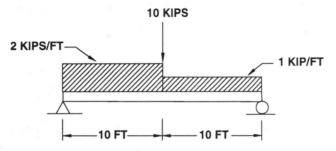

Begin by solving the external support reactions. This is accomplished by using the laws of equilibrium and we would find that the left reaction is 22.5 kips upwards, while the right reaction is 17.5 kips upwards. The free-body solution would be as follows:

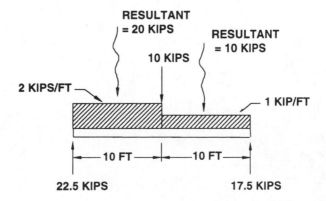

The shear diagram for this beam is a graphical representation of the shear force across the beam. Since the beam has concentrated loads (at the reactions and the 10 kip load in the center of the beam), the shear diagram will move up and down (vertically) under the application of each of these loads. The uniformly distributed loads will decrease by a rate equivalent to their intensity. This means the left-hand uniform load will decrease by 2 kips every foot or by a total of 20 kips over its 10 ft. length. Similarly, the right-hand uniform load will decrease by 1 kip every foot or by a total of 10 kips. Starting on the left side, these variations of shear could be catalogued as follows:

- The left reaction causes an increase of 22.5 kips, shear moves from zero to +22.5 kips
- Going to the right, the shear changes (decreases) at a constant rate of 2 kips/ft., dropping a total of 20 kips over the 10 ft. length of this uniform load. This decreases shear from +22.5 kips to +2.5 kips, as follows:

$$+22.5 \text{ kips} - (2 \text{ kips/ft.})(10 \text{ ft.}) = +2.5 \text{ kips}$$

- At this point the 10 kips concentrated load decreases the shear immediately. This decreases the shear from +2.5 kips to −7.5 kips.

$$+2.5 \text{ kips} - 10 \text{ kips} = -7.5 \text{ kips}$$

- Progressing to the right, the final uniform load decreases shear at a constant rate of 1 kip/ft., dropping a total of 10 kips over the 10 ft. length. This decreases shear to −17.5 kips.

$$-7.5 \text{ kips} - (1 \text{kip/ft.})(10 \text{ ft.}) = -17.5 \text{ kips}$$

- Finally, the right reaction applies a concentrated force of +17.5 kips upwards, bringing the sum of forces at that point to zero.

$$-17.5 \text{ kips} + 17.5 \text{ kips} = 0$$

These steps will be represented in the shear diagram as follows:

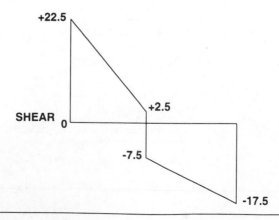

Bending Moment Diagrams

Moment diagrams are graphical representations of how the bending moment varies over the length of the beam. To best illustrate the relationship between the shear forces and the bending moments on a beam, the moment diagram is usually shown directly beneath the shear diagram.

It has already been stated that the bending moment at any point along the beam is simply the sum of all forces either to the left or right of that point, multiplied by their individual distances to that point. Stating this in a simplified manner, the bending moment at any point is the sum of the moments to the right or left of that point. Previously, it was illustrated that the sum of all forces at any point along the beam was represented by the construction of the shear diagram. How can the bending moment along the beam be similarly illustrated?

The answer is through understanding the relationship between shear forces and bending moments. Any force applied to a beam causes a change in the shear at that point along the beam. This change is represented by the movement of the shear diagram. The magnitude of the shear diagram is to represent the *sum* of all shear forces to the right or to the left of that point. Likewise, any force applied to a beam also creates bending moment, which is defined by the force multiplied by a perpendicular distance to the point under consideration. The moment diagram is supposed to represent the *sum* of all bending moments to the right (or to the left) of any point under consideration. The sum of all bending moments to the right (or left) of a point is simply the the sum of all the forces, multiplied by the distance to that point. Since the shear diagram already shows the magnitude of the sum of the forces, all that is necessary to find bending moment is the distance to the point of consideration. This means that *the bending moment at any point along the beam can be found by summing the areas contained under the shear diagram.* The area contained under the shear diagram represents moment because the magnitude of the shear diagram represents the sum of the forces and the length to a particular point represents the distance. Individual "areas" under the shear diagram can be calculated and algebraically added to obtain the bending moment up to any point along the beam.

This technique will prove to be much quicker than the tedious point-by-point calculation of bending moments, shown earlier in the chapter. There are a number of points to be made with the bending moment distribution along the beam and these are as follows:

- The bending moment diagram will start and end at zero. Any moment other than zero indicates a beam not in equilibrium under the applied loading.
- The location of zero shear will prove to be an important location along the beam. This will be a location of maximum or minimum moment.
- "Rectangular" or "square" areas under the shear diagrams will translate into straight, diagonal lines on the moment diagram. Each incremental area, when cumulatively added within these rectangles, would indicate that

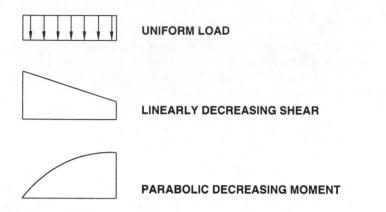

Figure 11-8   Relationship Between Uniform Load, Shear and Moment

the moment increases (or decreases) in a uniform manner. The slope of such a line will be dependent on the magnitude of shear force.

■ Triangular, trapezoidal, and other types of areas under the shear diagram will translate into curved (parabolic) lines on the moment diagram. When cumulatively adding each incremental piece of area in these shapes, the moment will increase (or decrease) at a nonuniform rate. The intended shape of these curves is outlined in Figure 11-8.

The following examples will demonstrate the construction of moment diagram based on the technique of adding areas under the shear diagram.

EXAMPLE 11.6

Construct a moment diagram for the following beam.

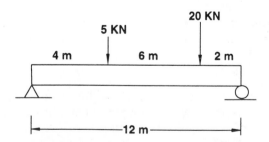

This problem is the same as example 11.3, which we have previously completed. Based on the results of example 11.3, we know the shear diagram is as follows:

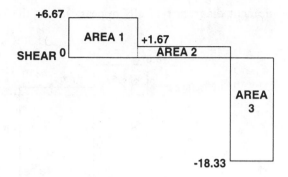

The moment diagram can be constructed utilizing the areas under the shear diagram for this beam. Begin by breaking the shear diagram into familiar geometric shapes (rectangles, triangles, etc.). In this problem, all the areas are rectangular as shown in the shear diagram and are calculated as follows:

$$\text{Area 1} = (+6.67 \text{ KN})(4 \text{ m}) = +26.67 \text{ KN-m}$$

$$\text{Area 2} = (+1.67 \text{ KN})(6 \text{ m}) = +10.00 \text{ KN-m}$$

$$\text{Area 3} = (-18.33 \text{ KN})(2 \text{ m}) = -36.67 \text{ KN-m}$$

The moment diagram is the graphical representation of bending moment along the beam and starting from the left side, the moment diagram could be defined by summing the aforementioned areas as follows:

■ Starting at zero, the moment increases linearly to a value of +26.67 KN-m after the first four feet. The straight line is due to the fact that all incremental pieces of area within this rectangle are the same. Therefore, the cumulative sum of these areas will yield a straight line.

■ From the 4 m mark, the moment diagram increases linearly to a value of +36.67 KN-m at the 10 m mark. This is obtained by adding area 1 and area 2. This is also a straight line, due to the fact that each incremental piece of area within the second area is the same.

$$+26.67 \text{ KN-m} + 10 \text{ KN-m} = 36.67 \text{ KN-m}$$

■ From the 10 m mark, the moment diagram decreases linearly to zero at the end of the beam. This is obtained by adding area 3 to the previous area sum of +36.67 KN-m. This appears accurate because the moment diagram should close to zero.

$$+36.67 \text{ KN-m} - 36.67 \text{ KN-m} = 0$$

The moment diagram typically appears under the shear diagram and is as follows:

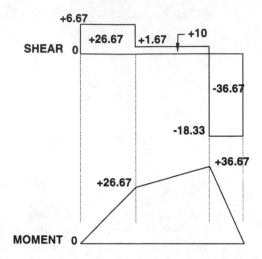

Notice that the maximum bending moment (+36.67 KN-m) in this example occurs at the 10 m mark. This coincides with the location where the shear diagram crosses zero.

The previous example illustrated the construction of a moment diagram when all forces on the beam were concentrated. The following examples will deal with the construction of moment diagrams for beams which also have uniformly distributed loaded applied to them. The technique used for uniformly distributed loads will be exactly the same as shown previously although the moment diagrams will have curved segments. This is because the cumulative addition of incremental areas is occurring at a constantly increasing or decreasing rate.

EXAMPLE 11.7

Construct the moment diagram for the beam shown below.

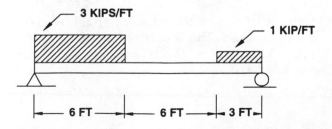

This problem uses the same beam as was used in example 11.4, for which we have already solved and constructed the shear diagram. The shear diagram for this beam is as follows:

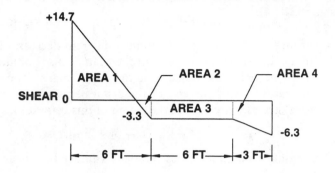

Because of the uniformly distributed loads, the areas under the shear diagram are not all rectangular as they were in the previous example. Still we proceed in a same manner—breaking the areas under the shear diagram into typical geometric shapes and calculating each area. This can be done as follows:

Area 1 = 1/2(4.9 ft.)(14.7 kips) = +36.02 kip-ft.

Area 2 = 1/2(1.1 ft.)(–3.3 kips) = –1.82 kip-ft.

Area 3 = (6 ft.)(–3.3 kips) = –19.8 kip-ft.

Area 4 = ((–3.3 kips + – 6.3 kips)/2)(3 ft.) = –14.4 kip-ft.

We have already stated that the moment diagram is the graphical representation of bending moment along the beam and starting from the left side the moment diagram could be defined by summing the aforementioned areas as follows:

- Starting at zero, the moment increases in a parabolic curve to a value of +36.02 kip-ft. after the first 4.9 feet. The concave parabolic line starts out with a steep slope which decreases constantly over the 4.9 ft. length. This due to the fact that all incremental pieces of area within this triangle get smaller throughout the area. The 4.9 ft. mark is the location of zero shear.
- From the 4.9 ft. mark, the moment diagram decreases to a value of +34.2 kip-ft. at the end of the 6 ft. mark. This is obtained by adding area 1 and area 2. This is again a parabolic line which is actually a continuation of the first curve. This line curves downward at an ever increasing rate due to the fact that each incremental piece of area within this small triangle increases in area.

+36.02 kip-ft. – 1.82 kip-ft. = 34.2 kip-ft.

- From the 6 ft. mark, the moment diagram decreases linearly to +14.4 kip-ft. at the 12 ft. mark by adding area 3 to the previous moment value of 34.2 kip-ft. This is a straight diagonal line because each incremental piece of area with area 3 will decrease the cumulative moment by the same amount.

+34.2 kip-ft. – 19.8 kip-ft. = +14.4 kip-ft.

- From the 12 ft. mark, the moment diagram decreases in a curved fashion to a value of zero at the end of the beam. This is obtained by adding area 4 to the previous moment at the 12 ft. mark. The slope of the curve increases toward the end of the beam because the incremental pieces of area increase throughout the length of this trapezoid.

+14.4 kip-ft. – 14.4 kip-ft. = 0

The moment diagram typically appears under the shear diagram and is as follows:

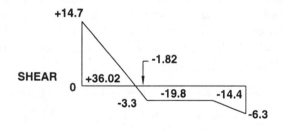

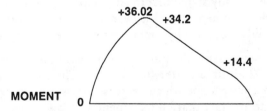

Notice that the maximum bending moment (+36.02 kip-ft.) in this example occurs at the 4.9 ft. mark. This marks the location where the shear diagram crosses zero, which is very important because this coincides with the location of maximum moment. Locations of zero shear will always be either local minimum or maximum values of moment.

EXAMPLE 11.8

Construct the moment diagram for the beam shown below.

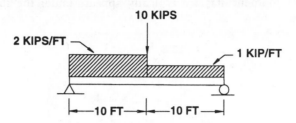

This problem uses the same beam as was used in Example 11.5 and for which we have already solved and contructed the shear diagram. The shear diagram for this beam is as follows:

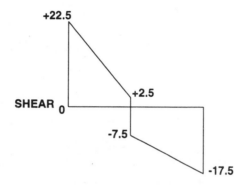

Because of the uniformly distributed loads, the areas under the shear diagram are trapezoidal in this case. We solve the moment diagram by calculating each area. This can be done as follows:

$$\text{Area 1} = ((+22.5 \text{ kips} + 2.5 \text{ kips})/2)(10 \text{ ft.}) = +125 \text{ kip-ft.}$$

$$\text{Area 2} = ((-17.5 \text{ kips} -17.5 \text{ kips})/2)(10 \text{ ft.}) = -125 \text{ kip-ft.}$$

Starting from the left side, the moment diagram could be defined by summing the aforementioned areas as follows:

- Starting at zero, the moment increases in a parabolic curve to a value of +125 kip-ft after the first 10 feet. The concave parabolic line starts out with a steep slope that continually decreases over the 10 ft. length of the first area. This due to the fact that all incremental pieces of area within this trapezoid get smaller throughout the area.
- From the 10 ft. mark, the moment diagram decreases to a value of zero at the end of the beam. This is obtained by adding area 2 to the previous value of moment (+125 kip-ft.). This is again another parabolic line that

curves downward at an ever increasing rate, due to the fact that each incremental piece of area within this trapezoid increases in area.

The moment diagram typically appears under the shear diagram and is as follows:

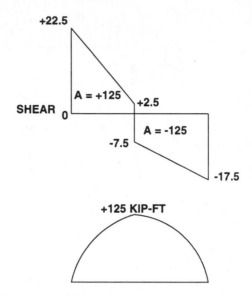

Notice that the maximum bending moment (+125 kip-ft.) in this example occurs at the 10 ft. mark, which coincides with the location of zero shear.

Relationship Between Load, Shear, and Moment

The shear and moment diagrams that were discussed previously are graphical representations of the magnitudes of shear and bending moments over the length of a beam. If a section of a beam is considered between points 1 and 2 on Figure 11-9, the vertical forces that act on that section can be defined by the following equilibrium equation:

$$\Sigma F_y = 0$$

$$+V_1 - w\Delta x - V_2 = 0$$

where

V_1 = shear force at section 1

$w\Delta x$ = additional weight in the incremental x distance

V_2 = shear force at section 2

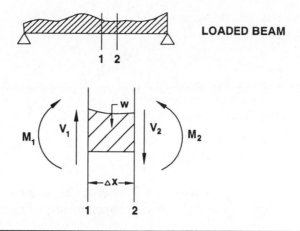

Figure 11-9 Shear Forces and Moments on Incremental Section

Rearranging the previous equation it can be seen that:

$$V_2 = + V_1 - w\Delta x$$

This equation states that the change of shear between any two points is equal to the external load applied between those two points.

Similarly, if moments are summed about the right-hand edge of the incremental section (1-2), the following moment equation can be obtained:

$$\Sigma M_{right} = 0$$

$$+V_1\Delta x + M_1 - w\Delta x(\Delta x / 2) - M_2 = 0$$

where

$$M_1 = \text{bending moment at section 1}$$

$$w\Delta x(\Delta x / 2) = \text{additional moment in the incremental } x \text{ distance}$$

$$M_2 = \text{shear force at section 2}$$

$$V_1 = \text{shear force at section 1}$$

Rearranging the previous equation it can be seen that:

$$M_2 = +V_1\Delta x + M_1 - w\Delta x(\Delta x / 2)$$

When Δx is small, the term $w\Delta x(\Delta x\,/\,2)$ becomes insignificant and the moment equation becomes:

$$M_2 = +V_1\Delta x + M_1$$

This equation states that the change of bending moment between any two points is equal to the moment at the beginning of that section plus the moment caused by the shear force over that section. This can be expressed as follows:

$$M_2 = +V_1\Delta x + M_1$$

$$\Delta M = M_2 - M_1 = V_1\Delta x$$

The aforementioned expression is important because the term $V_1\Delta x$ can also be descibed by the area under the shear diagram between points 1 and 2.

11.6   Moving Loads

In the previous discussion, the shear and bending moments on a beam were caused by loads that were in a fixed position. However, on many beams, such as those on bridges, the loads move from one end of the beam to the other. It is very important on these beams to be able to determine the magnitude and location of the maximum shear and bending moment, such that the beam can be adequately designed.

Moving loads are typically assumed to be concentrated loads set at a fixed distance apart that move in unison across the beam. Such would be the case with the axle loads of a truck or a train (Figure 11-10). Since the moving loads are made up of individual concentrated loads, *the maximum shear occurs as one of the loads sits directly on a support.* To establish the maximum value of shear, the individual loads may be placed over a support (one at a time) and the reactions calculated in each case. The distance between wheel or axle loads should remain constant throughout this process since typically they will be fixed.

The maximum moment caused by a moving load will generally occur under one of the concentrated wheel or axle loads. This is due to the fact that the shear diagram on a simple beam will have to cross zero under one of these loads. The maximum moment caused by any concentrated load (within the set which comprises the moving load) will occur when the center of the simple beam bisects the distance between that load and the resultant of the full set. This can be seen graphically in Figure 11-11.

The maximum moment caused by the complete moving load can be found by alternating the different wheel loads until the maximum is found. It should be noted that this rule is general by nature and should only be applied to simple beams.

The following example will demonstrate this technique of finding maximum moments and shears caused by a moving load.

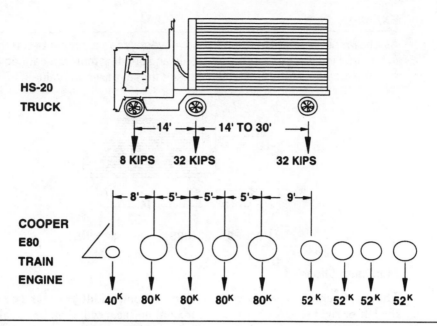

HS-20 TRUCK

COOPER E80 TRAIN ENGINE

Figure 11-10 Standard Truck and Train Loads

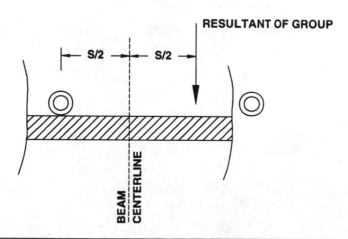

Figure 11-11 Condition of Wheel Load Causing Maximum Moment

EXAMPLE 11.9

As shown below, an HS-20 truck load moves across a simple beam that is 30 ft. in length. Find the maximum shear force and bending moment caused by this load and the location of the truck as it causes these maximum values.

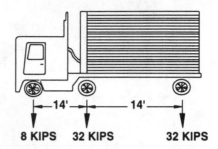

8 KIPS   32 KIPS   32 KIPS

Maximum Shear

The maximum shear will occur as either the front, middle, or rear axle load is over the left or right support because the maximum shear will then be equal to the support reaction. Therefore, let's "roll" the truck across the beam and calculate the support reactions at each location. This can be seen as follows:

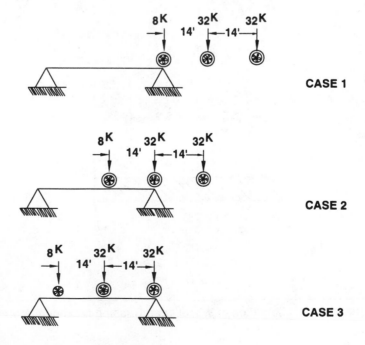

Case 1

R_r = 8 kips, R_l = 0

Case 2

R_r = 32 kips + 8 kips(16 ft. / 30 ft.) = 36.27 kips

R_l = 8 kips(14 ft. / 30 ft.) = 3.73 kips

Case 3

R_r = 32 kips + 32 kips(16 ft. / 30 ft.) + 8 kips(2 ft. / 30 ft.) = 49.6 kips

R_l = 22.4 kips

Therefore, the maximum shear occurs in Case 3 as the rear axle sits over the support and has a value of 49.6 kips.

Maximum Bending Moment

To determine the maximum bending moment caused by this truck load, we must first determine the location and magnitude of the resultant of the HS-20 truck. This is easily accomplished by using the principles learned in Chapter 6 (Centroids) and the resultant is 72 kips located at 18.67 ft. from the front axle, as shown below.

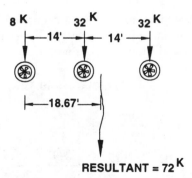

It was already stated that the maximum bending moment would be located under one of the concentrated "wheel" loads and would occur when the center of the beam was halfway between that wheel load and the resultant for the entire moving load. Since the HS-20 truck has three "wheel" loads, the maximum bending moment can be found by considering three separate cases—one with each of the wheel loads and the resultant straddling the center of the beam. These three cases can be seen as follows:

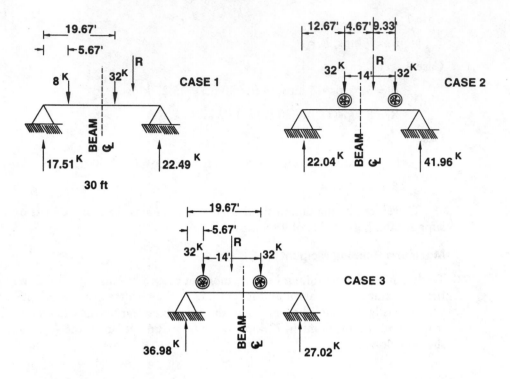

The maximum moment for Case 1 occurs under the middle axle and is calculated as follows by summing moments to either the left or the right of this axle. This is calculated and shown as follows:

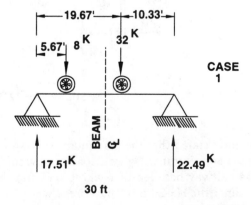

$$M_{\text{middle axle}} = 22.49 \text{ kips}(10.33 \text{ ft.}) = 232.4 \text{ kip-ft.}$$

The maximum moment for Case 2 occurs under the middle axle and is calculated as follows by summing moments to either the left or the right of this axle. This is calculated and shown as follows:

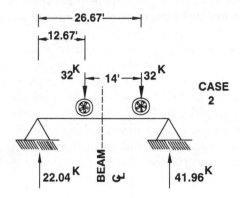

$$M_{\text{middle axle}} = 22.04 \text{ kips}(12.67 \text{ ft.}) = 279.2 \text{ kip-ft.}$$

The maximum moment for Case 3 occurs under the rear axle and is calculated as follows by summing moments to either the left or the right of this axle. This is calculated and shown as follows:

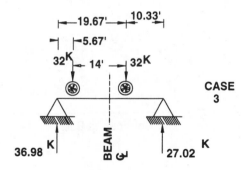

$$M_{\text{rear axle}} = 27.02 \text{ kips}(10.33 \text{ ft.}) = 279.2 \text{ kip-ft.}$$

The moment produced under Case 2 and Case 3 are the same and should be considered as causing the maximum moment for this load case.

11.7   Shear and Moment Diagrams for Continuous Beams

The shear and moments diagrams developed in this chapter have dealt with beams that generally spans between two supports, which is referred to as a simple beam.

Beams that span over more than two supports are referred to as continuous beams (Figure 11-12). These beams are frequently used in bridges and buildings because a bending moment is developed over the interior supports, thereby reducing the bending moment in the middle of the span. Since the bending moment is reduced in the middle of the span, a smaller beam section can be designed to adequately carry the load (Figure 11-13).

Continuous beams are more difficult to analyze because they are statically indeterminate— meaning that the support reactions cannot be solved using only the equations of equilibrium developed in Chapter 3. Methods for the solution of indeterminate structures is beyond the scope of this text, although one method is developed in Appendix F. There are, however, numerous beam tables that solve the reactions, shears, and moments for many types of continuous beams. Such beam tables are found in Appendix D. Once reactions are determined for continuous beams, the construction of shear and moment diagrams are developed using all the same principles as previously discussed.

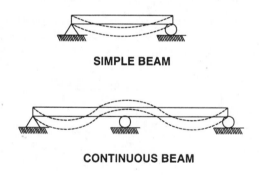

Figure 11-12 Simple and Continuous Beam

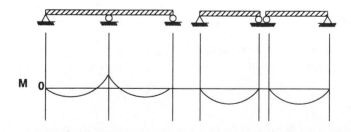

Figure 11-13 Comparison of Moment on Simple and Continuous Beams under Uniform Load

11.8   Summary

As loads are applied to a beam, the beam has to resist both shear forces and bending moments. Shear forces are those which tend to "rip" the beam apart, while bending moments tend to rotate or bend the beam. The magnitude of both shear forces and bending moments typically vary over the length of a beam, depending on the type and location of applied load. It is very important for the designer to have the ability to determine the variation of both shear and bending moment across the beam, as well as the magnitude of the maximum values for each.

The shear force at any location on a beam can simply be found by summing all the forces either to the right or left of the point under consideration. A graphical technique used to represent the shear force at every location across the beam is known as a shear diagram. This is a very useful tool for the designer because it visually locates the regions of high and low shear, which are imperative for design considerations. Similarly, the bending moment at any location across a beam can be found by summing all the moments that occur either to the right or to the left of that point. A graphical technique used to represent the bending moment at every point across the beam is known as a moment diagram. Moment diagrams can be found easily by summing the areas under the shear diagram to a particular point, since the area under a piece of the shear diagram represents force multiplied by distance. This technique is very fast and useful for the designer because it allows a quick determination of the regions of high and low moment as well as the magnitude and location of the maximum moment.

EXERCISES

1. Construct the shear and bending moment diagrams for the beam shown. List all pertinent values including location and magnitude of maximum shear and moment.

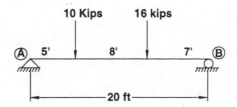

2. Construct the shear and bending moment diagrams for the beam shown. List all pertinent values including location and magnitude of maximum shear and moment.

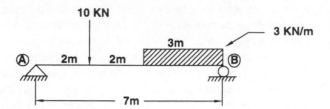

3. Construct the shear and bending moment diagrams for the beam shown. List all pertinent values including location and magnitude of maximum shear and moment.

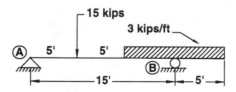

4. Construct the shear and bending moment diagrams for the beam shown. List all pertinent values including location and magnitude of maximum shear and moment.

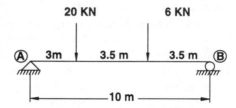

5. Construct the shear and bending moment diagrams for the beam shown. List all pertinent values including location and magnitude of maximum shear and moment.

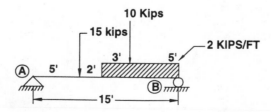

6. Construct the shear and bending moment diagrams for the beam shown. List all pertinent values including location and magnitude of maximum shear and moment.

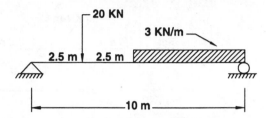

7. Construct the shear and bending moment diagrams for the beam shown. List all pertinent values including location and magnitude of maximum shear and moment.

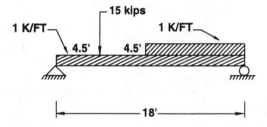

8. Construct the shear and bending moment diagrams for the beam shown. List all pertinent values including location and magnitude of maximum shear and moment.

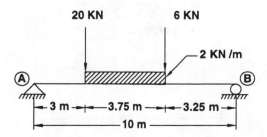

9. Construct the shear and bending moment diagrams for the beam shown. List all pertinent values including location and magnitude of maximum shear and moment.

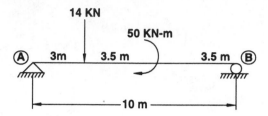

10. Construct the shear and bending moment diagrams for the beam shown. List all pertinent values including location and magnitude of maximum shear and moment.

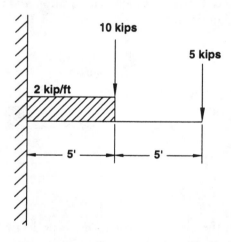

11. Construct the shear and bending moment diagrams for the beam shown. List all pertinent values including location and magnitude of maximum shear and moment.

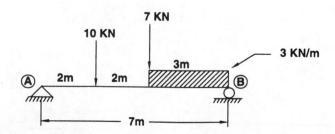

12. The moving load shown below travels across a simply supported beam of 20 meters. Determine the maximum shear and moment across the beam.

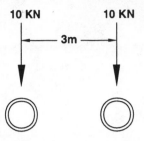

13. If the beam in the previous problem is only 15 m long, how does the maximum value of shear and moment change?

14. The moving load shown below travels across a simply supported 25-foot-long beam. Determine the magnitude of the maximum shear and bending moment and the position of the loads for each.

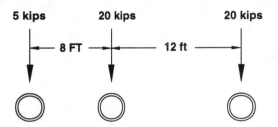

15. If the beam in the previous problem is 40 ft. long, how does the value of maximum shear and moment change?

CHAPTER

12

BEAM BEHAVIOR AND BENDING STRESS

12.1 Introduction

The previous chapter explored the shear forces and bending moments that were caused when forces were applied to a beam. Techniques were illustrated for determining the magnitude and location of shear forces and bending moments across the length of the beam. These shear forces and bending moments could be considered external because they result from the applied loads. To offset these external shears and bending moments, the beam must develop its own internal resistance. The beam's internal resistance to applied shear forces is referred to as **shear stress,** while its internal resistance to applied bending moment is referred to as bending or **flexural stress.** The primary topic of this chapter will be the development and calculation of bending and shear stresses within a beam. Shear stresses have been covered previously in Chapter 8 and will also be discussed at the end of this chapter.

Bending and shear stresses are very important to a designer since they may occur alone or in conjunction with other types of stresses that have been mentioned earlier. This chapter will study these stresses occurring alone, while Chapter 14 will explore the topic of combined stresses.

12.2 Behavior of a Beam in Bending

When a beam is subjected to a vertical force it responds by physically deflecting under that load. Consider the simply supported beam that is subjected to two con-

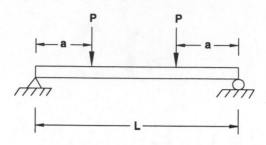

Figure 12-1 Beam Loaded with two Equal Loads.

centrated loads spaced an equal distance apart (Figure 12–1). This load will produce a section of the beam that is under uniform moment. It is also under zero
shear. The beam will deflect along its length and, if viewed more closely, stretch
along its bottom and compress along its top. This can be seen by viewing a pair
of points along the top (*A* and *B*) and a pair along the bottom (*C* and *D*). As the
beam bends, points *A* and *B* respond by being "squeezed" closer together, while
points *C* and *D* respond by stretching further apart (Figure 12-2).

This "squeezing together" on the top of the beam and "stretching apart" on
the bottom of the beam, when viewed in the context of the beam's overall length,
can be seen as flexural strain. It was previously discussed in Chapter 8 that tension
stress would cause a stretching of a member, while a compressive stress would
cause a shortening of a member. Thus is the case with a beam. Tension stress
develops on one side of the beam causing the elongation, while compressive
stresses on the other side of the beam cause the shortening of the member.

A closer look at Figure 12-2 would indicate that since there is tension on one
side and compression on the other, there is a location where the beam does not

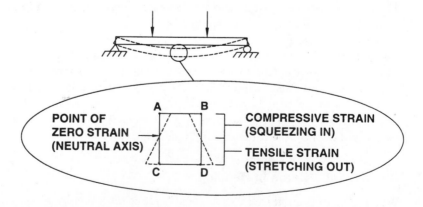

Figure 12-2 Bending Strains in a Simply Supported Beam

strain. This location is referred to as the **neutral axis**. The neutral axis is the plane within the beam that has zero strain. Because the neutral axis has zero strain, it must also correspondingly have zero bending stress. The cross-section of the beam can be assumed to be rotating or pivoting about this point. If the beam is homogeneous and symmetrical, the neutral axis will also coincide with the shape's centroidal axis. Another important feature of Figure 12-2 is the realization that this pivoting of strain about the beam's neutral axis occurs in a linear or straight line relationship. The importance of this item comes from the realization that the strains become greater as the distance away from the neutral axis increases. Therefore, the strain at the very outside of the beam (on both the top and the bottom) is the largest strain on the beam's cross-section. Consequently, the bending stresses at the very outside of the beam will likewise be the largest found over the cross-section of the beam.

These internally-developed bending stresses vary over the beam—from zero at the neutral axis to a maximum value at the extreme outside surface. It has also been determined that the stress on one side of the neutral axis will be in tension, while on the other side the stresses will be in compression. How exactly does this development of varying tension and compression stress resist the applied bending moment? The simple answer is: by developing a counteracting moment through the utilization of the bending stress. The bending stress, as previously mentioned, is distributed over the beam cross-section being tension on one side and compression on the other side. Instead of considering the distribution of stress, suppose one considered the resultant force, which, as we know from Chapter 8, is equal to the stress multiplied by the area (Figure 12-3). The resultant force on the tension side (T) would have to be equal to the resultant force on the compression side (C), due to horizontal equilibrium. This tension and compression resultant would be of equal magnitude and separated by some distance. This, as we have previously

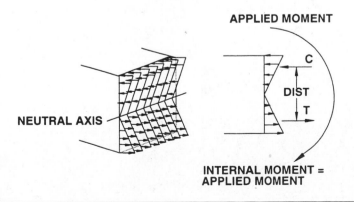

Figure 12-3   Bending Stress Distribution

discussed, is a case of a pure moment referred to as a couple. This *T-C couple* is the internal moment developed by the beam in response the externally applied moment. This internal moment must always be equal to the externally applied moment, for the beam to be in static equilibrium.

This mechanism is very revealing as to the true behavior of a beam with regard to bending moment. If a small amount of bending moment is applied to a beam, the bending stresses are similarly small, and a small internal moment is developed. However, as the moment on the beam becomes large, the bending stresses need to become large so the internal moment developed will be able to resist the applied moment. Should the stresses exceed some limiting amount which the material can accept, the material which the beam is made from will begin to break down. If this occurs, the equilibrium between externally applied moment and internal resisting moment will be lost and the beam will fail. The methods used to increase a beam's bending resistance would be (1) improve the material composition of the beam (change from wood to steel),(2) prevent a failure mechanism from occurring (brace a beam which is susceptible to buckling), and (3) increase the depth of the beam thereby "spreading out" the distance between the T and C resultants.

12.3 Bending Stress Formula

The beam behavior which we have considered up to this point has shown that the maximum strain occurs at the outermost fiber of the beam's cross-section. From Figure 12-4 the calculated flexural strain of B-B is one-half that of A-A. If the bending moment applied to the beam is low enough to keep the stresses under the material's proportional limit, then Hooke's law can be applied. Hooke's law

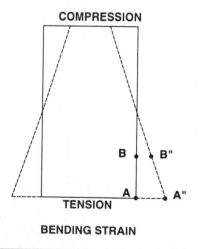

COMPRESSION

TENSION

BENDING STRAIN

Figure 12-4   Bending Strain in a Longitudinal Beam Section

would state that the flexural strains are related to the flexural stresses by a known material constant referred to as the **modulus of elasticity (E)**. This relationship can be expressed as follows;

$$\sigma_b = (E)\varepsilon_b$$

where

$$\sigma_b = \text{bending stress}$$

$$\varepsilon_b = \text{bending strain}$$

The aforementioned statement, regarding the proportionality of bending strains to bending stresses, has very important implications that must be grasped in order to understand the behavior of bending stresses caused by an applied moment. The most important implication is that bending stresses caused by a moment are proportional to the distance away from the neutral axis of the beam. This means that the "outer fibers" of the beam will have the highest bending stresses over the cross-section. The other implication is that an increase in area located away from the neutral axis will yield a section of increased resistance to bending.

When a moment, M, is applied to a beam, a counteracting moment must internally develop such that equilibrium is maintained. These bending stresses, which vary over the cross-section of the beam, will create the internal resisting moment.

From the discussion in Chapter 8, it is found that stress is equal to a force divided by some incremental unit of area ($\sigma = P / A$) or conversely, the force applied to an incremental piece of area is equal to the stress on that area multiplied by that area ($P = \sigma \times A$). If one would consider a small incremental piece of area somewhere over a cross-section of the beam (Figure 12-5), it could be seen that the bending stress on that area is proportional to the distance of that area from the neutral axis of the beam. This distance will be referred to as the distance c. The bending stress on this incremental area could then be described as follows:

$$\sigma_{\text{area}} = \sigma_b \, (y / c) \qquad \qquad \text{(Eq. 12-1)}$$

where

$\sigma_{\text{area}} = $ bending stress on the incremental area

$\sigma_b = $ maximum bending stress on the outside surface of the beam

$y = $ distance from the neutral axis to the incremental area being considered

$c = $ distance from neutral axis to the outside surface of the beam (1/2 beam depth)

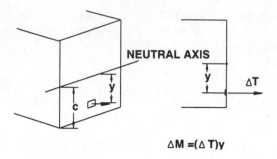

$$\Delta M = (\Delta T)y$$

Figure 12-5 Bending Stress on an Incremental Area of Beam

Each incremental piece of area will develop an incremental force on that area. Such incremental forces can be referred to as ΔT if they are tension forces and ΔC if they are compressive forces. To calculate the incremental amount of resisting bending force (ΔT or ΔC) developed by an incremental piece of area, we need only to take the bending stress developed in Equation 12-1 and multiply it by the area, as follows:

$$\text{incremental bending force} = \text{incremental stress } x \text{ area}$$

$$\Delta T \text{ (or } \Delta C) = (\sigma_{\text{area}}) (\Delta A)$$

or

$$\Delta T = [\sigma_b (y/c)] (\Delta A) \qquad\qquad \text{(Eq. 12-2)}$$

where

$$\Delta T = \text{incremental bending force (tension)}$$

$$\Delta A = \text{incremental piece of area}$$

This incremental bending force ΔT developed over ΔA is only one of many incremental bending forces that would be developed over the beam cross-section to resist the externally applied moment. However, forces alone do not resist a moment. An internal moment has to be generated in order to resist the externally applied moment. The internal resisting moment of the incremental area is calculated as the incremental force ΔT (or ΔC) multiplied by the distance away from the neutral axis y. This would make the internal resisting moment of the incremental area as follows:

$$\Delta M = (\Delta T) (y)$$

where

$$\Delta M = \text{incremental moment resistance over a piece of area}$$

$$y = \text{distance from the neutral axis to } \Delta A$$

By substituting in Equation 12-2 for the ΔT

$$\Delta M = ([\sigma_b\,(y\,/\,c)]\,[\Delta A])\,(y)$$

or

$$\Delta M = ([\sigma_b\,/\,c]\,[\Delta A y^2]) \qquad\qquad \text{(Eq. 12-3)}$$

Although the above expression is only for an increment of area, the total internal resistance of the beam can be thought of as the sum of all areas. This would change Equation 12-3 to the following:

$$\Sigma M = \Sigma(\sigma_b/\,c)\,(\Delta A y^2) \qquad\qquad \text{(Eq. 12-4)}$$

where

ΣM = total internal resistance of a beam to applied moment

From Chapter 7, it was learned that the term $\Sigma \Delta A y^2$ is a property of area referred to as the **moment of inertia, I.** This property helps to define a shape's resistance to bending moment, meaning that a high moment of inertia will reflect a larger resistance to bending moment. Substituting this value into Equation 12-4, we find:

$$\Sigma M = (\sigma_b\,/\,c)(I) \qquad\qquad \text{(Eq. 12-5)}$$

Since internal moment resistance and externally applied moment must be equal in cases of equilibrium, it can be stated that $\Sigma M = M$. Equation 12-5 can also be rearranged to solve for the bending stress at any location along the cross-section of the beam as:

$$\sigma_b = M(c)\,/\,I \qquad\qquad \text{(Eq. 12-6)}$$

where

σ_b = bending stress at the point under consideration on the beam's cross-section

M = applied bending moment to the cross-section

c = distance from the neutral axis to the point on the cross-section where stress is being considered

I = cross-section's moment of inertia

Equation 12-6 is referred to as the bending stress or flexural stress formula. It can be used in many applications to either calculate bending stresses on an existing beam or to design the necessary moment of inertia, I, which is required to resist a certain amount of applied moment. For such design situations, an allowable value of bending stress that the beam material can withstand would have to be given in order to provide a safe design.

The following examples will illustrate the use of the bending stress formula in a variety of applications.

EXAMPLE 12.1

The beam cross-section shown below has 150 kip-ft. of bending moment applied to it. Calculate the bending stress at the outside surfaces and also at 3 in. away from the neutral axis (consider bending takes place about the x-axis).

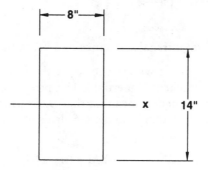

The cross-section is rectangular and therefore symmetrical about both the x-axis and the y-axis. Since bending is considered to take place about the x-axis, the neutral axis coincides with the centroidal x axis for the shape. From the discussion of moment of inertia in Chapter 7, the moment of inertia for a rectangular shape about its centroidal axis is:

$$I_x = 1/12 \; bh^3$$

$$I_x = 1/12 \;(8 \text{ in.})(14 \text{ in.})^3 = 1829.3 \text{ in.}^4$$

Solving for bending stress at the outside surface of the beam, we would find the following:

$$M = 150 \text{ kip-ft.} = 1800 \text{ kip-in.}$$

$$c = 1/2(14 \text{ in. }) = 7 \text{ in.}$$

$$I_x = 1829.3 \text{ in.}^4$$

$$\sigma_b = M(c) / I \qquad\qquad\qquad \text{(Eq. 12-6)}$$

$$\sigma_b = 1800 \text{ kip-in. (7 in.) / } 1829.3 \text{ in.}^4 = 6.89 \text{ ksi}$$

To calculate the bending stress at 3 in. from the neutral axis, the following calculation would be performed:

$$M = 150 \text{ kip-ft.} = 1800 \text{ kip-in.}$$

$$c = 3 \text{ in.}$$

$$I_x = 1829.3 \text{ in.}^4$$

$$\sigma_b = M(c) \, / \, I \qquad\qquad \text{(Eq. 12-6)}$$

$$\sigma_b = 1800 \text{ kip-in. } (3 \text{ in.}) \, / \, 1829.3 \text{ in.}^4 = 2.95 \text{ ksi}$$

The bending stress is much less near the neutral axis than at the outside surface. This is because the strains associated with the outside surface are greater than strains near the neutral axis. A secondary note should be made on the units involved. Notice that the units for the moment were changed to provide compatibility with those used for c and I. This is important so that the units of stress will be correct.

EXAMPLE 12.2

The beam shown below is used to span a distance of 10 m and carry a concentrated load of 10 KN at midspan. Calculate the maximum bending stress at the 1/4 points as well as at midspan.

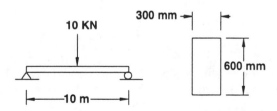

To begin this problem one must determine the applied moment that is generated at the 1/4 points and at midspan. This is easily accomplished by shear and moment diagrams, which were discussed in the previous chapter. The shear and moment diagrams for this particular beam would be as follows:

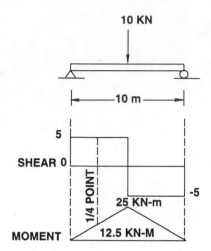

From the moment diagram, it is obvious that the maximum applied moment occurs at midspan and has a magnitude of 25 KN-m. The moment at each of the 1/4 points can be found by totaling the area under the shear diagrams up to those points. The moment at each of the 1/4 points is 12.5 KN-m.

The cross-section is rectangular and therefore symmetrical about both the x-axis and the y-axis. Since bending is considered to take place about the x-axis, the neutral axis coincides with the centroidal x axis for the shape. From the discussion of moment of inertia in Chapter 7, the moment of inertia for a rectangular shape about its centroidal axis is:

$$I_x = 1/12 \, bh^3$$

$$I_x = 1/12 \, (0.3 \text{ m})(0.6 \text{ m})^3 = 0.0054 \text{ m}^4$$

Solving for bending stress at the outside surface of the beam at midspan, we would find the following:

$$M = 25 \text{ KN-m}$$

$$c = 1/2(0.6 \text{ m}) = 0.3 \text{ m}$$

$$I_x = 0.0054 \text{ m}^4$$

$$\sigma_b = M(c) / I \qquad\qquad\qquad (\text{Eq. 12-6})$$

$$\sigma_b = 25 \text{ KN-m}(0.3 \text{ m}) / 0.0054 \text{ m}^4 = 1388.9 \text{ KN/m}^2 = 1388.9 \text{ KPa}$$

Solving for bending stress at the outside surface of the beam at the 1/4 points, we would find the following:

$M = 12.5$ KN-m

$c = 1/2(0.6$ m $) = 0.3$ m

$I_x = 0.0054$ m^4

$\sigma_b = M(c) / I$ (Eq. 12-6)

$\sigma_b = 12.5$ KN-m$(0.3$ m$) / 0.0054$ m$^4 = 694.4$ KN/m$^2 = 694.4$ KPa

The student should notice that the amount of bending stress developed by the beam is directly proportional to the amount of bending moment applied to the beam's cross-section.

EXAMPLE 12.3

The T-beam shown below is subjected to an applied moment of 200 kip-ft. Find the bending stress at the top and bottom of the beam. Assume the applied moment causes bending about the centroidal x axis.

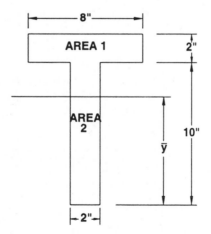

In this example the location of the beam's neutral axis is not known upon inspection because the beam is not symmetrical about the x-axis. Therefore, the first step is to locate the shape's centroidal x axis because this will coincide with the T-beam's neutral axis. The location of the centroidal axis is found by breaking the shape into simple geometric areas and proceeding with the process that was discussed in Chapter 6. This process would involve selecting a reference point and setting up a table as shown below. (The reference point that is used in this example is the bottom of the T-beam).

Shape	Area	x	ax	y	ay
1	16 in²	—	—	11 in²	176 in³
2	20 in²	—	—	5 in²	100 in³
	36 in²				276 in³

$$\bar{y} = \frac{276 \text{ in}^3}{36 \text{ in}^2} = 7.67 \text{ in}$$

$$\bar{y} = 7.67 \text{ in.}$$

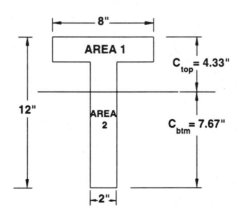

This means that the centroidal x-axis lies at a distance 7.67 in. up from the bottom of the beam. Knowing this information, the next step is to calculate the moment of inertia for the shape about this centroidal *x* axis. This procedure was discussed in Chapter 7 and involves using the parallel-axis theorem. The moment of inertia table for this shape would be as follows:

Shape	I_{XCNTRD}	A	d^2	AD^2	I_{XTOTAL}
1	5.33 in^4	16 in^2	3.33	177.4 in^4	182.7 in^4
2	166.7 in^4	20 in^2	2.67	177.4 in^4	309.3 in^4
	TOTAL		$\longrightarrow$		492 in^4

Therefore $I_x = 492$ in.^{4}

Knowing the moment of inertia, we can calculate the bending stress at the top and bottom of the T-beam. Recognize that the c distance in the bending stress equation is not the same for the top and bottom bending stress. This is typical for beams that are not symmetrical. The stresses can be calculated as follows:

$$M = 200 \text{ kip-ft.} = 2400 \text{ kip-in.}$$

$$c_{top} = 12 \text{ in.} -7.67 \text{ in.} = 4.33 \text{ in.}$$

$$c_{btm} = 7.67 \text{ in.}$$

$$I_x = 492 \text{ in.}^4$$

Top Stress:

$$\sigma_b = M(c) / I \qquad\qquad \text{(Eq. 12-6)}$$

$$\sigma_b = 2400 \text{ kip-in. } (4.33 \text{ in.}) / 492 \text{ in.}^4 = 21.1 \text{ ksi}$$

Bottom Stress:

$$\sigma_b = M(c) / I \qquad\qquad \text{(Eq. 12-6)}$$

$$\sigma_b = 2400 \text{ kip-in. } (7.67 \text{ in.}) / 492 \text{ in.}^4 = 37.4 \text{ ksi}$$

The next problem will focus on beam design, which is primarily a task of choosing the proper size of beam to resist an applied bending moment. The property of area that is to be designed for would be moment of inertia, I, or more likely the section modulus, S. The section modulus, S, is an elastic property of area, which is simply the moment of inertia divided by the distance from the neutral axis to the outside surface of the beam (I/c). For standard shaped beams, the section modulus is exclusively used—since the moment of inertia and the c distance are always a constant for a particular beam. The section modulus is commonly used for beams with standard dimensions, such as those made from steel or timber, and will be fully discussed in Chapter 15.

EXAMPLE 12.4

Calculate the width of a 12-inch-deep rectangular timber beam that is needed to resist an applied moment of 10 kip-ft. The allowable bending stress for this beam is 1.5 ksi.

As previously mentioned, the design of a beam typically involves solving for a property of area referred to as the section modulus, S. The section modulus on a standard timber beam is a ratio of moment of inertia to one-half the beam depth (I / c). The values of S can typically be looked up in a property table, such as those listed in Appendix E. Recognizing this, the standard bending stress formula can be rearranged as such:

$$\sigma_b = M(c) / I \qquad \text{(Eq. 12-6)}$$
$$\sigma_b = M / (I/c)$$

or

$$\sigma_b = M / S \qquad \text{(Eq. 12-7)}$$

Solving for the section modulus needed, Equation 12-7 can be rearranged as:

$$S = M / \sigma_b$$

And the section modulus solved for as follows:

$$S = M / \sigma_b$$
$$S = 10 \text{ kip-ft.}(12 \text{ in. / ft.}) / 1.5 \text{ ksi}$$
$$S = 80 \text{ in.}^3$$

The section modulus of a rectangular beam is:

$$S = bh^2 / 6$$

Using the depth (h) of the beam as 12 in., the width (b) can be solved for as follows:

$$b = S / (h^2 / 6)$$
$$b = 80 \text{ in.}^3 / (12^2 / 6) = 3.33 \text{ in.}$$

12.4   Shear Stresses in Beams

In addition to withstanding applied bending moments, beams also need to have the ability to resist shear forces. As discussed in Chapter 11, shear forces along a beam are also generated by applied loads and actually try to "rip the beam apart". These shear forces have to be resisted by a beam so that a failure does not occur. The mechanism used by a beam to resist the externally-applied shear forces is to internally create an equal and opposite resisting force. The mechanism used to create this internal resisting force is achieved by utilizing the creation of shear stresses (Figure 12-6).

The shear stresses that are developed vertically on a beam element must likewise develop on perpendicular planes such that equilibrium is maintained (Figure 12-7). This horizontal shearing stress resists the horizontal forces that tend to cause slippage of planes within the beam. This is the reason that individual planks, when stacked on top of each other, are not as strong as a beam of similar size. These planks would have a tendency to slide past one another, causing each plank to act independently. A solid beam will have fibers resisting this horizontal shear force and in essence, transfer the shear stress from plank to plank (Figure 12-8).

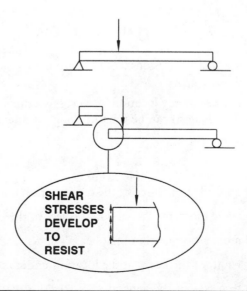

Figure 12-6   Vertical Shear Stresses in a Beam

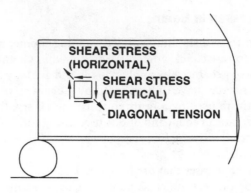

Figure 12-7   Development of Vertical and Horizontal Shear Stresses

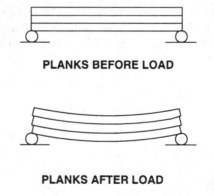

PLANKS BEFORE LOAD

PLANKS AFTER LOAD

Figure 12-8   Effect of Horizontal Shear Stresses

The shear stress at any location on a cross-section of a beam (and at any point along the beam's length) can be found through the general shear stress equation.

$$\sigma_v = V\,Q\,/\,I\,b \qquad\qquad\qquad\text{(Eq. 12-8)}$$

where

V = applied shear force at the cross-section under consideration

Q = static moment of area with respect to the centroidal axis

I = moment of inertia of the beam's cross section

b = width of the beam at the location under consideration

Of all the pieces in the general shear stress equation, the static moment of area, Q, probably requires some further explanation. The static moment of area is the first moment of area of the beam lying above the plane where shear stress is

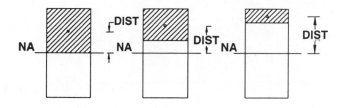

Figure 12-9   Illustrations of the Static Moment of Area, Q

being calculated on the cross-section. This first moment of area is calculated by taking the area above the plane where shear stress is being calculated and multiplying by the distance from the neutral axis to the centroid of that area. Properly calculating the static moment of area is perhaps the most critical procedure in using the general shear stress equation. Figure 12-9 illustrates three different values of Q for the same beam. Notice that the calculation of the static moment is always an area multiplied by a distance to the centroid of that area.

Should the area outside the plane where shear stress is being calculated be complex, the area should be broken down into a number of simple areas (i.e rectangles, squares, etc.) and the static moment for each individual area (Q_i) found. After calculating the static moment of areas for each individual area, the total static moment of area can be found by summing together the individual static moments of area. This can be represented as follows:

$$Q = Q_1 + Q_2 + Q_3 + ...Q_n$$

where

$$Q = \text{total static moment of area}$$

$$Q_1, Q_2, \text{etc.} = \text{static moment of area for the individual pieces}$$

The following examples will illustrate the calculation of shear stress over the cross-section of a beam.

EXAMPLE 12.5

Calculate the shear stress for the beam shown below; at the neutral axis, 3 in. away from the neutral axis, and at the outside surface. The shear force on the beam is 50,000 pounds.

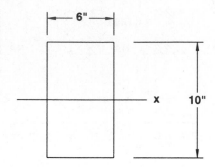

The general shear stress equation can be applied at each of the sections, with the following the same in all cases:

$$V = 50,000 \text{ lbs}$$

$$I = 1/12 \, bh^3 = 1/12(6 \text{ in.})(10 \text{ in.})^3 = 500 \text{ in.}^4$$

$$b = 6 \text{ in.}$$

The only item that will be different for the three locations is the static moment of area, Q. This can be calculated as follows.

Shear Stress at the Neutral Axis

The figure below illustrates the area involved with the determination of Q and the static moment of area, calculated as follows:

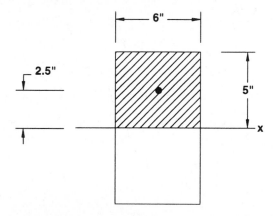

$$Q = (\text{area})(\text{distance}) = (30 \text{ in.}^2)(2.5 \text{ in.}) = 75 \text{ in.}^3$$

The general shear stress equation can then be calculated as follows:

$$\sigma_v = V \, Q / I \, b \qquad\qquad \text{(Eq. 12-8)}$$

$$\sigma_v = 50,000 \text{ lbs}(75 \text{ in.}^3) / 500 \text{ in.}^4 (6 \text{ in.}) = 1250 \text{ psi}$$

Shear Stress at 3 inches away from Neutral Axis

The figure below illustrates the area involved with the determination of Q and the static moment of area, is calculated as follows:

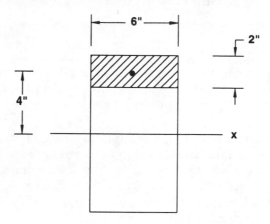

$$Q = (\text{area})(\text{distance}) = (12 \text{ in.}^2)(4 \text{ in.}) = 48 \text{ in.}^3$$

The general shear stress equation can then be calculated as follows:

$$\sigma_v = V\,Q\,/\,I\,b \qquad\qquad \text{(Eq. 12-8)}$$

$$\sigma_v = 50{,}000 \text{ lbs}(48 \text{ in.}^3)\,/\,500 \text{ in.}^4\ (6 \text{ in.}) = 800 \text{ psi}$$

Shear Stress at the Outside Surface

Above the outside surface there is no area and therefore the calculation of Q is simply zero. The general shear stress equation likewise would be zero as the following calculation indicates:

$$\sigma_v = V\,Q\,/\,I\,b \qquad\qquad \text{(Eq. 12-8)}$$

$$\sigma_v = 50{,}000 \text{ lbs}(0 \text{ in.}^3)\,/\,500 \text{ in.}^4\ (6 \text{ in.}) = 0 \text{ psi}$$

The previous example illustrates that shear stresses along a beam's cross-section are greatest at the neutral axis and vary to a minimum of zero at the outside surface. This distribution is opposite of the bending stress distribution, which was zero at the neutral axis and a maximum at the outside surface. The importance of this fact is crucial in the inspection of existing structures that have suffered some type of structural distress. The inspector must be technically competent to understand that flexural distress will typically become evident by large amounts of deflection accompanied by cracking on the outside surface. Shear distress will usually occur near the neutral axis and may not be accompanied by large amounts of deflection or displacement. Vertical shear stresses may also combine with their horizontal counterparts to form diagonal tension resultants that may lead to diagonal cracking in brittle materials such as concrete.

EXAMPLE 12.6

Calculate the shear stresses at the neutral axis and at the midpoint of the flange in a W 16 × 77, if the shear force is 100 kips (444.8 KN). Use the general shear stress equation for the stress calculation at both locations.

To calculate the shear stresses using the general formula, the majority of work will go into the calculation of the static moment of area, Q, since moment of inertia, I_x, and the width of the web and the flange can readily be taken from the appropriate section table in the Appendix A. Remember that Q is simply the area outside the plane under consideration multiplied by the distance from the neutral axis to the centroid of that area.

Shear Stress at the Neutral Axis

For shear stress at the neutral axis, Q can be calculated by representing the area outside the neutral axis as two rectangles. The Q can then be readily found by summing the areas multiplied by their respective distances, as illustrated below. The thicknesses and depths of the various member components can be found in the section tables in Appendix A. The static moment of area is calculated as follows:

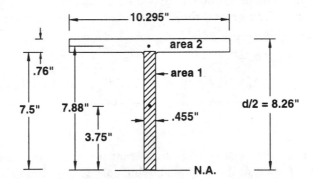

$$Q = Q_1 + Q_2$$

$$Q = (7.5 \text{ in.} \times .455 \text{ in.})(7.5 \text{ in.} / 2) + (10.295 \text{ in.} \times .76 \text{ in.})$$
$$(7.88 \text{ in.}) = 74.45 \text{ in.}^3$$

The general shear stress equation can then be calculated as follows:

$$\sigma_v = V Q / I b \qquad \text{(Eq. 12-8)}$$

with

$V = 100$ kips

$Q = 74.45$ in.^{3} (from above)

$I = 1110$ in.^{4} (from Appendix A)

$\sigma_v = 100$ kips$(74.45$ in.$^3) / 1110$ in.^{4} $(.455$ in.$) = 14.74$ ksi

Shear Stress at the Midpoint of Flange

For shear stress at the midpoint of the flange, Q can be calculated by representing the area outside this plane as a rectangle, shown below. Values for flange width and thickness can be found in the section tables in Appendix A. Q can be readily found by multiplying this area by the distance from the neutral axis to the centroid of this area as follows:

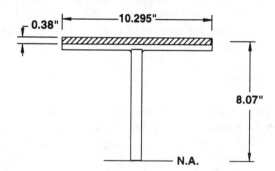

$$Q = (10.295 \text{ in.} \times .38 \text{ in.})(8.07 \text{ in.}) = 31.57 \text{ in.}^3$$

The general shear stress equation can then be calculated for the midpoint of the flange, as follows:

$$\sigma_v = V Q / I b \qquad \text{(Eq. 12-8)}$$

$\sigma_v = 100$ kips$(31.57$ in.$^3) / 1110$ in.^{4} $(10.29$ in.$) = .27$ ksi

As you can see, the shear stresses at the neutral axis are much greater than at other parts of the section. It should be pointed out that the shear stress at the junction of the flange and web is much higher than at the midpoint of the flange. Also note that the "rounded corners", or fillets, at the junction of the flange and web were excluded in order to simplify this example. If he has access to the WT 8×38.5 section properties, the static moment of area relative to the neutral axis can be calculated as 74.92 in.^{3} with the fillets included. This represents a difference of approximately 0.6%.

12.5 Summary

When loads are applied to beams, the two primary load effects that occur are bending moments and shear forces. These external effects must be resisted by the beam or the beam will experience distress and possibly failure. A bending moment will try to rotate or excessively bend a beam until failure occurs, while shear forces try to displace or "rip" the beam along a given section. From the previous chapter, it was shown that shear forces generally are highest around the supports or concentrated loads, while bending moments are typically high near the middle of the beam.

To resist applied bending moments, the beam develops internal bending stresses that create an internal moment to counteract the externally applied bending moment. The bending stress is a tensile stress on one side of the neutral axis and a compressive stress on the other. The bending stress varies from zero at the neutral axis to a maximum value at both the tensile and compressive side of the beam. The shear forces are resisted by the beam developing internal shearing stresses perpendicular to the shear force. Shearing forces also occur along horizontal planes and the planes may actually cause a splitting failure in materials such as timber. Shear stresses are typically a maximum at the neutral axis and decrease to zero at the outside surface of the beam.

EXERCISES

1. A rectangular beam measuring 10 in. wide and 18 in. deep has an applied bending moment of 120 kip-ft. acting upon it. Determine the maximum tensile and compressive bending stresses on the beam and also determine the bending stress at a distance 3 in. above the neutral axis.

2. A rectangular beam measuring 100 mm wide and 300 mm deep has an applied bending moment of 100 KN-m acting upon it. Determine the maximum tensile and compressive bending stresses on the beam and also determine the bending stress at a distance 35 mm above the neutral axis.

3. A rectangular beam measuring 8 in. wide and 16 in. deep has a maximum bending stress of 11 ksi on its outside fiber. Determine the magnitude of applied moment on the beam to cause this stress.

4. A rectangular beam measuring 80 mm wide and 180 mm deep has a maximum bending stress of 30,000 KPa on its outside fiber. Determine the magnitude of applied moment on the beam to cause this stress.

5. A rectangular tube has an applied moment of 1500 kip-in. acting upon it. If the tube is 4 in. wide and 8 in. deep with a wall thickness of 0.25 in., determine the maximum bending stress developed in the beam. Also determine the bending stress 2 in. below the neutral axis.

6. A square tube has an applied moment of 150 KN-m acting upon it. If the tube is 150 mm square with a wall thickness of 25 mm, determine the

maximum bending stress developed in the beam. Also determine the bending stress 20 mm below the neutral axis.

7. For the beam shown below, calculate the maximum bending stress developed at locations measuring 5 ft., 10 ft., and 15 ft. from the left support. The beam is rectangular, measuring 10 in. wide and 20 in. deep.

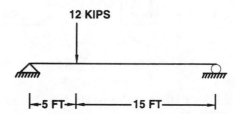

8. For the beam shown below, calculate the maximum bending stress developed at 3 ft. from the left support and at midspan. The beam is a W 12 × 50.

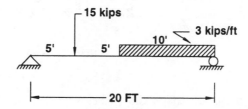

9. For the beam shown below, calculate the maximum bending stress developed at 3 ft. from the left support and at midspan. The beam is a W 10 × 30.

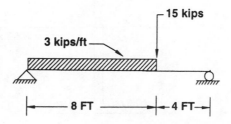

10. For the beam shown below, calculate the maximum bending stress developed at 2 m, 4 m, and 6 m from the left support. The beam is a W 360 × 110.

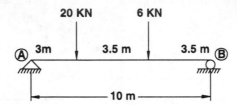

11. For the beam shown below, calculate the maximum bending stress developed if the beam is a rectangular tube measuring 80 mm wide and 250 mm deep. The wall thickness is 20 mm.

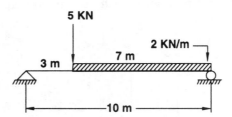

12. For the beam shape shown below, calculate the maximum tensile and compressive bending stress if the beam is subjected to an applied moment of 200 kip-ft.

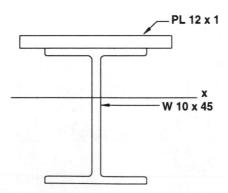

13. For the beam shape shown below, calculate the maximum tensile and compressive bending stress if the beam is subjected to an applied moment of 210,000 lb-ft.

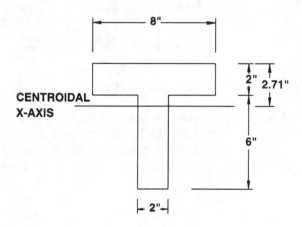

14. For the beam shape shown below, calculate the maximum tensile and compressive bending stress if the beam is subjected to an applied moment of 150 KN-m.

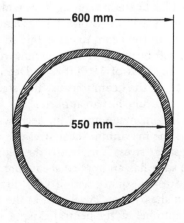

15. Determine the maximum tensile and compressive bending stress for the beam section shown in exercise 12 if the beam has loadings applied, as shown below.

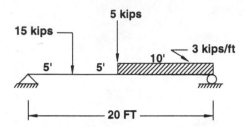

16. A solid square beam is to be design to carry an applied moment of 30 kip-ft. If the maximum allowable bending stress that can be developed is 2000 psi, determine the dimensions of the beam.

17. A 8-inch-square tube beam is to be designed to carry an applied moment of 45 kip-ft. If the maximum allowable bending stress which can be developed is 12 ksi, determine the proper wall thickness for the beam

18. A 250-mm-square beam is to be design to carry an applied moment of 50 KN-m. If the maximum allowable bending stress which can be developed is 55000 KPa, determine the proper wall thickness for the beam.

19. A rectangular beam measuring 10 in. wide and 18 in. deep has an applied shear force of 80,000 lbs acting upon it. Determine the maximum shear stress on the beam and also determine the shear stress at a distance 3 in. above the neutral axis.

20. A rectangular beam measuring 100 mm wide and 300 mm deep has an applied shear force of 40 KN acting upon it. Determine the maximum shear stress on the beam and also determine the shear stress at a distance 35 mm above the neutral axis.

21. A rectangular beam measuring 8 in. wide and 16 in. deep has developed a maximum shear stress of 6.5 ksi. Determine the magnitude of applied shear force on the beam to cause this stress.

22. A rectangular beam measuring 80 mm wide and 180 mm deep has a maximum shear stress of 18,000 KPa. Determine the magnitude of applied shear force on the beam to cause this stress.

23. A rectangular tube has an applied shear force of 55 kips acting upon it. If the tube is 4 in. wide and 8 in. deep with a wall thickness of 0.25 in., determine the maximum shear stress developed in the beam. Also determine the shear stress 2 in. below the neutral axis.

24. A square tube has an applied shear force of 3 MN acting upon it. If the tube is 150 mm square with a wall thickness of 25 mm, determine the maximum shear stress developed in the beam. Also determine the shear stress 20 mm below the neutral axis.

25. For the beam shown below, calculate the maximum shear stress developed at 3 ft.and 6 ft. from the left support. The beam is a W 12 × 50.

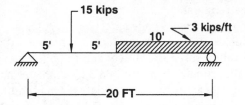

26. For the beam shown below, calculate the maximum shear stress developed at 3 ft. and 10 ft. from the left support. The beam is a W 10 × 30.

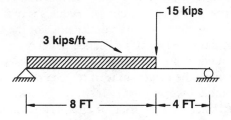

27. For the beam shown below, calculate the maximum shear stress developed at 2 m, 4 m, and 6 m from the left support. The beam is a W 360 × 110.

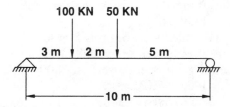

28. For the beam shown below, calculate the maximum bending stress developed if the beam is a rectangular tube measuring 80 mm wide and 250 mm deep. The wall thickness is 20 mm.

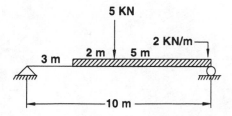

29. For a W12 × 87 rolled steel beam, calculate the shear stress at the centroidal x-axis and at the junction between the web and the flange. The shear force is 40 kips and the fillets can be neglected (assume the web and flange to be rectangles).

30. For a W 310 × 86 rolled steel beam, calculate the shear stress at the centroidal x-axis and at the junction between the web and the flange. The solid shear force is 150 KN and the fillets can be neglected

31. A solid, circular beam with a 3 in. diameter has an applied shear force of 50,000 lbs acting on it. Calculate the maximum shear stress on the section.

32. A solid, circular beam with a 300 mm diameter has an applied shear force on 175 KN acting on it. Calculate the maximum shear stress on the section.

33. A solid square beam is to be designed to carry an applied shear force of 25 kips. If the maximum allowable shear stress which can be developed is 200 psi, determine the dimensions of the beam.

34. A solid square beam is designed to carry an applied moment of 80 KN. If the maximum allowable shear stress that can be developed is 1.5 KPa, determine the dimensions of the beam.

35. A standard wood joist is 14 ft. long and supports a uniform load of 225 lbs/ft. under the flooring of an apartment building. Select a size that will be adequate for both shear and bending if the allowable shear stress is 100 psi and the allowable bending stress is 1200 psi.

36. A standard wood joist is 14 ft. long and supports a concentrated load at midspan of 5 kips. Select a size that will be adequate for both shear and bending if the allowable shear stress is 90 psi and the allowable bending stress is 1350 psi.

37. A W 310 rolled steel beam is 8 m long and carries a uniform load of 8 KN/ m. Select a size which will be adequate for both shear and bending if the allowable shear stress is 70,000 Kpa and the allowable bending stress is 140,000 KPa.

38. A W 10 × 33 rolled steel beam is 15 ft. long and has an allowable shear stress of 14 ksi and an allowable bending stress of 20 ksi. Calculate the maximum uniform load that this beam can safely carry.

CHAPTER

13

BEAM DEFLECTION

13.1 Introduction

As a beam is loaded it develops internal stresses to resist these loads. As these internal stresses are developed, the beam undergoes a vertical displacement which is referred to as deflection (Figure 13.1). If the deflection of a beam becomes excessive it can cause a beam to behave in an unsatisfactory manner. Such unwanted behavior may manifest itself through the unsightly cracking of ceilings and walls, perceptible vibration of floor systems, or poor performance of adjacent building components such as doors and windows. Such problems all project a poor appearance to the casual observer and should be prevented. These items are not strength capacity failures of the beams but rather failures of their serviceability requirements. The American Institute of Steel Construction's (AISC) specification defines **serviceability** as "a state in which the function of a building, its appearance, maintainability, durability, and comfort of its occupants are preserved under normal usage".[1]

The computation of beam deflection is necessary to ensure the serviceability of the structure. Usually, in buildings and other structures, the deflection of beams is relatively small compared to the overall dimensions of the section and its span length. However, since deflection criterion may control in some instances, the structural designer needs to be able to calculate the anticipated deflection in order to compare this value to the limiting values as set forth in the applicable building code.

A beam that has excessive amounts of deflection will not function as its occupants wish and can be considered to be a serviceability-related failure. A serviceability failure can be as much of a problem as a strength-based failure, because the

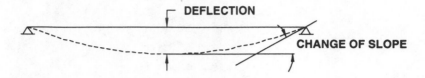

Figure 13-1   Deflection of a Simply-Supported Beam

structure may not be used in the manner intended. Therefore, a designer must be able to compute the deflection that a beam will undergo such that it may be judged against the appropriate building code requirements. Methods to calculate beam deflection will be the focus of this chapter. It should also be mentioned that excessive deflections may also lead to strength failures as in the case of ponded water on flexible roof systems. This is an extremely important consideration for the design of roof systems and should be thoroughly investigated in any design.

13.2   The Mechanics of Deflection

As a beam is subjected to load, it will deflect. Deflection, δ, can simply be thought of as the distance which the beam displaces from its original horizontal position. This displacement occurs as the straight beam is deflected along a smooth curve (Figure 13-2). Different points along the beam displace or deflect by various amounts. The curve formed at the location of the neutral axis is referred to as the displacement curve and this curve can be considered as a circular arc when viewed in very small segments. Within a small segment of the beam , the edges of the circular arc intersect at a common point located a certain distance away. This distance is referred to as the **radius of curvature, ρ**. If a small segment of the beam is con-

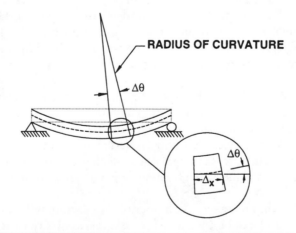

Figure 13-2   Relationship of Radius of Curvature to Deflection

sidered, it is assumed that the distance along the arc is approximated by the horizontal distance, Δx. Using right triangle trigonometry, the following can be stated:

$$\tan \Delta\theta = \Delta x\ /\ \rho$$

or

$$1\ /\ \rho = \tan \Delta\theta\ /\ \Delta x$$

At very small angles, $\tan \Delta\theta$ is approximately equal to $\Delta\theta$ and therefore:

$$1\ /\ \rho = \Delta\theta\ /\ \Delta x \qquad\qquad \text{(Eq. 13-1)}$$

Equation 13-1 is important because it states that the relative rotation a segment experiences ($\Delta\theta\ /\ \Delta x$) is inversely proportional to the radius of curvature. Therefore, a segment that has a large amount of deflection will have a large relative rotation between segments, and a smaller radius of curvature. This can be coupled with the realization that the maximum amount of bending strain (ε_b) that a segment undergoes is proportional to the distance from the neutral axis to the outside surface of the beam *(c)*, and inversely proportional to the length of the radius of curvature (Figure 13-3). This can be expressed as follows:

$$\varepsilon_b = c\ /\ \rho \qquad\qquad \text{(Eq. 13-2)}$$

From Hooke's law it is known that stress is simply modulus of elasticity multiplied by the strain, and therefore Equation 13-2 can be used to determine bending stress(σ_b) as follows:

$$\sigma_b = \varepsilon_b\ E$$

$$\sigma_b = (c\ /\ \rho)E \qquad\qquad \text{(Eq. 13-3)}$$

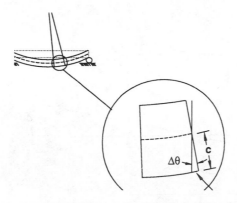

Figure 13-3   Bending Strain = $(\tan \Delta\theta)(c)\ /\ (\tan \Delta\theta)(\rho) = c\ /\ \rho$

Equating Equation 13-3 with the bending stress formula developed in Chapter 12 (Eq. 12-6) we can see the following:

$$(c / \rho)E = Mc / I$$

Rearranging we can find the following:

$$1 / \rho = M / EI \qquad \text{(Eq. 13-4)}$$

Equation 13-4 states that a larger bending moment applied to a beam segment will result in a shorter radius of curvature, the consequence thereof being more deflection. In a similar manner, a larger stiffness of the beam segment (EI) will result in a longer radius of curvature, the consequence being less deflection.

If Equation 13-1 and Equation 13-4 are equated, the following can be seen:

$$\Delta\theta / \Delta x = M / EI$$

or

$$\Delta\theta = (M / EI)\Delta x \qquad \text{(Eq. 13-5)}$$

Using Equation 13-5 we can determine that the change of rotation between the edges of a beam segment is equal to $M/EI(\Delta x)$. The change of rotation ($\Delta\theta$) between the two edges of a beam segment also corresponds to the vertical displacement (Δy) between the two edges also. This displacement is the result of the change of rotation or slope ($\Delta\theta$) which occurs between different points along a beam (Figure 13-4). In fact, deflection at any point along the beam can be found by summing up all the vertical displacements of all beam segments to that particular point. This can be stated as:

$$\delta = \Sigma\Delta y$$

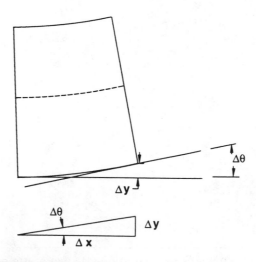

Figure 13-4   Deflection of an Incremental Segment of Beam

where

δ = total deflection at a point

Δy = incremental vertical displacements of each beam segment

The incremental pieces of deflection can also be defined as the *sum of incremental changes in rotation along a beam*. The amount of rotational change that a beam undergoes between two points can be approximated by the sum of all rotational changes ($\Sigma\Delta\theta$). This can be expressed by modifying Equation 13-5 to give the following:

$$\Sigma\Delta\theta = \Sigma(M / EI)\Delta x \qquad \text{(Eq. 13-6)}$$

and since

$$\delta = \Sigma\Delta y$$

and

$$\Sigma\Delta y = \Sigma\Delta\theta$$

Equation 13-6 can be modified to describe deflection as follows:

$$\delta = \Sigma(M / EI)\Delta x \qquad \text{(Eq. 13-7)}$$

This discussion of deflection mechanics will serve as the foundation for a discussion of particular methods that will be used throughout the chapter. Using this information, how is beam deflection calculated? Actually, there are many methods that are available based on the relationships that exist between the bending moment on a beam of known rigidity and the corresponding rotation (slope). The following sections will present some of these methods.

13.3   The Moment-Area Method

One of the most popular methods to calculate the deflection at any point along a beam is the moment-area method. The moment-area method for calculating deflection is based on the principle that the deflection between any two points on a deflection curve is equal to the change of rotation between those points. This principle was developed in the previous section and can be summarized by Equation 13-6:

$$\Sigma\Delta\theta = \Sigma(M / EI)\Delta x \qquad \text{(Eq. 13-6)}$$

The moment diagram of a beam can be changed into a *M/EI* diagram for beams having constant values of *E* and *I* (which is typically the case). If incremental pieces of area under the *M/EI* diagram are utilized, the area will approximate a rectangle. Therefore, the summation of the rotational change between two points ($\Sigma\Delta\theta$) is simply the area under the *M/EI* diagram (Figure 13-5). The rotational change between two points should be measured in radians and the following conclusion stated:

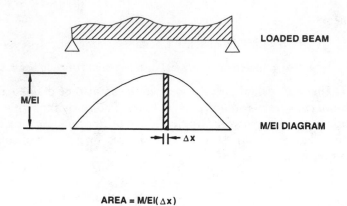

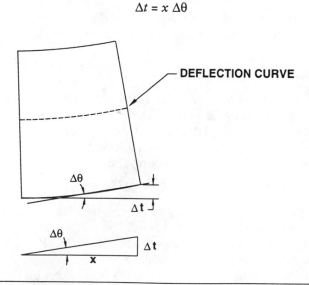

Figure 13-5   Rotational Change Related to the Area under the M/EI Diagram

The rotational change between two points on the deflection curve is equal to the area under the *M/EI* diagram between those points.

It can also be seen from Figure 13-6 that the vertical displacement (Δt) between two points on the deflection curve is a function of the rotational change and the length (x) between the two points. From the previous discussion, it was found that the rotational change between two points is equal to the area under the *M/EI* diagram, such that the vertical displacement between two points can be expressed as follows:

$$\Delta t = x \, \Delta\theta$$

Figure 13-6   Elevation Difference as a Function of $\Delta\theta$

However, from Equation 13-6, the rotational change ($\Delta\theta$) can be replaced to yield:

$$\Delta t = x \, (M \, / \, EI)\Delta x \qquad\qquad \text{(Eq. 13-8)}$$

This small vertical difference Δt is the elevation difference between the tangents of the two points located on each side of some segment of the deflection curve. These differences in elevations over small segments can be summed up to give elevation differences of points spaced much further apart. This larger elevation difference is referred to as the **tangent deviation** of one point relative to another. In Figure 13 -7, the tangent deviation of point B relative to point A is denoted by $t_{B/A}$. Since the area between the points in question is large, the x distance used in Equation 13-8 should be the x distance to the centroid of the area under the M/EI diagram. The sum of all the incremental elevation differences will be the total tangent deviation and this yields the following:

$$\Sigma\Delta t = \Sigma x \, (M \, / \, EI)\Delta x \qquad\qquad \text{(Eq. 13-9)}$$

or using Figure 13-7:

$$t_{B/A} = x \, (M \, / \, EI)\Delta x$$

where

x = distance from B to the centroid of the area under the M/EI diagram between A and B

$(M \, / \, EI)\Delta x$ = area under the M/EI diagram between A and B

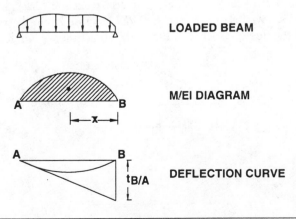

LOADED BEAM

M/EI DIAGRAM

DEFLECTION CURVE

Figure 13-7   Tangent Deviation of Point B relative to Point A

The following can be stated about the tangent deviation between points:

> The tangent deviation of point B relative to point A is the elevation difference found by drawing a vertical line from B and intersecting that with a tangent of the deflection curve drawn from point A. This can mathematically be found by multiplying the area under the M/EI diagram between A and B, with the x distance from point B to the centroid of that area.

These tangent deviations can be manipulated to find the deflection at various points along a beam. The following examples will illustrate how the moment-area technique is utilized.

EXAMPLE 13.1

Using the moment-area method, determine the deflection at the midpoint of the simply supported beam shown below. Consider the beam to be made of steel ($E = 29 \times 10^6$ psi) and to have a moment of inertia, I_x, of 500 in^4.

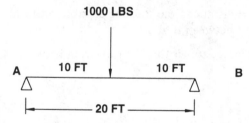

The first order of business is to construct the M/EI diagram for such a case, which is shown below. Notice the M/EI diagram is simply the moment diagram divided by the flexural stiffness, although at this time we will leave E and I as variables.

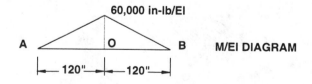

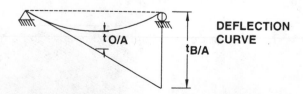

The deflection at the beam's center, the maximum deflection, is desired. The deflection at this point can be considered to be as follows:

$$1/2\ t_{B/A} - t_{O/A}$$

The tangent deviation of B relative to A ($t_{B/A}$) can be found by taking the area under the M/EI diagram between A and B (which can be viewed as two triangles) and multiply this area by the x distance from B to the centroid of the area. The area under the M/EI diagram can be calculated as:

$$A = 2(1/2)(120\ \text{in.})(60,000/EI) = 7,200,000\ \text{in.}^2\text{-lb}/EI$$

The x distance from B to the centroid of this area is simply 120 in., and therefore the tangent deviation from B to A is as follows, from Equation 13-9:

$$\Sigma\Delta t = \Sigma x\ (M/EI)\Delta x \qquad\qquad\qquad \text{(Eq. 13-9)}$$

$$t_{B/A} = x\ (M/EI)\Delta x$$

$$t_{B/A} = 120\ \text{in.}\ (7,200,000\ \text{in.}^2\text{-lb}/EI) = 864 \times 10^6\ \text{in.}^3\text{-lb}/EI$$

Therefore $1/2\ t_{B/A}$ is equal to:

$$1/2\ t_{B/A} = 432 \times 10^6\ \text{in.}^3\text{-lb}/EI$$

The tangent deviation of O relative to A ($t_{O/A}$) can be found by taking the area under the M/EI diagram between O and A (which can be viewed as a triangle) and multiply this area by the x distance from O to the centroid of the area. The area under the M/EI diagram can be calculated as:

$$A = (1/2)(120\ \text{in.})(60,000/EI) = 3,600,000\ \text{in.}^2\text{-lb}/EI$$

The x distance from O to the centroid of this area is simply 40 in., and therefore the tangent deviation from O to A is as follows, from Equation 13-9:

$$\Sigma\Delta t = \Sigma x\ (M/EI)\Delta x \qquad\qquad\qquad \text{(Eq. 13-9)}$$

$$t_{O/A} = x\ (M/EI)\Delta x$$

$$t_{O/A} = 40\ \text{in.}\ (3,600,000/EI) = 144 \times 10^6\ \text{in.}^3\text{-lb}/EI$$

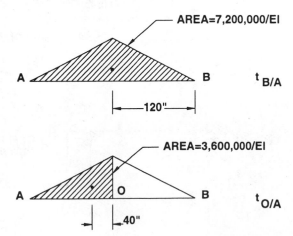

The deflection at the midpoint can be found as follows:

$$\delta = 1/2 \, t_{B/A} - t_{O/A}$$

$$\delta = 432 \times 10^6 \text{ in.}^2\text{-lb}/EI - 144 \times 10^6 \text{ in.}^3\text{-lb}/EI = 288 \times 10^6 \text{ in.}^3\text{-lb}/EI$$

Substituting in the values for E and I we find the following:

$$\delta = 288 \times 10^6 \text{ in.}^3\text{-lb} / (29 \times 10^6 \text{ psi})(500 \text{ in.}^4) = .02 \text{ in.}$$

Therefore the beam will deflect a total of .02 in. at its midpoint.

Particular attention must be paid to the units used throughout a problem when using the moment-area technique. Notice that the length units were changed to inches in the previous example to maintain dimensional continuity.

EXAMPLE 13.2

Using the beam from the previous example, find the deflection of the beam at the quarter point. Use the same E and I values.

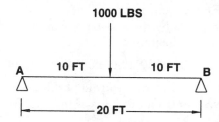

This problem is worked in a similar fashion to the first example and begins by constructing the M/EI diagram. This is shown below.

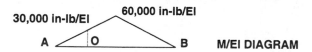

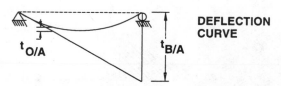

The deflection at the quarter point can be considered to be as follows:

$$\delta = 1/4 \, t_{B/A} - t_{O/A}$$

The tangent deviation of B relative to A ($t_{B/A}$) can be found by taking the area under the M/EI diagram between A and B (which can be viewed as two triangles) and then multiplying this area by the x distance from B to the centroid of the area. The area under the M/EI diagram can be calculated as:

$$A = 2(1/2)(120 \text{ in.})(60{,}000/EI) = 7{,}200{,}000 \text{ in.}^2\text{-lb}/EI$$

The x distance from B to the centroid of this area is simply 120 in., and therefore the tangent deviation from B to A is as follows, from Equation 13-9:

$$\Sigma \Delta t = \Sigma x \, (M \, / \, EI) \Delta x \qquad\qquad \text{(Eq. 13-9)}$$

$$t_{B/A} = x \, (M \, / \, EI) \Delta x$$

$$t_{B/A} = 120 \text{ in. } (7{,}200{,}000/EI) = 864 \times 10^6 \text{ in.}^3\text{-lb}/EI$$

Therefore $1/4 t_{B/A}$ is equal to:

$$1/4 \, t_{B/A} = 216 \times 10^6 \text{ in.}^3\text{-lb}/EI$$

The tangent deviation of O relative to A ($t_{O/A}$) can be found by taking the area under the M/EI diagram between O and A (which can be viewed as a triangle) and then multipling this area by the x distance from O to the centroid of the area. The area under the M/EI diagram can be calculated as:

$$A = (1/2)(60 \text{ in.})(30{,}000/EI) = 900{,}000 \text{ in.}^2\text{-lb}/EI$$

The x distance from O to the centroid of this area is 20 in., and therefore the tangent deviation from O to A is as follows:

$$\Sigma \Delta t = \Sigma x \, (M \, / \, EI) \Delta x \qquad\qquad \text{(Eq. 13-9)}$$

$$t_{O/A} = x \, (M \, / \, EI) \Delta x$$

$$t_{O/A} = 20 \text{ in. } (900{,}000 \text{ in.}^2\text{-lb/EI}) = 18 \times 10^6 \text{ in.}^3\text{-lb}/EI$$

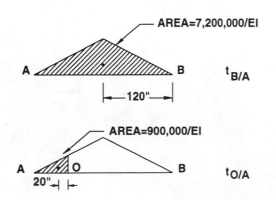

The deflection at the midpoint can be found as follows:

$$\delta = 1/4\ t_{B/A} - t_{O/A}$$

$$\delta = 216 \times 10^6\ \text{in.}^2\text{-lb}/EI - 18 \times 10^6\ \text{in.}^3\text{-lb}/EI = 198 \times 10^6\ \text{in.}^3\text{-lb}/EI$$

Substituting in the values for E and I, we find the following:

$$\delta = 198 \times 10^6\ \text{in.}^3\text{-lb} / (29 \times 10^6\ \text{psi})(500\ \text{in.}^4) = .013\ \text{in.}$$

Therefore the beam will deflect a total of .013 in. at its quarter point.

EXAMPLE 13.3

For the cantilever beam shown below, find the deflection at the free end. Consider the beam to have an $E = 200 \times 10^9$ Pa (200×10^9 N/m^2) and $I = 500 \times 10^6$ mm^4.

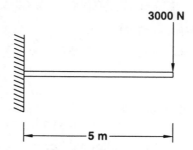

The problem again starts by constructing the M/EI diagram, which is shown as follows:

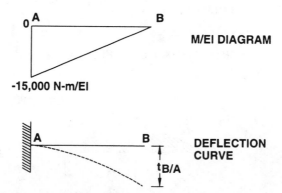

The deflection at the beam's free end is desired. The deflection at this point is as follows:

$$\delta = t_{B/A}$$

The tangent deviation of B relative to A ($t_{B/A}$) can be found by taking the area under the M/EI diagram between A and B (which can be viewed as a triangle) and then multiplying this area by the x distance from B to the centroid of the area. The area under the M/EI diagram can be calculated as:

$$A = (1/2)(5 \text{ m})(-15{,}000 \text{ N-m}/EI) = -37{,}500 \text{ N-m}^2/EI$$

The x distance from B to the centroid of this area is simply two-thirds of the 5 m length, which is 3.33 m, and therefore the tangent deviation from B to A is as follows:

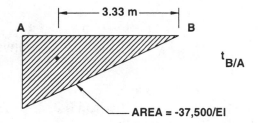

$$t_{B/A} = x \, (M \,/\, EI)\Delta x$$

$$t_{B/A} = 3.33 \text{ m} \, (-37{,}500 \text{ N-m}^2/EI) = -124{,}875 \text{ N-m}^3/EI$$

The deflection at the free end is equal to $t_{B/A}$, and substituting in the values for E and I, we find the following:

$$\delta = -124{,}875 \text{ N-m}^3 \,/\, (200 \times 10^9 \text{ N/m}^2)(.0005 \text{ m}^4) = -0.00125 \text{ m}$$

Therefore the beam will deflect a total of .00125 m downwards at its free end.

The application of uniform loads on a beam will result in the M/EI diagram being a second degree parabola. Although curved M/EI diagrams may be a bit more difficult to work with, the moment-area procedure remains the same. The following example will illustrate such a case.

EXAMPLE 13.4

For the cantilever beam shown below, find the deflection at the free end. The beam is a W 12 × 50. Consider $E = 29 \times 10^6$ psi and $I_x = 394$ in.[4].

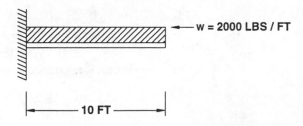

The problem again starts by constructing the M/EI diagram, which is shown as follows:

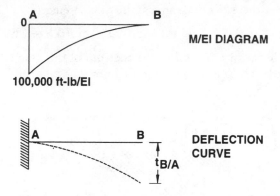

M/EI DIAGRAM

100,000 ft-lb/EI

DEFLECTION CURVE

$t_{B/A}$

The deflection at the beam's free end is desired. The deflection at this point can be considered to be as follows:

$$\delta = t_{B/A}$$

The tangent deviation of B relative to A ($t_{B/A}$) can be found by taking the area under the M/EI diagram between A and B. This area is a second degree parabola, which needs to be multiplied by the x distance from B to the centroid of the area. The area under the parabolic M/EI diagram can be calculated as:

$A = 1/3(bh)$

$A = (1/3)(120 \text{ in.})(-1,200,000 \text{ in.-lb}/EI) = -48,000,000 \text{ in.}^2\text{-lb}/EI$

The centroid of a second order parabola is located at an x distance of 3/4 of the base dimension. Therefore, the x distance from B to the centroid of this area is simply three-fourths of the 120 in. length (which is 7.5 ft., or 90 in.) The tangent deviation from B to A is as follows:

$t_{B/A} = x \, (M \, / \, EI)\Delta x$ $\qquad$ (Eq. 13-9)

$t_{B/A} = 90 \text{ in. } (-48,000,000 \text{ in.}^2\text{-lb}/EI) = -4.32 \times 10^9 \text{ in.}^3\text{-lb}/EI$

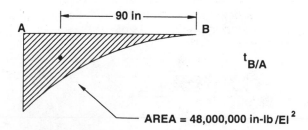

90 in

$t_{B/A}$

AREA = 48,000,000 in-lb/EI^2

The deflection at the free end is equal to $t_{B/A}$ and substituting in the values for E and I, we find the following:

$$\delta = -4.32 \times 10^9 \text{ in.}^3\text{-lb}/EI / (29 \times 10^6 \text{ psi})(394 \text{ in.}^4) = -0.378 \text{ in.}$$

Therefore the beam will deflect a total of .378 in. downward at its free end.

The moment-area method for finding the deflections along a beam is very useful. It is limited because the areas under the M/EI diagram must be simple standard shapes for the method to be relatively straight-forward. It also must be emphasized that the moment-area method is a solution for *relative* vertical displacements between two points. This relative displacement may not always correspond exactly with the actual deflection when deflection other than the maximum is desired.

13.4   Standard Deflection Formulas and Superpositioning

In addition to the moment-area technique that was discussed in the previous section, many formulas exist for the calculation of beam deflection under standard load cases. Formulas may be written to calculate the maximum deflection of a beam or the deflection at a required point. Most of the time, designers are concerned with the maximum beam deflection. Such formulas are found in many references, such as the AISC Manual of Steel Construction[1]. For a quick reference, a list of beam deflection formulas for some of the more common beam loadings is shown in Table 13-1.

In the formulas, the student can see that the magnitude of deflection is inversely proportional to the beam stiffness, EI. This simply means that the larger this term (EI) becomes, the smaller the magnitude of deflection will be. It should be noted that the magnitude of deflection is typically small when using normal beam lengths and therefore units should probably be measured in inches or millimeters. It is very important to keep consistent units when using these formulas . Units must be expressed in dimensionally correct forms and therefore the reader should make sure all uniform loads are in units of pounds per inch, kips per inch, kilonewtons per meter, or other (depending on what units the modulus of elasticity is expressed in). Similarly, the span lengths should be typically listed in units that match the modulus of elasticity.

The usage of standard formulas can also be expanded when a beam is subjected to more than one common load type. Through the **principle of superpositioning,** the deflection caused on a beam at a particular point by two different types of loads (i.e. concentrated load and uniform load) can be viewed as being the **sum** of the individual deflections caused by each individual load (Figure 13-8). As shown in Figure 13-8, the deflection at the midpoint of the beam is found by summing the deflection caused by the concentrated load and the deflection caused

Table 13-1 Common Deflection Formulas

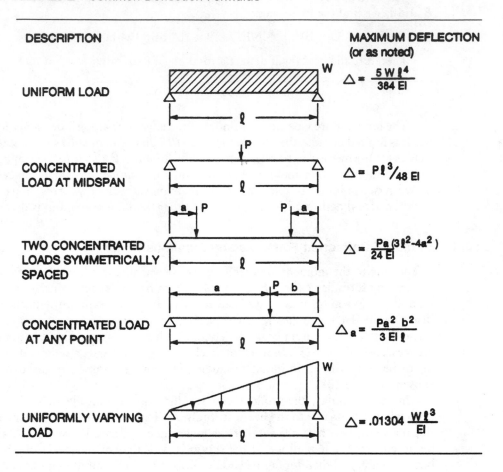

DESCRIPTION	MAXIMUM DEFLECTION (or as noted)
UNIFORM LOAD	$\triangle = \dfrac{5\,W\,\ell^4}{384\,EI}$
CONCENTRATED LOAD AT MIDSPAN	$\triangle = P\ell^3/48\,EI$
TWO CONCENTRATED LOADS SYMMETRICALLY SPACED	$\triangle = \dfrac{Pa\,(3\ell^2 - 4a^2)}{24\,EI}$
CONCENTRATED LOAD AT ANY POINT	$\triangle_a = \dfrac{Pa^2\,b^2}{3\,EI\,\ell}$
UNIFORMLY VARYING LOAD	$\triangle = .01304\,\dfrac{W\ell^3}{EI}$

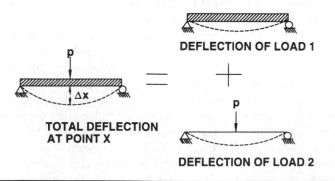

DEFLECTION OF LOAD 1

TOTAL DEFLECTION
AT POINT X

DEFLECTION OF LOAD 2

Figure 13-8 Principle of Superpositioning

by the uniform load. It should be noted that the maximum deflection calculated by some of the standard beam formulas are not necessarily located at the same point along the beam. Superpositioning will only work correctly if the deflections for each individual load effect are calculated at the point in question. This may mean that in some cases a deflection that is less than the maximum possible under a particular load effect will have to be calculated. In this situation, the reader may choose to revert back to the moment-area technique for finding deflections at a particular point.

The following examples will illustrate the usage of the standard beam formulas and the principle of superpositioning.

EXAMPLE 13.5

Calculate the live load deflection of the 20-foot-long W 12 × 50 under a uniform live load of 3 kips/ft. over its entire length.

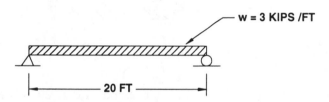

In this problem, the deflection is calculated from the formula found in Table 13-1. The variables needed for this calculation are as follows:

w = 3 kips/ft. = (3 kips/ft.) / (12 in. / ft.) or .25 kips/in.

l = 20 ft. = 240 in.

E = 29,000 ksi

I = 394 in.^{4} (W 12 × 50 about its strong axis)

Calculating deflection from the formula given in Table 13-1:

$\delta = 5wl^4 / 384EI$

$\delta = 5(.25 \text{ kips/in.})(240 \text{ in.})^4 / 384(29000 \text{ ksi})(394 \text{in.}^4)$

$\delta = .945$ in.

To include dead load deflection caused by the beam's self weight, one would simply consider an additional 50 lbs/ft. of loading (or 4.17 lb/in.). To calculate the additional deflection due to self weight, the process would be;

Calculating deflection from the formula given in Table 13-1:

$\delta = 5wl^4 / 384EI$

$\delta = 5(.00417 \text{ kips/in.})(240 \text{ in.})^4 / 384(29000 \text{ ksi})(394 \text{in.}^4)$

$\delta = .015 \text{ in.}$

This self weight would cause an additional .015 in. of deflection.

EXAMPLE 13.6

Calculate the deflection at midspan on the beam shown below. The beam is a W 12 × 50.

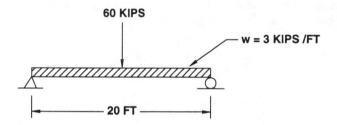

Upon reviewing the load case, one notices that there is no formula for this exact case. However, because both the uniform load case and the concentrated load case produced maximum deflections at the beam's midspan, the deflection at midspan can be determined through superpositioning. We can simply sum the deflections from each individual load case to get the combined deflection. Obtaining the section properties for the W 12 × 50 from Appendix A, the procedure for this method can be illustrated as follows:

$$l = 20 \text{ ft.} = 240 \text{ in.}$$

$$w = 3 \text{ kips/ft.} = .25 \text{ kips/in.}$$

$$P = 60 \text{ kips}$$

$$I = 394 \text{ in.}^4$$

$$E = 29000 \text{ ksi}$$

Using the individual formulas taken from Table 13-1, a "superimposed" formula is created as follows:

$$\delta_{total} = 5wl^4 / 384EI + Pl^3 / 48 \, EI$$

$$\delta_{total} = .945 \text{ in.} + 1.51 \text{ in.} = 2.455 \text{ in.}$$

EXAMPLE 13.7

Calculate the deflection at midspan on the beam shown below. The beam is a W 12 × 50.

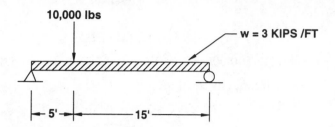

Upon reviewing the load case, superpositioning would again seem to be the likely choice. However, when considering the two load effects separately, one notices that (unlike the previous example) the maximum deflection does not occur at midspan for the concentrated load case. Therefore, either a deflection formula must be found for the deflection at midspan for the concentrated load or another deflection method must be used. This example will proceed as if there is not a formula available for the midspan deflection caused by the concentrated load (although it can be easily found) and the moment-area method will be used. Obtaining the section properties from Appendix A, the procedure for this method can be illustrated as follows:

Deflection at midspan by a uniform load using standard formula

$$l = 20 \text{ ft.} = 240 \text{ in.}$$

$$w = 3 \text{ kips/ft.} = .25 \text{ kips/in.}$$

$$I = 394 \text{ in.}^4$$

$$E = 29000 \text{ ksi}$$

From Table 13-1, the deflection formula can be calculated as follows:

$$\delta_{uniform} = 5wl^4 / 384EI$$

$$\delta_{uniform} = 5(.25 \text{ k/in.})(240 \text{ in.})^4 / 384(29000 \text{ ksi})(394 \text{ in.}^4) = .95 \text{ in.}$$

For the concentrated load at the 5 foot point, the moment area method will be used to calculate the deflection at midspan.

The deflection at this point can be considered to be as follows:

$$\delta = 1/2 \, t_{A/B} - t_{O/B}$$

The tangent deviation of A relative to B ($t_{A/B}$) can be found by taking the area of the M/EI diagram for each triangle which can be calculated as:

$$A_1 = 1/2(60 \text{ in.})(450,000/EI) = 13,500,000 \text{in.}^2\text{-lb}/EI$$

$$A_2 = 1/2(180 \text{ in})(450,000/EI) = 40,500,000 \text{in.}^2\text{-lb}/EI$$

The x distance from A to the centroid of each area is simply 120 in. for area 2, and 40 in. for area 1. Therefore, the tangent deviation of A relative to B ($t_{A/B}$) is as follows:

$$t_{A/B} = x \, (M \, / \, EI)\Delta x$$

$$t_{A/B} = 40 \text{ in.}(13,500,000/EI) + 120 \text{ in. }(40,500,000/EI) = 5.4 \times 10^9 \text{ in.}^3\text{-lb}/EI$$

Therefore $1/2 \, t_{A/B}$ is equal to:

$$1/2 \, t_{A/B} = 2.7 \times 10^9 \text{ in.}^3\text{-lb}/EI$$

The tangent deviation of O relative to B ($t_{O/B}$) can be found by taking the area under the M/EI diagram between O and B (which can be viewed as a triangle) and then multipling this area by the x distance from O to the centroid of the area. The area under the M/EI diagram can be calculated as:

$$A_3 = 1/2(120 \text{ in.})(300,000/EI) = 18,000,000 \text{ in.}^2\text{-lb}/EI$$

The x distance from O to the centroid of this area is simply 40 in., and therefore the tangent deviation from O to B is as follows:

$$t_{O/B} = x \, (M \, / \, EI)\Delta x$$

$$t_{O/A} = 40 \text{ in. }(18,000,000/\text{in.}^2\text{-lb } EI) = 720 \times 10^6 \text{ in.}^3\text{-lb}/EI$$

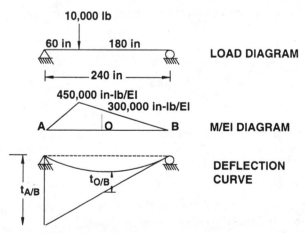

The deflection caused by the concentrated load at the 5 foot location can be found as follows:

$$\delta = 1/2 \, t_{A/B} - t_{O/B}$$

$$\delta = 2.7 \times 10^9 \text{ in.}^3\text{-lb}/EI - .720 \times 10^9 \text{ in.}^3\text{-lb}/EI = 1.98 \times 10^9 \text{in.}^3\text{-lb}/EI$$

Substituting in the values for E and I, we find the following:

$$\delta_{concentrated} = 1.98 \times 10^9 \text{in.}^3\text{-lb} / (29 \times 10^6 \text{ psi})(394 \text{ in.}^4) = .173 \text{ in.}$$

Therefore, the total deflection at the midpoint of the beam is:

$$\delta_{total} = \delta_{uniform} + \delta_{concentrated}$$

$$\delta_{total} = .95 \text{ in.} + .173 \text{ in.} = 1.12 \text{ in.}$$

A limitation of superpositioning is that the maximum deflection caused by separate loadings may not always coincide at the same point along the beam. Therefore, the exact magnitude and location of maximum deflection could not be found directly from this method. In such cases, a more analytical approach would have to be utilized.

Previously, the discussion has focused on possible methods to calculate deflections—but what are the limiting values of deflections that are used in standard practice? Usually, building codes and specifications will limit deflections to a fraction of a beam's span length. Most building codes, including the AISC specification, limit the live load deflection on beams under service load conditions to 1/360th of the span length (or $l/360$) in plastered construction. When dealing with beams that are a part of unplastered construction, this requirement is relaxed to 1/180th of the span length (or $l/180$) in roof assemblies and to 1/240th of the span length in floor assemblies. In comparison, the AASHTO specification[2] limits the live load deflection on steel bridge girders to 1/800th of the span length (or $l/800$) to maintain clearance height under typically heavy live loads.

The student should remember that the aforementioned deflection limits are recommended maximums, which are meant to provide guidance (not a guarantee) in achieving a serviceable design. A designer must always use his personal judgment, taking into account the goals of a specific project. These values may be adjusted, either upward or downward, should that individual project warrant such a decision.

13.5 Deflection Strength Failure: Ponding

Typically beam deflection is solely a serviceability concern. However, there is one notable exception where deflection problems can actually be the catalyst behind a destructive strength-based failure. This phenomena is referred to as **ponding**, the results of which can be devastating.

Ponding refers to the retention of water on a flat or semi-flat roof during periods of heavy rainfall or snow melt. As the water accumulates on the roof faster than it can drain off, the roof deflects under the weight of the water, forming a bowl-shaped profile that enables the roof to retain even more water (Figure 13-9). The roof will keep deflecting, thereby holding more water, until it finally collapses.

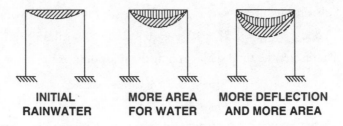

INITIAL RAINWATER	**MORE AREA FOR WATER**	**MORE DEFLECTION AND MORE AREA**

Figure 13-9   Successive Deflections caused by Ponding

Ponding has become more of a problem with the advent of flexible or "light" roof framing systems. With the roof beams and girders being more flexible, there is a greater chance that ponding may occur in a severe rainstorm. Therefore, the AISC requires that specific ponding criteria be met to ensure adequate stiffness for primary and secondary roof members. Some lessons learned from ponding failures are more subtle in nature and more practically oriented than the aforementioned stiffness specifications. The easiest solution to ponding is simply to have the roof sloping at an adequate pitch, typically a minimum of 1/4 in. per foot. Other solutions can be simple lessons such as locating roof drains away from column lines. This is because the columns do not deflect nearly as much as the roof framing, and therefore, the areas adjacent to columns remain "high" in regards to roof profile. Ponding will take place between columns, and therefore, the optimum location for roof drains would be at midspan between columns (if feasible). [3]

13.6   Summary

The ability to determine the magnitude of deflection for a beam is very important to a structural designer. Excessive deflections can lead to serviceability problems—such as cracking in walls and ceilings, improper functioning of building hardware, or poor appearance. All of these problems cast doubt upon the structural integrity of the building. Therefore, it is important to be able to calculate deflections and compare these with appropriate levels that are typically found in a building code.

Many techniques can be used to calculate deflections—including standard deflection formulas that are found in many references. When the loadings on a beam are non-standard, other techniques must be utilized. The moment-area method for finding deflections is also used many times because of its simplicity. Although deflection is usually considered to affect only the intended function of a building, it can also lead to catastrophic failure—such as a structure's collapse due to ponding of rainwater or snowmelt. The designer must be aware of the serviceability aspects of a particular structure and adjust specification requirements accordingly.

EXERCISES

1. Calculate the live load deflection caused by two 50 kip live loads, each located at 5 ft. in from their respective supports on a 25 ft. simply-supported beam. The beam is a W 12 × 120 and the self weight is neglected. Would this deflection be acceptable under service floor load conditions using plastered construction?

2. Calculate the live load deflection for a W 310 × 97 that is 10 meters long and has a 20 KN force located at midspan.

3. Calculate the total deflection for the beam in Exercise 1 if dead load from the beam's self weight is to be included.

4. Calculate the total deflection for the beam in Exercise 2 if dead load from the beam's self weight is to be included.

5. Using the moment-area method, calculate the deflection at the midpoint for a simply supported beam which is 20 ft. long and has 5 kip loads applied at the 5 ft. and 15 ft. locations. The beam is a W 12 × 40 with $E = 29 \times 10^6$ psi.

6. Using the moment-area method, calculate the deflection at the midpoint for a simply supported beam shown below. The beam is a W 10 × 33 with $E = 29 \times 10^6$ psi.

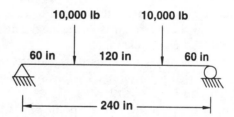

7. Using the moment-area method, calculate the deflection at the midpoint, for the cantilevered beam shown below. The beam is a W 310 × 86 with $E = 200 \times 10^3$ MPa.

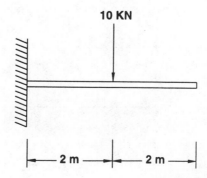

8. For the beam used in Exercise 7, calculate the deflection at the free end using the moment-area method also.

9. For the beam shown below, calculate the deflection at midspan using the principle of superpositioning. The beam is a W 10×45 with $E = 29000$ ksi.

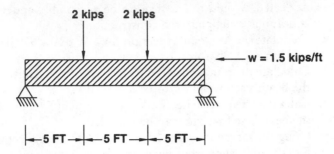

10. For the beam used in Exercise 9, calculate the increase needed in the magnitude of the concentrated load to reach a deflection of 1.25 in.

11. Calculate the concentrated live load, P, which, when applied to midspan of a W 250×101, causes a deflection of 65 mm. The beam is 7.5 m long.

12. Calculate the magnitude of uniform load needed to cause a 2×10 floor joist to deflect 0.50 in. The floor joist spans 16 ft. and the $E = 1.4 \times 10^6$ psi.

13. Calculate the concentrated live load to cause a 0.50 in. live load deflection in a W 12×14 that is 20 ft. long. The live load is to be applied at midspan and $E = 29000$ ksi.

14. Calculate the deflection caused by a uniform load of 2 KN/m over a W 250×101 that is 9 m long. The beam also has a 10 KN load applied at midspan and $E = 200 \times 10^3$ MPa.

15. Calculate the deflection at the point of loading caused by a 8 KN load applied at the third point on a W 310×179 that is 9 m long.

16. Using the moment-area method, derive the deflection formula at the 1/4 point for a beam with a concentrated load, P, at its center. Consider the span length to be L.

17. Calculate the deflection under the 10k load for the beam shown below. The beam is made from two 2×10's and has an $E = 1.6 \times 10^6$ psi.

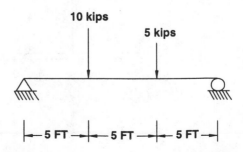

18. A 20-foot-long, simply-supported beam holds a uniform load of 3 kips/ft. across its full length. Calculate the required moment of inertia for the beam deflection to be limited to 0.25 in. $E = 29,000$ ksi.

19. A 10-meter-long, simply-supported beam holds a uniform load of 4 KN/m across its full length. Calculate the required moment of inertia for the beam deflection to be limited to 20 mm. $E = 200,000$ MPa.

20. Calculate the minimum diameter required for a solid round beam that is 10 ft. long and cantilevered out from a wall. The beam is made from steel ($E = 29000$ ksi) and the maximum allowable deflection is 0.20 in. The beam is under a uniform load of 2 kips/ft.

21. A W 8 × 35 spans 15 ft. and carries a 7 kips load at each third point. Using the moment-area method, calculate the deflection at midspan. Consider $E = 29 \times 10^6$ psi.

REFERENCES

1. *Manual of Steel Construction,* Ninth Edition, American Institute of Steel Construction, Chicago, 1989.

2. *Standard Specifications for Highway Bridges,* 14th Edition, American Association of State Highway and Transportation Officials, Washington, DC , 1989.

3. Kaminetsky, Dov "Design and Construction Failures, Lessons Learned from Forensic Investigations", McGraw-Hill Publishing, New York, 1991, p. 249.

CHAPTER

14

COMBINED STRESSES

14.1 Introduction

In the previous chapters we have discussed members being subjected to axial, bending, and shear stresses in a separate and individual manner. The analysis of these members was direct because the stresses were one of the aforementioned types. However, many times a member will be subjected to more than one type of stress and therefore it will be important to determine the behavior of different stresses acting in combination. A typical example is that of a standard beam. When a beam is acted upon by bending moment it is also being acted upon by shear stresses as well. The tension or compression components caused by bending stress will combine with the shear stress at various locations to potentially produce a critical combined stress. The designer must have the ability to recognize and to calculate the maximum stress effect caused by such a condition.

When two or more types of stress are applied to a given member, it will become necessary to consider breaking these stresses into components that act either normal to the plane in question or parallel to the plane in question. Those stresses acting normal to the plane are referred to as **normal stresses** and those acting parallel to the plane in question are referred to as **shear stresses**. By considering the components of individual stress types, it will be easier to calculate the resultant effect of some typical combined stress scenarios.

14.2 Combined Axial and Bending Stress

One of the most common combined stress situations is that of a member subjected to axial stresses as well as bending stresses. This combination occurs in many everyday applications; such as prestressed beams, eccentrically loaded columns,

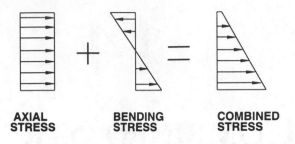

AXIAL STRESS   **BENDING STRESS**   **COMBINED STRESS**

Figure 14-1   Principle of Super Positioning with Combined Stress

and foundations subjected to lateral loads. The actual stress that is developed in such members is the result of an axial load effect being coupled with the bending effect produced from an applied moment.

The combined stress effect produced by an axial stress coupled with a bending stress can be calculated using the principle of superpositioning (Figure 14-1). Superpositioning can be used because both the axial effect and bending effect produce stresses that are collinear, or along the same line, at a given location on a member cross-section. Since bending stress produces both tension and compression stresses, these add directly to the tension or compression stress produced from the axial load. The combined axial and bending stress formula can be written as follows:

$$\sigma_c = P / A + Mc / I \qquad \text{(Eq. 14-1)}$$

where

$P =$ applied axial force

$A =$ cross-sectional area resisting axial force

$M =$ applied bending moment to the cross-section under consideration

$c =$ distance from neutral axis to location bending stress is to be computed

$I =$ moment of inertia about axis which moment is applied

Several items should be noted with regard to the use of Equation 14-1. It must be remembered that bending stress produces tensile and compressive stresses that vary from zero at the neutral axis to a maximum at the outside surface of the member. Typically, designers are most interested with the maximum and minimum combined stresses on the cross-section. Because the bending stress component will vary from zero at the neutral axis to a maximum at the outside surface, the maximum or minimum combined stress will occur at the outside of the member. If the maximum or minimum combined stress is desired, the distance "c" will be the distance from the neutral axis to the outside surface of the section (on a symmetrical section, c will therefore be 1/2 the depth of the section). Secondly, it should be noted that similar axial stresses will be summed, while dissimilar axial stresses will be subtracted. Because the magnitude of collinear vectors add in a numeric fashion, the tensile stress caused by an axial load will add to the tensile portion of the bending stress. Conversely, a tensile stress due to axial loads will subtract from the compressive portions of the bending stress. Because tension stress can only "add" to tension stress (and compression to compression), it may be easier to assign a sign convention—such as tension stress is always positive while compressive stress is always negative. This will help to eliminate the possibility of inadvertently adding dissimilar stresses.

The following example will demonstrate the use of combined stress principles regarding members simultaneously subjected to axial and bending stresses.

EXAMPLE 14.1

The beam shown is 10 in. wide and 16 in. deep. An axial tension force of 50,000 lbs is applied at the centroid of the cross-section. Calculate the minimum and maximum combined stress that occurs on the beam.

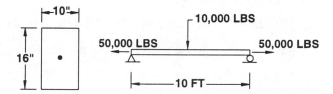

The minimum and maximum combined effect will occur where the bending stress component and the axial stress component are maximized. Since the axial stress component is constant throughout the member, the maximum bending stress will occur where the applied moment, M, is the highest and on the outside surface of the beam (when c is equal to 1/2 depth). The maximum moment can be found by creating a shear and moment diagram as shown below.

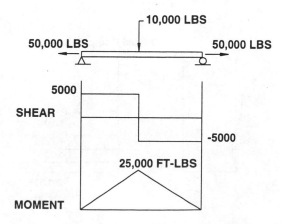

From the moment diagram, it can be seen that the maximum bending moment on the beam occurs at midspan and has a value of 25,000 lb-ft. This moment will produce a bending stress that results in tension on the bottom of the beam and compression on the top. Recognizing that the axial stress is also tension, this means that the maximum combined stress effect will occur on the bottom of the beam, while the minimum combined stress will occur on the top. Calculating Equation 14-1:

$$\sigma_c = P / A + Mc / I \qquad \text{(Eq. 14-1)}$$

with

$P = 50,000$ lbs

$A = 160$ in.^{2}

$M = 25,000$ lb-ft.

$c = 1/2(16$ in.$) = 8$ in.

$I = 1/12bh^3 = 1/12(10$ in.$)(16$ in.$)3 = 3413$ in.4

$\sigma_{cmax} = +(50,000$ lb $/ 160$ in.$^2) + [(25,000$ lb-ft.$)(12$ in. $/$ ft.$)(8$ in.$) / 3413$ in.$^4]$

$\sigma_{cmax} = + 312.5$ psi $+ 703.2$ psi $= + 1015.7$ psi (tension)

$\sigma_{cmin} = +(50,000$ lb $/ 160$ in.$^2) - [(25,000$ lb-ft.$)(12$ in. $/$ ft.$)(8$ in.$) / 3413$ in.$^4]$

$\sigma_{cmin} = + 312.5$ psi $- 703.2$ psi $= - 390.7$ psi (compression)

The distribution of stress across the cross-section of the beam at midspan would be as follows:

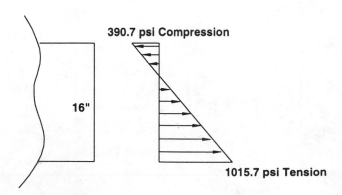

EXAMPLE 14.2

The concrete footing shown below holds a column that is subjected to an axial load of 20 KN and a lateral load of 3 KN. Calculate the maximum and minimum bearing stresses that are produced on the footing.

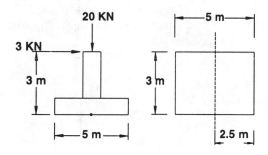

The bending stress on the footing occurs about its centroidal axis and can be calculated as follows:

$$M = (3 \text{ KN})(3 \text{ m}) = 9 \text{ KN-m}$$

The applied moment will bend the footing about this centroidal (neutral) axis, producing variable compressive stresses in the front part of the footing and variable tensile stresses in its rear portion. Since the axial stress component is constant throughout the member, the maximum bending stress will occur at the outside surface of the footing (front and rear). Recognizing that the axial stress is also compressive, this means that the maximum combined stress effect will occur in the front of the footing, while the minimum combined stress will occur in the rear. Calculating Equation 14-1:

$$\sigma_c = P / A + Mc / I \qquad \text{(Eq. 14-1)}$$

with

$P = 20 \text{ KN}$

$A = 15 \text{ m}^2$

$M = 9 \text{ KN-m}$

$c = 1/2(5 \text{ m}) = 2.5 \text{ m}$

$I = 1/12bh^3 = 1/12(3 \text{ m})(5 \text{ m})^3 = 31.25 \text{ m}^4$

Therefore

$$\sigma_{cmax} = -(20 \text{ KN} / 15 \text{ m}^2) - [(9 \text{ KN-m})(2.5 \text{ m}) / 31.25 \text{ m}^4]$$

$$\sigma_{cmax} = -1.33 \text{ KN/m}^2 - .72 \text{ KN/m}^2 = -2.05 \text{ KN/m}^2 \text{ or } 2.05 \text{ KPa (compression)}$$

$$\sigma_{cmin} = -(20 \text{ KN} / 15 \text{ m}^2) + [(9 \text{ KN-m})(2.5 \text{ m}) / 31.25 \text{ m}^4]$$

$$\sigma_{cmin} = -1.33 \text{ KN/m}^2 + .72 \text{ KN/m}^2 = -0.61 \text{ KN/m}^2 \text{ or } 0.61 \text{ KPa (compression)}$$

The distribution of combined stress under the footing would be completely in compression, which is typical for a footing. The distribution of this combined stress would be as follows:

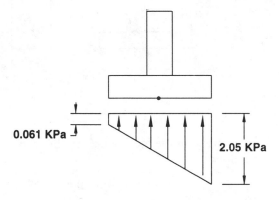

0.061 KPa   **2.05 KPa**

EXAMPLE 14.3

A rectangular beam measuring 12 in. wide by 20 in. deep is prestressed using steel cables located at the centroidal axis and delivering a compressive force of 80 kips to the end of the beam. If the beam is loaded as shown below, calculate the combined stress distribution in the beam.

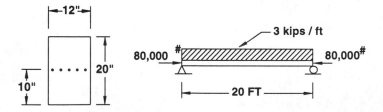

The minimum and maximum combined effect will occur where the bending stress component and the axial stress component are maximized. Since the axial stress component is constant throughout the member, the maximum bending stress will occur where the applied moment, M, is the highest and on the outside surface of the beam (when c is equal to 1/2 depth). The maximum moment can be found by creating a shear and moment diagram as is shown below.

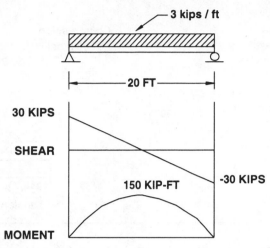

From the moment diagram it can be seen that the maximum bending moment on the beam occurs at midspan and has a value of 150 kip-ft. This moment will produce a bending stress that results in tension on the bottom of the beam and compression on the top. Recognizing that the axial stress is also compression, this means that the maximum combined stress effect will occur on the top of the beam, while the minimum combined stress will occur on the bottom. Calculating Equation 14-1:

$$\sigma_c = P / A + Mc / I \qquad \text{(Eq. 14-1)}$$

with

$P = 80{,}000$ lbs

$A = 240$ in.^{2}

$M = 150$ kip–ft. $= 150{,}000$ lb-ft.

$c = 1/2(20 \text{ in.}) = 10$ in.

$I = 1/12bh^3 = 1/12(12 \text{ in.})(20 \text{ in.})^3 = 8000$ in.^{4}

Calculating Equation 14-1:

$\sigma_{cmax} = -(80{,}000 \text{ lb} / 240 \text{ in.}^2) - [(150{,}000 \text{ lb-ft.})(12 \text{ in.} / \text{ft.})(10 \text{ in.}) / 8000 \text{ in.}^4]$

$\sigma_{cmax} = -333.3 \text{ psi} - 2250 \text{ psi} = -2583.3$ psi (compression on top)

$\sigma_{cmin} = -(80{,}000 \text{ lb} / 240 \text{ in.}^2) + [(150{,}000 \text{ lb-ft.})(12 \text{ in.} / \text{ft.})(10 \text{ in.}) / 8000 \text{ in.}^4$

$\sigma_{cmin} = -333.3 \text{ psi} + 2250 \text{ psi} = +1916.7$ psi (tension in bottom)

The distribution of stress across the cross-section of the beam at midspan would be as follows:

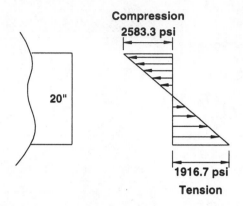

In the preceding problem it appears that compressive stresses are actually increased in the top portion of the beam, however in real life the prestressing force is offset from the neutral axis. This produces a bending effect from the prestressing which actually works oppositely of the applied bending moment. We will discuss this further in the upcoming section.

14.3 Combined Stresses Due to Eccentric Loads

Another case of combined stress due to axial and bending effects arises when an axial load acts away from the centroidal axis of a member. Such a load is therefore referred to as an **eccentric load**. Eccentric loads, by their nature, subject the member to axial stress and to a bending stress. The bending stress develops from the moment caused by the load, multiplied by the eccentric distance or eccentricity.

A common case of combined stress due to eccentric loads is the case of an eccentricly loaded column (Figure 14-2). This column carries an axial force, P, that must be resisted by the column developing a compressive stress. The compressive stress would be uniform over the entire resisting area and calculated using the direct stress formula from Chapter 8:

$$\sigma = P / A \qquad \text{(Eq. 8-1)}$$

Because the load is not applied through the center of the column, it will cause bending about the shape's neutral axis—typically the centroidal axis. The moment that causes this bending is as follows:

$$M = (P)(e)$$

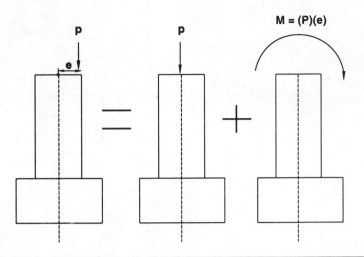

Figure 14-2   Column Behavior Under Eccentric Load

where

$$e = \text{eccentric distance or eccentricity}$$

The bending stress is then calculated based on the bending stress formula which was developed in Chapter 12:

$$\sigma = Mc / I \qquad \text{(Eq. 12-6)}$$

Using this information, the combined stress can be found by superpositioning the stresses from both the axial and bending behaviors and summing their effects. It should be noted that similar stress effects will be summed, while dissimilar stress effects will be subtracted. This would mean that a column under axial compression would have a higher combined stress at the location under compression due to bending. Consequently, the portion that was under tension due to bending would have a lower net combined stress.

The following examples will demonstrate the use of combined stress principles due to eccentric loads.

EXAMPLE 14.4

Calculate the maximum and minimum combined stress effect on the column shown below. The eccentricity about the x-axis is 6 inches.

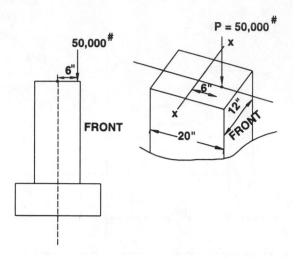

The minimum and maximum combined effect will occur where the bending stress component and the axial stress component are maximized. Since the column is in axial compression, the maximum effect will occur where the bending effect creates compression also. Since the axial load is on the front part of the column, the rotation about the centroidal axis will "squeeze" this part of the column closer together, indicating that the front part of the column is under compression due to bending. The maximum bending stress will therefore occur on the front side of this column. The maximum moment can be found as follows:

$$M = (P)(e)$$

$$M = 50{,}000 \text{ lbs}(6 \text{ in.}) = 300{,}000 \text{ lb-in.}$$

Calculating Equation 14-1:

$$\sigma_c = P / A + Mc / I \qquad\qquad\qquad \text{(Eq. 14-1)}$$

with

$$P = 50{,}000 \text{ lbs}$$

$$A = 240 \text{ in.}^2$$

$$M = 300{,}000 \text{ lb-in.}$$

$$c = 1/2(20 \text{ in.}) = 10 \text{ in.}$$

$$I = 1/12bh^3 = 1/12(12 \text{ in.})(20 \text{ in.})^3 = 8000 \text{ in.}^4$$

Calculating Equation 14-1:

$$\sigma_{cmax} = -(50{,}000 \text{ lb} / 240 \text{ in.}^2) - [(300{,}000 \text{ lb-in.})(10 \text{ in.}) / 8000 \text{ in.}^4]$$

$$\sigma_{cmax} = -208.3 \text{ psi} - 375 \text{ psi} = -583.3 \text{ psi (compression in front)}$$

$$\sigma_{cmin} = -(50{,}000 \text{ lb} / 240 \text{ in.}^2) + [(300{,}000 \text{ lb-in.})(10 \text{ in.}) / 8000 \text{ in.}^4$$

$$\sigma_{cmin} = -208.3 \text{ psi} + 375 \text{ psi} = +166.7 \text{ psi (tension in back)}$$

The distribution of stress across the cross-section of the column would be as follows:

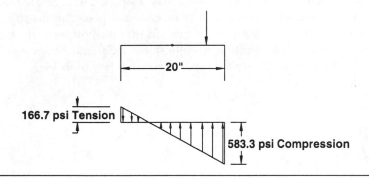

A common mistake made in the previous example is to mistake the eccentricity, e, used in obtaining the moment with the "c" distance used in the bending stress formula. Remember c is the distance from the neutral axis to where the bending stress is calculated. If a maximum or minimum stress is required, c will be the distance from the neutral axis to the outside of the member.

EXAMPLE 14.5

A rectangular beam measuring 12 in. wide by 20 in. deep is prestressed using steel cables located 4 in. below the centroidal axis and delivering a compressive force of 80 kips to the end of the beam. If the beam is loaded as shown below, calculate the combined stress distribution in the beam.

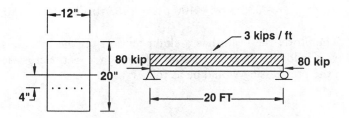

In this example, the prestressing force will cause a combined axial and bending stress effect by causing bending about the beam's neutral axis. The prestressing force is a compressive force that also causes bending because of its eccentricity. The combined effect of just the prestressing will be calculated first and then added to the bending stress effect caused by the loading.

Prestressing Effect

The minimum and maximum combined effect from the prestressing will occur where the bending stress component and the axial stress component are maximized. Since the axial prestressing load is situated 4 in. below the neutral axis, the bending stress will occur such that tension is placed in the top of the beam and compression in the bottom. This means that the bottom of the beam will cumulatively add the compression from both the axial and bending effects, while these effects are subtracted in the top. Calculating Equation 14-1:

$$\sigma_c = P/A + Mc/I \qquad \text{(Eq. 14-1)}$$

with

$$P = 80,000 \text{ lbs}$$

$$A = 240 \text{ in.}^2$$

$$M = (P)(e) = 80,000 \text{ lb}(4 \text{ in.}) = 320,000 \text{ lb-in.}$$

$$c = 1/2(20 \text{ in.}) = 10 \text{ in.}$$

$$I = 1/12bh^3 = 1/12(12 \text{ in.})(20 \text{ in.})^3 = 8000 \text{ in.}^4$$

Calculating:

$$\sigma_{cmax} = -(80,000 \text{ lb}/240 \text{ in.}^2) - [(320,000 \text{ lb-in.})(10 \text{ in.})/8000 \text{ in.}^4]$$

$$\sigma_{cmax} = -333.3 \text{ psi} - 400 \text{ psi} = -733.3 \text{ psi (compression in bottom)}$$

$$\sigma_{cmin} = -(80,000 \text{ lb}/240 \text{ in.}^2) + [(320,000 \text{ lb-ft.})(10 \text{ in.})/8000 \text{ in.}^4$$

$$\sigma_{cmin} = -333.3 \text{ psi} + 400 \text{ psi} = +66.7 \text{ psi (tension in top)}$$

Therefore, the beam has a slight tension on the outside surface of the top before it is loaded. The distribution of stress across the cross-section of the beam from the prestressing would be as follows:

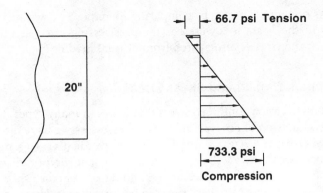

PRESTRESSING STRESS

Load Effect

The bending moment stress caused by the uniform load would be maximum at midspan and reach a value of 150,000 lb-ft. The bending stress caused by this moment would be calculated from Equation 12-6 as follows:

$$\sigma = Mc / I \qquad\qquad \text{(Eq. 12-6)}$$

$$\sigma = 150{,}000 \text{ lb-ft.}(12 \text{ in./ft.})(10 \text{ in.}) / 8000 \text{ in.}^4 = 2250 \text{ psi}$$

This stress would be a tension stress along the bottom surface of the beam and a compressive stress along the top surface of the beam. Adding the load effect to the prestressing effect we would find the following net effect:

$$\sigma_{\text{net-top}} = +66.7 \text{ psi} - 2250 \text{ psi} = -2183.3 \text{ psi (compression)}$$

$$\sigma_{\text{net-btm}} = -733.3 \text{ psi} + 2250 \text{ psi} = 1516.7 \text{ psi (tension)}$$

The distribution of net stress would be as follows:

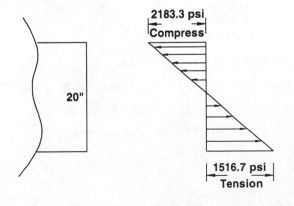

NET STRESS

As one would notice by comparing Example 14.5 with Example 14.3, the net stress in the beam is lowered as the prestressing effect is utilized to counteract some of the stresses anticipated from normal loading.

14.4 Combined Normal and Shear Stress

Another common combined stress that arises in many types of members is due to the combination of normal and shear stresses. As we have discussed previously, a normal stress is a stress acting normal (perpendicular) to a member or a portion of a member that we will refer to as an "element". Normal stresses will either be tension or compression stresses, depending on whether they are trying to "pull apart" or "squeeze together" the member or element. Shear stresses act parallel to a plane within a member or an element and try to rip the member or element apart (Figure 14-3).

Many types of members or elements within such members will be subjected to varying combinations of normal and shear stresses. These combinations that exist on the faces of the element also create many combinations of stress on various planes within that element. These stresses on the other planes may in fact be higher, and therefore more critical, than the actual applied normal and shear stresses. An example of this might be a member that is placed in vertical compression (Figure 14-4). Up to now, it has conveniently been assumed that such a member should have its compressive stress calculated by the direct stress formula, $\sigma = P / A$. This stress is the one developed on the plane perpendicular or normal to the applied load. This, however, is not the only plane of stress that is present within this member. In reality, the applied compressive force will set up a combination of normal and shear stresses on throughout this member. As one can see in

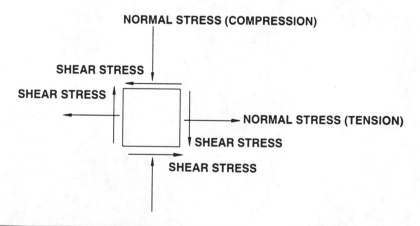

Figure 14-3 Normal and Shear Stresses Acting in Combination on an Element

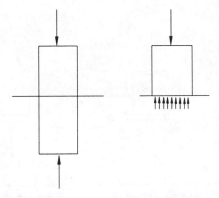

Figure 14.4   Compressive Stress on a Plane Normal to an Axial Load

Figure 14-5, the *oblique planes* (not perpendicular to the applied force), when subjected to the compressive force, will have a component of the force acting perpendicular to that plane (normal component) and another acting parallel to that plane (shear component).

Using the laws of equilibrium, we can derive mathematical equations for determining the normal stress acting on any plane within a body or an element. Consider an element of a body that has normal stresses acting on its x face (σ_x) and its y face (σ_y), along with shear stress τ_{xy} (shear acting on an x face in the y direction), and τ_{yx} (shear acting on a y face in the x direction). If a plane making an angle of θ is cut through the element, a free-body can be constructed (Figure 14-6). Knowing that equilibrium must be maintained, we can write equations by

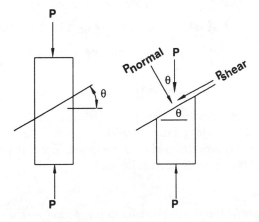

Figure 14-5   Normal and Shear forces Acting on Oblique Planes

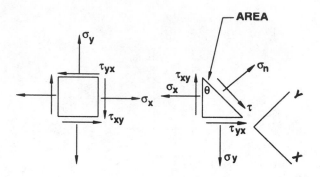

Figure 14-6   Stresses Acting Within an Element at Angle θ

breaking the stresses down into x and y components that are relative to the angle θ and multiplying them by their respective areas. This can be accomplished as follows:

$$\Sigma F_y = 0$$

$\sigma_n A$ – (normal force components in the y axis) – (shear force components in y axis) = 0

$\sigma_n A$ = (normal components) + (shear components)

Realizing the area along the x face is $A\cos\theta$ and that the area along the y face is $A\sin\theta$, the following can be written:

$$\sigma_n A = [(\sigma_x\, A\cos\theta)\cos\theta + (\sigma_y\, A\sin\theta)\sin\theta] + [(\tau_{xy}\, A\cos\theta)(\sin\theta) + [(\tau_{yx}\, A\sin\theta)(\cos\theta)]$$

Knowing that $\tau_{xy} = \tau_{yx,}$ this can be arranged as:

$$\sigma_n A = [\sigma_x\, A\cos^2\theta + \sigma_y\, A\sin^2\theta] + [2\tau_{xy}\, A\sin\theta\cos\theta] \qquad \text{(Eq. 14-2)}$$

Dividing through by A and utilizing standard trigonometric relationships, Equation 14-2 can become:

$$\sigma_n = [((\sigma_x + \sigma_y)/ 2) + ((\sigma_x - \sigma_y)/ 2)\cos 2\theta] + \tau_{xy}\sin 2\theta \qquad \text{(Eq. 14-3)}$$

This is an equation to calculate the normal stress on any plane throughout the element. To determine the shear stress (τ_p) on any plane throughout the element a similar process can be followed by summing forces in the x direction parallel to the plane. This can be done as follows:

$$\Sigma F_x = 0$$

$\tau_p\, A$ – (normal force components in the x-axis) – (shear force components in x-axis) = 0

$\tau_p A$ = (normal components) + (shear components)

Realizing the area along the x face is $A\cos\theta$ and that the area along the y face is $A\sin\theta$, the following can be written:

$$\tau_p A = [(-\sigma_x A\cos\theta)\sin\theta + (\sigma_y A\sin\theta)\cos\theta] + [-(\tau_{xy} A\sin\theta)(\sin\theta) + (\tau_{yx} A\cos\theta)(\cos\theta)]$$

Knowing that $\tau_{xy} = \tau_{yx}$ this can be arranged as:

$$\tau_p A = [(-\sigma_x + \sigma_y)(A\sin\theta\cos\theta)] + [(\tau_{xy}A)\cos2\theta - \sin2\theta] \qquad \text{(Eq. 14-4)}$$

Dividing through by A and utilizing standard trigonometric relationships, Equation 14-4 can become:

$$\tau_p = [(-(\sigma_x - \sigma_y)/2)\sin2\theta] + \tau_{xy}\cos2\theta \qquad \text{(Eq. 14-5)}$$

Therefore Equation 14-3 can be used to calculate the normal stress at any plane throughout the element, while Equation 14-5 can be used to calculate the associated shear stress on those planes. This gives the designer a tool for the analysis of member that are subjected to a combination of normal and shear stresses. The following examples will illustrate the use of these formulas.

EXAMPLE 14.6

The element below is taken from a beam and is subjected to the normal and shear stresses as shown. Determine the normal and shear stresses on planes within the element of 10°, 30°, and 45°.

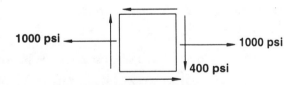

The determination of normal and shear stresses on different planes throughout the element can be found using Equations 14-3 and 14-5, which were derived previously.

At $\theta = 10°$, the normal and shear stresses can be found as follows:

$$\sigma_n = [(\sigma_x + \sigma_y/2) + (\sigma_x - \sigma_y/2)\cos2\theta] + t_{xy}\sin2\theta \qquad \text{(Eq. 14-3)}$$

with

$\sigma_x = +1000$ psi (the positive sign indicating tension)

$\sigma_y = 0$

$\tau_{xy} = +400$ psi (the positive sign indicating clockwise)

Substituting into Equation 14-3, we find:

$$\sigma_n = [(1000 + 0)/2 + (1000-0)/2 \cos 20°] + 400 \sin 20° = +1106.6 \text{ psi}$$

The shear stress can be found as follows:

$$\tau_p = [-(\sigma_x - \sigma_y / 2) \sin 2\theta] + \tau_{xy} \cos 2\theta \qquad \text{(Eq. 14-5)}$$

Substituting in the stress values we find:

$$\tau_p = [-(1000 - 0/ 2) \sin 20°] + 400 \cos 20° = + 204.8 \text{ psi}$$

At $\theta = 30°$, the normal and shear stresses can be found as follows:

$$\sigma_n = [(\sigma_x + \sigma_y / 2) + (\sigma_x - \sigma_y / 2) \cos 2\theta] + \tau_{xy} \sin 2\theta \qquad \text{(Eq. 14-3)}$$

with

$\sigma_x = +1000$ psi (the positive sign indicating tension)

$\sigma_y = 0$

$\tau_{xy} = + 400$ psi (the positive sign indicating clockwise)

Substituting into Equation 14-3, we find:

$$\sigma_n = [(1000 + 0)/2 + (1000 - 0)/2 \cos 60°] + 400 \sin 60° = -1096.4 \text{ psi}$$

The shear stress can be found as follows:

$$\tau_p = [-(\sigma_x - \sigma_y / 2) \sin 2\theta] + \tau_{xy} \cos 2\theta \qquad \text{(Eq. 14-5)}$$

Substituting in the stress values, we find:

$$\tau_p = [-(1000 - 0/ 2) \sin 60°] + 400 \cos 60° = -233 \text{ psi}$$

At $\theta = 45°$, the normal and shear stresses can be found as follows:

$$\sigma_n = [(\sigma_x + \sigma_y / 2) + (\sigma_x - \sigma_y / 2) \cos 2\theta] + \tau_{xy} \sin 2\theta \qquad \text{(Eq. 14-3)}$$

with

$\sigma_x = +1000$ psi (the positive sign indicating tension)

$\sigma_y = 0$

$\tau_{xy} = +400$ psi (the positive sign indicating clockwise)

Substituting into Equation 14-3, we find:

$$\sigma_n = [(1000 + 0)/2 + (1000-0)/2 \cos 90°] + 400 \sin 90° = + 900 \text{ psi}$$

The shear stress can be found as follows:

$$\tau_p = [-(\sigma_x - \sigma_y / 2) \sin 2\theta] + \tau_{xy} \cos 2\theta \qquad \text{(Eq. 14-5)}$$

Substituting in the stress values, we find:

$$\tau_p = [-(1000 - 0/\ 2)\ \sin 90°] + 400 \cos 90° = -500 \text{ psi}$$

The stresses on these three planes could be illustrated as follows:

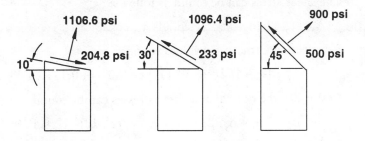

EXAMPLE 14.7

The element below is taken from a beam and is subjected to the normal and shear stresses as shown. Determine the normal and shear stresses on planes within the element of 15° and 40°.

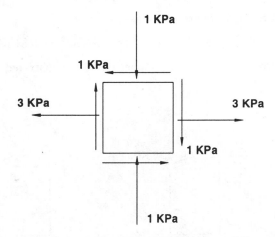

The determination of normal and shear stresses on different planes throughout the element can be found using Equations 14-3 and 14-5, which were derived previously.

At $\theta = 15°$, the normal and shear stresses can be found as follows:

$$\sigma_n = [(\sigma_x + \sigma_y /\ 2) + (\sigma_x - \sigma_y /\ 2) \cos 2\theta] + \tau_{xy} \sin 2\theta \qquad \text{(Eq. 14-3)}$$

with

$\sigma_x = +3$ KPa (the positive sign indicating tension)

$\sigma_y = -1$ KPa (indicating compression)

$\tau_{xy} = +1$ KPa (the positive sign indicating clockwise)

Substituting into Equation 14-3, we find:

$$\sigma_n = [(3-1)/2 + (3-(-1))/2 \cos 30°] + 1 \sin 30° = +3.23 \text{ KPa}$$

The shear stress can be found as follows:

$$\tau_p = [-(\sigma_x - \sigma_y / 2) \sin 2\theta] + \tau_{xy} \cos 2\theta \qquad \text{(Eq. 14-5)}$$

Substituting in the stress values we find:

$$\tau_p = [-(3-1/2) \sin 30°] + 1 \cos 30° = +0.36 \text{ KPa}$$

At $\theta = 40°$, the normal and shear stresses can be found as follows:

$$\sigma_n = [(\sigma_x + \sigma_y / 2) + (\sigma_x - \sigma_y / 2) \cos 2\theta] + \tau_{xy} \sin 2\theta \qquad \text{(Eq. 14-3)}$$

Substituting into Equation 14-3, we find:

$$\sigma_n = [(3-1)/2 + (3-(-1))/2 \cos 80°] + 1 \sin 80° = +2.33 \text{ KPa}$$

The shear stress can be found as follows:

$$\tau_p = [-(\sigma_x - \sigma_y / 2) \sin 2\theta] + \tau_{xy} \cos 2\theta \qquad \text{(Eq. 14-5)}$$

Substituting in the stress values, we find:

$$\tau_p = [-(3-1/2) \sin 80°] + 1 \cos 80° = -0.81 \text{ KPa}$$

The stresses on these two planes could be illustrated as follows:

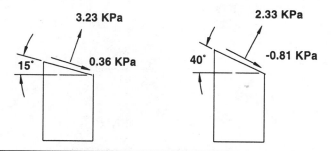

14.5 Maximum, Normal and Shear Stresses

In the previous section, equations were developed for determining the magnitude of normal stress (σ_n) and the magnitude of shear stress (τ_p) that occurred on any plane throughout an element subjected to normal and shear stress. These equations were as follows:

$$\sigma_n = [(\sigma_x + \sigma_y / 2) + (\sigma_x - \sigma_y / 2) \cos 2\theta] + \tau_{xy} \sin 2\theta \qquad \text{(Eq. 14-3)}$$

$$\tau_p = [-(\sigma_x - \sigma_y / 2) \sin 2\theta] + \tau_{xy} \cos 2\theta \qquad \text{(Eq. 14-5)}$$

The designer may, however, be interested in determining only the location of the planes that maximum values of normal and shear stress occur on, for a given state of stress. Therefore it is imperative that the maximum values of normal and shear stress be found, along with the planes on which they act.

The maximum normal stress is often referred to as the **principal stress** and occurs on a plane known as the **principal plane**. By differentiating Equation 14-3 with respect to θ, the maximum normal stress occurs when:

$$\tan 2\theta = 2\tau_{xy} / \sigma_x - \sigma_y$$

or

$$\tan 2\theta = \tau_{xy} / (\sigma_x - \sigma_y / 2) \qquad \text{(Eq. 14-6)}$$

Similarly the following relationship can be determined:

$$\sin 2\theta = \tau_{xy} / \sqrt{(\sigma_x - \sigma_y/2)^2 + \tau_{xy}^2} \qquad \text{(Eq. 14-7)}$$

$$\cos 2\theta = (\sigma_x - \sigma_y / 2) / \sqrt{(\sigma_x - \sigma_y/2)^2 + \tau_{xy}^2} \qquad \text{(Eq. 14-8)}$$

This relationship can be viewed in trigonometric terms seen in the Figure 14-7 and from this, the following can be determined:

$$X = (\sigma_x - \sigma_{y/}2)$$

$$Y = \tau_{xy}$$

$$H = \sqrt{(\sigma_x - \sigma_y/2)^2 + \tau_{xy}^2}$$

The solution for θ could actually have two solutions—the other occurring 180° away from the first. There will be a maximum value of normal stress and a minimum value of normal stress. By substituting Equations 14-6, 14-7, and 14-8, into Equation 14-3; the following solution for maximum normal stress is found:

$$\sigma_{nmax} = [(\sigma_x + \sigma_y / 2) + (X)(X / H)] + (Y)(Y / H) \qquad \text{(Eq. 14-9)}$$

where

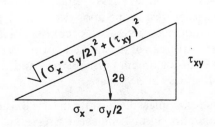

Figure 14-7   Relationship of Normal and Shear Stresses for Determination of Principle Stress

$$X = (\sigma_x - \sigma_y / 2)$$

$$Y = \tau_{xy}$$

$$X / H = \cos 2\theta$$

$$Y / H = \sin 2\theta$$

Consolidating Equation 14-9, we find:

$$\sigma_{n\max} = (\sigma_x + \sigma_y / 2) + (X^2 / H) + (Y^2 / H) \qquad \text{(Eq. 14-10)}$$

Realizing that the hypotenuse, H, is simply $\sqrt{X^2 + Y^2}$, Equation 14-10 becomes:

$$\sigma_{n\max} = (\sigma x + \sigma y / 2) + \sqrt{(X^2 + Y^2)}$$

and resubstituting the values for X and Y back into this equation, we find:

$$\sigma_{n\max} = (\sigma_x + \sigma_y / 2) + \sqrt{(\sigma_x - \sigma_y/2)^2 + \tau_{xy}^2} \qquad \text{(Eq. 14-11)}$$

The solution for the value and location of maximum shear stress can be found in a similar manner[1] and is found to be:

$$\tau_{\max} = +\sqrt{(\sigma_x - \sigma_y/2)^2 + \tau_{xy}^2} \qquad \text{(Eq. 14-12)}$$

while the angle on which it is located is given by:

$$\tan 2\theta_s = - (\sigma_x - \sigma_y / 2) / \tau_{xy} \qquad \text{(Eq. 14-13)}$$

The two previous equations can be used to determine the maximum normal and shear stresses that occur on an element under a certain condition of stress. By utilizing Equation 14-6 (or 14-7, or 14-8), the location of the plane of these stresses can likewise be found. Upon further manipulation of the previous formulas, the following relationship can be stated:

- The maximum normal stress (principle stress) will occur on a plane that has zero shear stress acting upon it.
- The maximum shear stress will occur on a plane which has a normal stress equal to the average of the applied normal stresses.
- The planes of maximum normal stress and the plane of maximum shear stress will always be 45° apart.

The following example will illustrate the use of these equations in determining the values of maximum normal and shear stresses.

EXAMPLE 14.8

The element below is the same as that used in example 14.6. Determine the magnitude of the maximum normal and shear stresses and the plane on which they act.

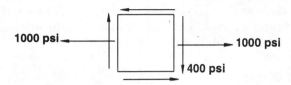

The determination of the maximum normal and shear stresses and the planes on which these act can be found using Equations 14-11 and 14-12, which were derived previously. The maximum normal stress can be found as follows:

$$\sigma_{nmax} = (\sigma_x + \sigma_y/2) + \sqrt{(\sigma_x - \sigma_y/2)^2 + \tau_{xy}^2} \qquad \text{(Eq. 14-11)}$$

With

$$\sigma_x = +1000 \text{ psi}$$
$$\sigma_y = 0$$
$$\tau_{xy} = +400 \text{ psi}$$

The equation becomes:

$$\sigma_{nmax} = (1000 + 0)/2 + \sqrt{(1000 - 0/2)^2 + (400)^2} = 1140.3 \text{ psi}$$

This stress acts at an angle which is computed as follows:

$$\tan 2\theta = \tau_{xy} / (\sigma_x - \sigma_y/2) \qquad \text{(Eq. 14-6)}$$
$$\tan 2\theta = 400 / 500$$
$$2\theta = \tan^{-1}(400 / 500) = 38.66°$$
$$\theta = 19.33°$$

The maximum shear stress can be determined as follows:

$$\tau_{max} = +\sqrt{(\sigma_x - \sigma_y/2)^2 + \tau_{xy}^2} \qquad \text{(Eq. 14-12)}$$

Substituting the values we find;

$$\tau_{max} = \sqrt{(1000 - 0/2)2 + (400)^2} = 640.3 \text{ psi}$$

This stress acts at an angle calculated as follows:

$$\tan 2\theta_s = -\,(\sigma_x - \sigma_y\,/\,2)\,/\,\tau_{xy}\,/\, = \qquad\qquad \text{(Eq. 14-13)}$$

$$\tan 2\theta_s = -\,(500\,/\,400)$$

$$2\theta_s = \tan^{-1}(-\,500\,/\,400) = -51.34°$$

$$\theta_s = -\,25.67°$$

The positive angle on the maximum normal stress indicates that this stress on a plane is located at a clockwise 19.33° from the original element, while a negative sign indicates an angle counterclockwise from the original element. Notice again how the planes are 45° apart.

14.6   Mohr's Circle

A graphical technique used to determine the maximum normal and shear stresses on a element stressed under combined normal and shear stresses, was developed in the late nineteenth century by a German engineer named Otto Mohr[2]. This graphical technique is based on the equations that were developed in the previous section and can be used to evaluate the stresses on all planes throughout an element, in addition to the maximum stresses.

Mohr noticed that the equations for finding normal stress and shear stress could be combined into the equation of a circle, and as such, the plotted circle represented a state of stress on planes throughout the element. Mohr's circle will plot normal stresses along the x-axis—with tension stresses given a positive sign and compression stresses given a negative sign. Shear stresses will be plotted along the y-axis—with a positive value given for a shear tending to rotate the element clockwise, while a negative sign is for shear tending to rotate an element counterclockwise (Figure 14-8).

The steps for constructing Mohr's circle can be seen in Figure 14-9 and are as follows:

1. Using the x face of the element, plot the x and y coordinates of stress. The normal stress is the x coordinate and the shear stress is the y coordinate (point A). Repeat this with the stresses found on the element's y face (point B).

2. Draw a line connecting the two points. This line represents the state of stress on the element as shown. Refer to this condition as the original element.

3. Locate the center point (C) of this line. This point is the center point of the circle and has to fall on the x-axis. The coordinates of this point are $(\sigma_x + \sigma_y\,/\,2,\ 0)$.

4. The maximum and minimum normal stresses are located at points Q and R, respectively. Notice that points Q and R lie a "radius" away from the center point of the circle, on the x axis. Therefore, the coordinates of the maximum normal stress are $(\sigma_x + \sigma_y\,/\,2 + R,\ 0)$, while those for the minimum normal stress are $(\sigma_x + \sigma_y\,/\,2 - R,\ 0)$.

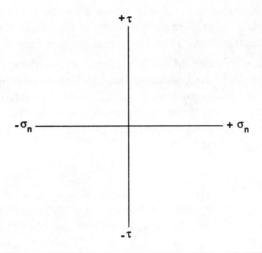

Figure 14-8 Orientation of Axes for Mohr's Circle

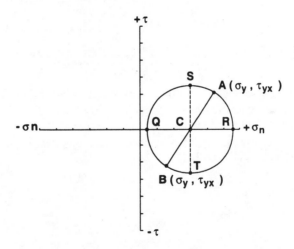

Figure 14-9 Construction of Mohr's Circle

5. The maximum and minimum shear stresses are located at points S and T respectively. Notice that points S and T lie a "radius" away from the center point of the circle, on the y axis. Therefore, the coordinates of the maximum normal stress are $(\sigma_x + \sigma_y / 2, + R)$ while those for the minimum normal stress are $(\sigma_x + \sigma_y / 2, - R)$.

6. The location of the planes on which the maximum stresses act can be found by calculating the angle from the original element to the plane in question. The angle from the original element to the plane of maximum

normal stress is $\angle ACR$ while the angle from the original element to the plane of maximum shear stress is $\angle ACS$. Remember these angles are expressed as 2θ and the actual angle is only θ.

The construction of Mohr's circle can yield very fast and accurate results when carefully plotted, although it is much more common to combine the plotting of the circle with trigonometric calculations. The calculation of the radius is typically computed using the Pythagorean theorem, while the angles are likely to be determined using the tangent function. The following example will illustrate the use of Mohr's circle.

EXAMPLE 14.9

For the element shown below, which is the same element used in Example 14.8, find the maximum normal and shear stresses and the planes on which they act, using Mohr's circle.

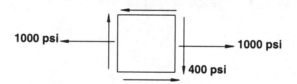

The first step is to establish (x,y) coordinates of an x face and a y face for the element. The coordinates for the x face are (+1000, +400), because the normal stress is 1000 psi tension and the shear stress is 400 psi clockwise. The coordinates for the y face are (0, – 400). The line connecting these point represents the original element, as can be seen below.

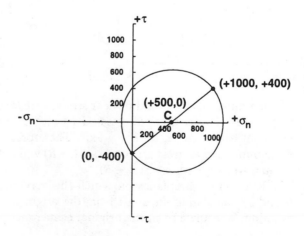

The middle of this line is the centerpoint of the circle. The coordinates of the point are ($\sigma_x + \sigma_y$ / 2, 0) or (500,0). A circle can be drawn using the centerpoint and the point for the original element.

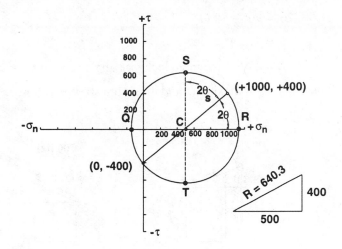

The radius of the circle can be determined, using the Pythagorean theorem as follows:

$$R = \sqrt{500^2 + 400^2} = 640.3 \text{ psi}$$

The maximum and minimum normal stress can be found by adding (or subtracting) the radius to the center point of the circle, as shown by point R and Q below:

$$\sigma_{nmax} = +500 \text{ psi} + 640.3 \text{ psi} = +1140.3 \text{ psi (tension)}, 0 \text{ psi shear}$$

$$\sigma_{nmin} = +500 \text{ psi-}640.3 \text{ psi} = -140.3 \text{ psi (compression)}, 0 \text{ psi shear}$$

The location of the plane of maximum and minimum normal stress can be found by calculating the angle 2θ as follow:

$$\tan 2\theta = 400 / 500$$

$$2\theta = \tan^{-1}(400 / 500) = 38.66°$$

$$\theta = 19.33° \text{ (positive angle indicates clockwise to horizontal axis)}$$

The maximum and minimum shear stress can be found be extending the radius vertically from the center point of the circle, as shown by point S and T below:

$$\tau_{max} = +500 \text{ psi (tension)}, +640.3 \text{ psi shear}$$

$$\tau_{min} = +500 \text{ psi (tension)}, -640.3 \text{ psi shear}$$

The location of the plane of maximum and minimum shear stress can be found by calculating the angle $2\theta_s$ as follows:

$$2\theta_s = 90° - 2\theta$$

$$2\theta_s = 90° - 38.66°$$

$$2\theta_s = 51.34°$$

$$\theta = 25.67° \text{ (counterclockwise angle to vertical axis)}$$

These planes of maximum and minimum stresses can be expressed as follows:

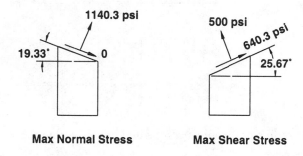

Max Normal Stress   **Max Shear Stress**

14.7   Summary

In many structures, different types of stresses will combined to create variations of stress that must be known such that proper design may be accomplished. Many different types of combinations may occur, however, the most common are the combinations of axial stress with bending stress and the combination of normal stress with shear stress.

The combination of axial and bending stress is calculated through the combined stress formula, which superimposes the effects of the axial stress (tension or compression) with the effects of bending stress (tension and compression). This results is the axial effect adding to the similar bending stress, thereby creating a larger combined stress, while simultaneously subtracting from the opposite bending stress, thus creating a minimum combined effect.

The combination of normal and shear stresses can be calculated through equations that are derived, based on the equilibrium of an element subjected to stress. The designer must know the magnitude of the maximum normal stress (sometimes referred to as the principle stress) and the maximum shear stress. Also important to the designer is the location of the planes on which the stresses reach their maximum value. In addition to the equations, a graphical technique known as Mohr's circle can be used to quickly and easily find these maximum normal and shear stresses.

EXERCISES

1. Calculate the minimum and maximum combined stress in the beam shown below. The axial load is applied at the centroid of the section. Locate the point of zero stress within the cross section.

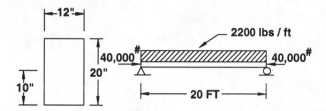

2. Calculate the minimum and maximum combined stress in the beam shown below. The axial load is applied at the centroid of the section. Locate the point of zero stress within the cross-section.

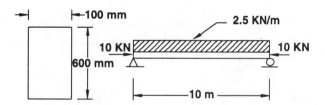

3. Calculate the axial load required in Exercise 1 that would create a combined stress of 0 psi in the bottom of the beam.

4. Calculate the axial load required in Exercise 2 that would create a combined stress of 6000 KPa compression in the bottom of the beam.

5. The bracket shown below is rigidly fixed to the table as shown. Calculate the maximum and minimum combined stress at section *A-A*. The bracket cross-section is 1 in. × 1in. square.

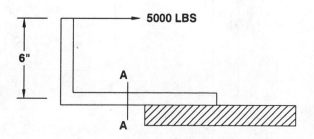

6. The C-clamp shown is tightened until the force P is equal to 1000 pounds. Calculate the maximum and minimum combined stress at section A-A if the clamp's cross-section is circular, with a 1 in. diameter.

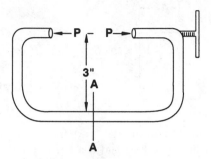

7. The column shown below is subjected to a load P equal to 15 KN. Calculate the maximum and minimum combined stress in the front and the rear of the column.

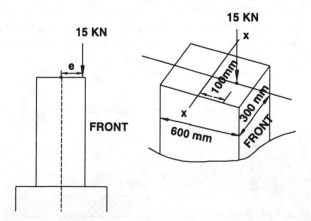

8. Calculate the force P that would be necessary to cause the stress in the rear of the column in Exercise 7 to be 10 KPa tension.

9. The column shown below is loaded as shown. Calculate the stresses at each of the four corners and determine whether they are in tension or compression.

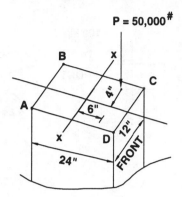

10. The beam shown below is prestressed with a force of P. Calculate the prestressing force that is necessary to create zero stress in the bottom of the beam at midspan after the uniform load is applied.

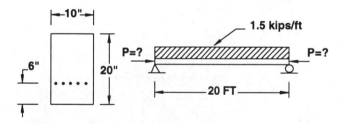

11. A rectangular beam which is 10 inches wide has an axial tensile load of 20,000 lbs applied at its ends. If the beam is 24 ft. long and loaded with a uniform load of 1000 lbs/ft. over its entire length, calculate the depth of the beam to limit the combined tensile stress to 1200 psi.

12. A rectangular beam that is 700 mm wide has an axial tensile load of 20,000 N applied at its ends. If the beam is 8 m long and loaded with a uniform load of 700 N/m over its entire length, calculate the depth of the beam to limit the combined tensile stress to 5 KPa.

13. A circular beam has an axial tensile load of 10 kips applied at its ends. If the beam is 15 ft. long and loaded with a concentrated load of 1 kip at its center, calculate the diameter of the beam to limit the maximum combined stress to 2 ksi.

14. The footing shown below has a compressive load of 40 kips applied to it. Calculate the maximum eccentricity, e, that can be allowed and still have the maximum combined stress be less than 4 ksf.

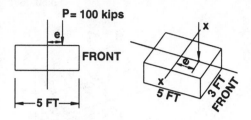

15. The industrial crane hook shown has a circular cross-section with a diameter of 25 mm. Calculate the maximum and minimum combined stress at section A-A.

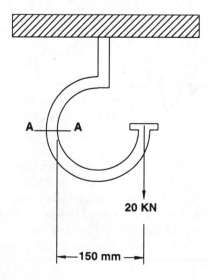

16. For the elements shown below, calculate the normal and shear stresses acting at planes of (CW) 10°, 20°, and 30°, using the stress formulas and also using Mohr's circle.

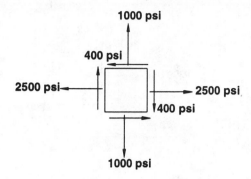

17. For the elements shown below, calculate the normal and shear stresses acting at planes of (CW) 10°, 20°, and 30°, using the stress formulas and also using Mohr's circle.

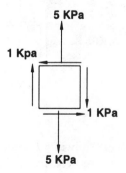

18. For the elements shown below, calculate the normal and shear stresses acting at planes of (CW) 10°, 20°, and 30°, using the stress formulas and also using Mohr's circle.

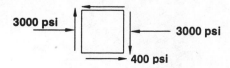

19. For the elements shown below, calculate the normal and shear stresses acting at planes of (CW) 10°, 20°, and 30°, using the stress formulas and also using Mohr's circle.

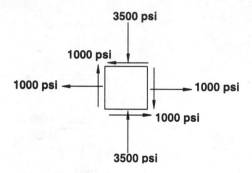

20. For the elements shown below, calculate the normal and shear stresses acting at planes of (CW) 10°, 20°, and 30°, using the stress formulas and also using Mohr's circle.

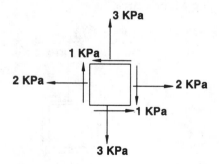

21.—25. For the elements shown in Exercises 16—20, calculate the maximum normal and shear stresses and locate the plane on which these act.

REFERENCES

1. Beer, Ferdinand P., and Johnston, E. Russell, *Mechanics of Materials*, McGraw-Hill Co., New York, 1981, P.292-95.
2. Straub, Hans, *A History of Civil Engineering*, Cambridge, MA, 1964, p.197-202.

15

BEAM DESIGN

15.1 Introduction and Review of Beam Theory

This chapter will attempt to explore the fundamental concepts of beam design with three very common building materials—steel, reinforced concrete, and timber. Steel beam design will focus on rolled-beams and will be presented in two parts to reflect the two prominent philosophies of steel design in use today. The current design methods used in steel are the allowable stress design (ASD) method and the load and resistance factor design (LRFD) method. The material regarding steel design in this chapter will be based on the most current American Institute of Steel Construction (AISC) specifications.[1,2] Design of timber members is based on an allowable stress methodology and all material contained within this chapter will be based on the American Institute of Timber Construction (AITC) specification.[3] Design of reinforced concrete beams is based on a philosophy known as Strength Design, and all material contained within this chapter will be based on the American Concrete Institute (ACI) code.[4]

A firm grasp of the fundamentals is needed before any discussion can occur on more involved design topics, and therefore the student is urged to review the fundamentals of beam mechanics. Beams, girders, stringers, purlins, and joists are all terms that describe members that may support loads applied perpendicular to their longitudinal axis. These members are integral pieces of all structures; from the steel beams that support the floor framing of buildings to the large steel girders supporting major interstate bridges. (Figure 15.1).

In Chapter 12, it was discussed how a beam subjected to an applied loading develops two primary stresses in order to maintain its integrity. In order to keep from ripping apart, the beam develops shear stresses to resist the shear force from

Figure 15-1 Steel Girder Construction on the Passiac River Bridge. (Courtesy Bethlehem Steel Corporation).

the applied loads. Also, in order to keep from deflecting or rotating excessively, the beam must develop internal bending stresses to resist the applied bending moments caused by the loads (Figure 15-2). If the beam cannot develop resisting stresses to offset the stresses that result from the applied loading, failure will occur.

The actual shear stress, f_v, developed by a beam has to be equal to the applied shear stress on the beam. This shear stress can be calculated from the following "exact" formula:

$$f_v = VQ \, / \, Ib$$

where

f_v = shear stress on beam at that location, ksi or psi

V = shear force at point under consideration, kips or lbs.

Q = Static moment of area, in.^{3}

I = Moment of inertia of beam, in.^{4}

b = Width of beam at point under consideration, in.

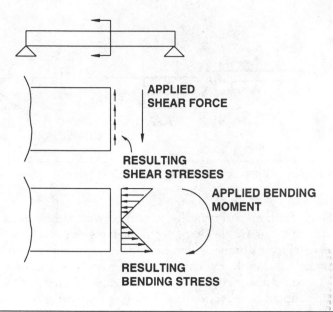

Figure 15-2   Development of Internal Resisting Shear and Moments in a Beam

Remember that the shear stress over the beam's cross-section will always be the maximum at the neutral axis. It should also be noted, that practically all design specifications refer to actual stress by the letter "f" instead of the Greek letter sigma (σ). Although this is different than what has been used throughout this text, the author chooses to make this change to remain consistent with the various design specifications.

The beam also has to resist the applied bending moments that are applied to the beam. The stress that the beam develops in response to applied bending moment is termed flexural stress or bending stress. The formula which we use to calculate actual bending stress, f_b, is as follows:

$$f_b = Mc\,/\,I$$

where

f_b = bending stress at location under consideration, ksi or psi

M = applied moment at point under consideration (usually the maximum moment along the beam), kip-ft. or kip-in.

c = distance from neutral axis to point on beam cross-section desired (usually outer fiber for design), in.

I = moment of inertia about bending axis, in.[4]

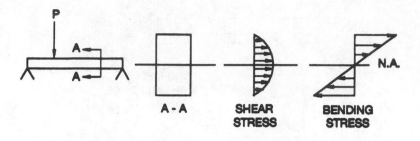

Figure 15-3   Shear and Bending Stress Distibution

Since designers are typically concerned with maximum stress, the value c is typically taken as 1/2 the beam depth. Bending stress is assumed to vary proportionally with the distance c in a homogeneous member, and therefore reaches a maximum value at the extreme outside (top and bottom) of a section.

The stress distributions over a rectangular beam can be seen in Figure 15-3. Again, it should be reiterated that *shear is maximum at the neutral axis, while the bending stress is zero at this point. Conversely, flexural stress is maximum at the outer fiber of the beam, while shear stress is zero.*

15.2   Potential Modes of Beam Failure

The behavior of a beam is a mixture of the two types of members which we have discussed in the book up to this point—tension and compression members. Therefore, it only stands to reason that the types of failure that a beam can undergo are essentially a mixture of the failure modes that we have studied in previous chapters.

The tension side of a beam experiences tensile stresses, which are highest at the extreme outside fiber. The mode of failure that we are concerned about on the tension side of the beam would involve excessive deformation, or stretching of the fibers. Because the stresses and strains over the beam's cross section are assumed to vary linearly, the compression side of the beam would also be subject to the same stresses. However, the major concern on the compression side of the beam is its ability to remain stable under these stresses. A failure or limit state can occur due to instability on the compression side of the beam in two potential manners— local buckling and lateral-torsional-buckling.

Local buckling refers to an instability of the individual elements of a beam (the flanges and the web). These could potentially fail under compressive stress before the beam is able to reach its full moment capacity. This instability is a function of the shape of the beam. In steel design, we will refer to the ability of a steel shape to reach yield stress across its entire cross sectional area before failure as "compactness". If a beam is compact, its individual elements will remain stable and exhibit no local buckling until the section reaches its full moment capacity.

The AISC specification will measure this ability to prevent local buckling by referring to a beam shape as being either **compact** or **noncompact**. This will be discussed further in Section 15.3.

The other potential limit state affecting the development of a beam's resistance to bending stress reflects the condition of lateral support along its compressive flange. This column-type instability in beams is referred to as **lateral-torsional-buckling or lateral buckling** because as the slender compression flange begins to buckle out of plane, the beam undergoes a torsional component due to the downward forces along the top flange (Figure 15-4).

Lateral buckling is a function of lateral bracing of the compression flange. A beam that has more bracing along its compression flange will be less likely to fail in this manner. A beam can have many conditions of lateral support along its com-

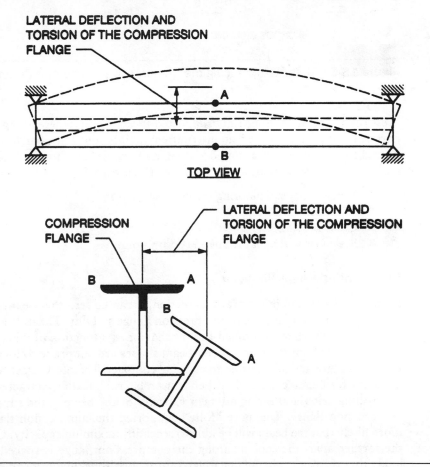

Figure 15-4 Lateral Torsional Buckling

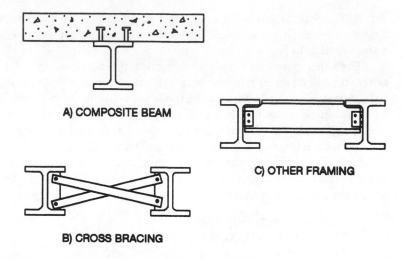

Figure 15-5 Methods of Bracing the Compressive Flange of a Beam

pression flange. It could be fully supported (braced everywhere along its flange), partially supported (braced at intermittent points along its flange), or unsupported (having no external support along its flange). Some methods that achieve a degree of lateral support are listed below (Figure 15-5).

- Non-Composite Flooring (at least by friction)
- Composite Flooring
- Cross Bracing
- Beams or struts that frame in certain points

15.3 Steel Rolled-Beam Behavior

From our discussion in the previous section, it can be seen that beam strength in bending is largely a function of compression flange stability. The following discussion will try to relate steel rolled-beam behavior as a function of this criteria.

The vast majority of steel rolled-beam shapes are not prone to local buckling types of failure and are referred to as being compact. Therefore, steel rolled-beam behavior for bending is largely dependent on lateral buckling characteristics. The controlling criteria affecting moment capacity of the beam is the support of the compression flange. The more "fully" supported the compression flange is, the more likely that the beam will be able to reach its maximum capacity. Conversely, the farther apart the support along the compression flange is spaced, the more likely that the beam will fail due to lateral-torsional-buckling prior to reaching its full capacity. Lateral-torsional-buckling can occur in two potential manners— inelastic and elastic.

Inelastic lateral-torsional-buckling takes place when the distance between braced points is short enough that it allows the beam to develop higher stresses that happen to be in the material's inelastic region. Elastic lateral-torsional-buckling occurs when the distance between lateral support of the compression flange becomes large, so that the stresses developed in the beam at failure are relatively low—below the material's elastic limit.

Formulas have been developed for beam capacity and allowable stress based on a beam's unbraced length. Basically, a beam's capacity or allowable stress can fall into one of three categories based on its unbraced length. If the beam's unbraced length is less than some defined value, the beam is viewed as not being able to fail by lateral-torsional-buckling, and is therefore rewarded by a high value of allowable or critical stress. Should a beam's length be greater than this defined value, lateral-torsional-buckling will be the mode of failure and the allowable stress or critical stress values will be reduced. The calculated bending capacity of the beam will be reduced as lateral-unbraced-length becomes greater and greater.

Shear behavior in steel rolled-beams is typically not a concern, because steel has a very high resistance to shearing failures. Only if the beams are subjected to extreme loads or have high concentrated point loads, will shear behavior be a factor. As steel beams become deeper (i.e. plate girders), the web does become more susceptible to shear failure and therefore will need to be reinforced. The following two sections will discuss steel rolled-beam design using both the ASD and LRFD techniques. The ASD method has been used since the beginning of this century and has been successful, although possibly conservative in some instances. The LRFD method was adopted in 1986 by the AISC to provide a more rational approach to steel design and is gaining more and more acceptance. Students interested in only one method may skip to the pertinent section.

15.4   AISC Rolled-Beam Philosophy Using LRFD

The most basic beam design requirement in the LRFD method, is that design moment capacity, $\phi_b M_n$, must be greater than or equal to the required (factored) flexural strength, M_u. The required flexural strength may be referred to by some as the "ultimate moment". This most basic requirement can be simply stated as follows:

$$\phi_b M_n \geq M_u$$

where

$$\phi_b = .90 \text{ for bending}$$

$$M_n = \text{nominal moment capacity}$$

$$M_u = \text{required (factored) moment}$$

As discussed earlier, bending strength is most affected by the unbraced length of the compression flange. Since rolled shapes are primarily bent about the strong axis and compact, we will restrict our evaluation of the LRFD requirements to those that focus on the unbraced length of the compression flange.

When considering the unbraced length of the beam's compression flange, l_b, there are two limiting lengths that we must compare this against. These values are referred to as L_p and L_r. The L_p limit is the dividing line under which a beam can actually reach the full capacity called the plastic moment, M_p, and is given as follows:

$$L_p = 300 r_y / \sqrt{F_y} \qquad \text{(Eq. 15-1)}$$

where

r_y = radius of gyration, in.

F_y = specified yield stress of steel, ksi

The L_r limit is the dividing line between elastic and inelastic buckling of the compression flange and is given in the LRFD manual as follows:

$$L_r = \frac{r_y X_1}{(F_{yf} - F_r)} \sqrt{1 + \sqrt{1 + X_2 (F_{yf} - F_r)^2}} \qquad \text{(Eq. 15-2)}$$

where

r_y = radius of gyration, in.

F_{yf} = specified yield stress of the flange steel, ksi (generally the same as flange steel)

X_1, X_2 = beam buckling factors, found in section tables

F_r = compressive residual stress, generally taken as 10 ksi for rolled-steel beams

The beam buckling factors, X_1 and X_2, are listed with the section properties in the AISC section tables (found in Appendix A) or they can be calculated from equations given in Chapter F of the AISC specification.

For compact, steel rolled-beams, the LRFD design philosophy for calculating nominal moment capacity, M_n, can be illustrated in Figure 15-6. This curve can be explained by looking at the three separate categories stated below that correspond to the three cases on the figure. All rolled-beams will be assumed to be compact in this discussion.

Category #1 Beams with $l_b \le L_p$

Beams in this region can achieve full plastic moment capacity, and therefore the nominal moment capacity in this region is the plastic moment, M_p. The plastic moment capacity is calculated as follows:

$$M_p = Z_x F_y. \qquad \text{(Eq. 15-3)}$$

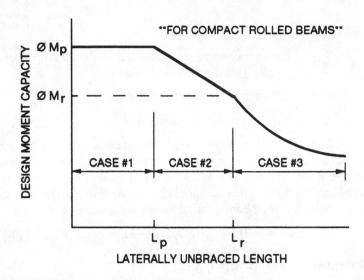

Figure 15-6   Effect of Laterally Unbraced Length on Design Chart. (Courtesy American Institute of Steel Construction)

Therefore, in this category a steel rolled-beam's nominal capacity, M_n is ;

$$M_n = M_p = Z_x F_y$$

Therefore the beam's design moment capacity is:

$$\phi_b M_n = 0.90(Z_x F_y)$$

Category #2 Beams with $L_p < l_b \le L_r$

Beams in this region exhibit inelastic lateral torsional buckling as the limit state before being able to achieve M_p. The LRFD specification gives the nominal moment capacity as follows:

$$M_n = C_b [M_p - (M_p - M_r)[(l_b - L_p)/(L_r - L_p)]] \le M_p \qquad \text{(Eq. 15-4)}$$

This equation is simply a linear interpolation between M_p and M_r, based on the unbraced length of beam. The moment capacity M_r requires some explanation at this point. M_r is the moment capacity of the beam as it reaches first yield, in this case due to lateral torsional buckling.

The student should be advised that residual stresses affect only the value of M_r and not the value of M_p, since plastic moment capacity considers yield over the entire area and residual stresses must be in equilibrium before loads are applied. The M_r value is calculated as follows;

$$M_r = (F_y - F_r)S_x \qquad \text{(Eq. 15-5)}$$

where

F_r = compressive residual stress in flange considered to be 10 ksi for rolled-beams

F_{yf} = specified yield stress of flange

S_x = elastic section modulus

The value of C_b in Equation 15-4 is referred to as the bending coefficient and is meant to account for the effect of moment gradient to the moment capacity of the beam. In our discussions, the value of C_b will be conservatively taken as 1.0, although the AISC reference should be explored to gain further insight as to its use.

Therefore, the beam's design moment capacity is:

$$\phi_b \, M_n = 0.90 \, C_b \, [M_p - (M_p - M_r)[(l_b - L_p)/(L_r - L_p)]] \leq M_p$$

Category #3 Beams with $l_b > L_r$

This region contains beams that fail by exhibiting the behavior of elastic lateral-torsional-buckling. The beams in this region have long, slender unsupported lengths of the compression flange and failure occurs before the sections reach yield. The nominal moment capacity of a beam in this region is equal to M_{cr}. The LRFD formula for this is shown below:

$$M_{cr} = \frac{C_b S_x X_1 \sqrt{2}}{l_b/r_y} \sqrt{1 + \frac{X_1^2 X_2}{2 \, (l_b/r_y)^2}} \leq M_p \qquad \text{(Eq. 15-6)}$$

Therefore the nominal moment capacity is;

$$M_n = M_{cr}$$

The design moment capacity is:

$$\phi_b \, M_n = \phi_b \, M_{cr}$$

The following examples will illustrate the evaluation of steel rolled-beams using the LRFD philosophy.

EXAMPLE 15.1

Determine the adequacy ($\phi_b \, M_n \geq M_u$) of the W 12 x 87 shown below if the compression flange is braced only at the ends. The factored bending moment is equal to 121.5 kip-ft. Steel is A36 and assume $C_b = 1.0$

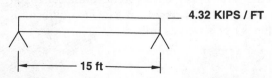

A W 12 × 87 is a compact section, so therefore, lateral buckling is the primary concern as far as capacity. Check unbraced length criteria to determine what case it falls into:

Unbraced length, $l_b = 15$ ft

From section tables in Appendix A:

$$r_y = 3.07 \text{ in}$$

$$X_1 = 3880 \text{ ksi}$$

$$X_2 = .000586$$

$$F_y = 36 \text{ ksi}$$

Calculating L_p and L_r using Eq.15-1 and 15-2;

$$L_p = 300(3.07 \text{ in.}) / \sqrt{36} \text{ ksi} = 153.5 \text{in. or } 12.8 \text{ ft.} \qquad \text{(Eq. 15-1)}$$

$$L_r = (3.07(3880 \text{ ksi}) / (36 \text{ ksi} - 10 \text{ ksi})) \left[\sqrt{1 + \sqrt{1 + 0.000586 \, (36 \text{ ksi} - 10 \text{ ksi})^2}} \right]$$

$$= 676.6 \text{ in. or } 56.4 \text{ ft.} \qquad \text{(Eq. 15-2)}$$

Therefore, since $L_p < l_b < L_r$, we can use Category #2 to determine nominal moment capacity. Calculate M_p and M_r, which are then used in Equation 15-4 as follows:

$$M_p = Z_x F_{yf} = 36 \text{ ksi}(132 \text{ in.}^3) / 12 \text{ in./ft.} = 396 \text{ kip-ft.}$$

$$M_r = (F_{yf} - F_r)S_x = (36 - 10 \text{ ksi})(118 \text{ in.}^3)/ 12 \text{ in./ft.} = 255.7 \text{ kip-ft.}$$

Calculate Equation 15-4 for M_n:

$$M_n = 1.0[396 - (396 - 255.7)(15 - 12.8/56.4 - 12.8)] = 388.9 \text{ kip-ft.} \quad \text{(Eq.15-4)}$$

Calculating design capacity:

$$\phi_b \, M_n = .90(388.9 \text{ kip-ft.}) = 350 \text{ kip-ft.} > 121.5 \text{ kip-ft.}$$

Therefore, the section is very adequate, since $\phi_b \, M_n \geq M_u$.

EXAMPLE 15.2

Recalculate the design moment capacity of the beam in the previous example if the W 12 × 87 has its compression flange braced at midspan.

In this case, the unbraced length, l_b, is reduced to 7.5 ft., which is less than L_p (L_p was calculated to be 12.8 ft. in Example 15.1). Therefore, this case falls into category #1 (which has $l_b < L_p$) and in this category the nominal moment capacity is calculated as follows:

$$M_n = M_p = Z_x F_y \qquad \text{(Eq. 15-3)}$$

$$M_n = (132 \text{ in}^3)(36 \text{ ksi}) = 4752 \text{ kip-in. or } 396 \text{ kip-ft.}$$

The design capacity would now be:

$$\phi_b M_n = (0.90)(396 \text{ kip-ft.}) = 356.4 \text{ kip-ft.}$$

Therefore, the bracing increases the section's capacity in this example slightly because the beam is controlled by category #1.

The design of steel rolled-beams is a bit more difficult than evaluation of existing beams because the value of nominal moment capacity changes as a beam falls into the aforementioned categories. There are many charts and aids that are available to a steel designer. One of the most commonly used charts to help in the design of rolled steel beams is published in Part 2 of the LRFD manual . A sample of these charts is provided in Figure 15-7.

These charts have plotted the design moment ($\phi_b M_n$) that wide flange sections can carry based on their unbraced length, l_b. The charts given in the manual can be used for the aforementioned shapes made from steels with $F_y = 36$ ksi and $F_y = 50$ ksi. These charts are graphical representations utilizing the three cases equations mentioned earlier in this chapter and are to be used for cases when the bending coefficient, C_b, is equal to 1.0.

To use the AISC charts, a designer can simply enter the unbraced length, l_b, on the bottom scale and intersect that with the moment needed to be resisted on the vertical scale. Any beam listed above and to the right of this intersection will satisfy the necessary requirements (considering that $C_b = 1.0$). Solid lines on the graph indicate the most economical section by weight in a given region, whereas the dashed line indicates that a lighter section will satisfy the strength requirement. The author would encourage the readers to use this as a preliminary design step because the fundamentals of beam behavior must still be fully understood. Items such as compactness and the bending coefficient must still be checked, in order to ensure proper usage of these charts. The open circles that appear on the line for an individual section indicate the value of L_r; over which the beam begins elastic behavior. The closed circles that may appear on the line for an individual section indicate that section's value of L_p.

If such charts are not available, the following is a basic format that can be used to help the beginner in the design of rolled beams using LRFD. From section 15-3, we know that compact rolled beams fall into one of three categories based on the unbraced length of the compression flange, l_b. Since l_b will be given or assumed in any problem, the task begins by looking at the dividing lines between plastic moment capacity and inelastic buckling, L_p; and inelastic and elastic buckling, L_r. Should the sections under consideration fall into category #1 (in which beams can

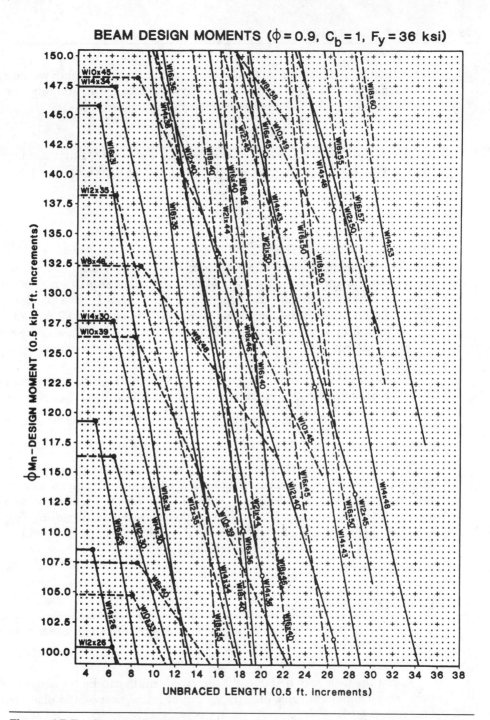

Figure 15-7 Standard LRFD Steel Rolled-Beam Design Chart. (Courtesy American Institute of Steel Construction)

reach their full plastic moment), then the section's unbraced length would have to be less than or equal to L_p. Using this strategy we can set $l_b = L_p$ and solve for a minimum value of r_y, as shown below by rearranging Equation 15-1:

$$r_{y\,min} = l_b \times \sqrt{F_y} / 300$$

If a majority of your beam sections meet the minimum requirement for r_y, you stand a very good chance of being in category #1 and therefore:

$$M_n = M_p = Z_x F_y$$

Solving for Z_x yields:

$$Z_{xreq'd} = M_n / F_y \qquad \text{(Eq. 15-7)}$$

Should the majority of beam sections not meet this requirement, it would seem likely that the beam chosen would stand a good chance of being in category #2 or #3. The nominal moment capacity decreases as $l_b > L_p$. We can use logic to help us in our assumptions, when our beam falls into category #2 and #3.

A beam is designed according to category #2 if $L_p < l_b < L_r$, and the nominal moment in category #2 ranges from M_p (when the unbraced length is equal to L_p) to M_r (when the unbraced length is equal to L_r).

Therefore, if the design fails to meet case #1 requirements, a designer can try to gauge the upper and lower limits of category #2 by solving the following:

$$Z_{x\,req'd} = M_n / F_y \qquad \text{(Upper limit of category \# 2)}$$

$$S_{x\,req'd} = M_n / (F_y - F_r) \qquad \text{(Lower limit of category \# 2)}$$

Realizing that the first assumption of a beam to fall in category #1 has given the section's value of L_p and L_r, try to gauge where the considered section might fall. If a beam's unbraced length, l_b, is barely over L_p, then select a section using $Z_{x\,req'd}$. If your l_b is barely greater than L_r, then choose a section using $S_{x\,req'd}$.

The following examples will illustrate this technique in the design of a rolled-beam section, using the aforementioned logic.

EXAMPLE 15.3

Design a W 12 section to hold a factored moment of 400 kip-ft. Steel is A36 and beam is 12 ft. long and unbraced ($C_b = 1.0$)

Using the logic stated above, the unbraced length can be set equal to the L_p value and a minimum value of r_y ($r_{y\,min}$) can be found that would place a section into category #1.

$$l_b = 12 \text{ ft.} = 144 \text{ in.}$$

Setting $l_b = L_p$ we can solve the following;

$$r_{y \, min} = 144 \times 6/300 = 2.88 \text{ in.}$$

Looking at the W 12 charts found in Appendix A, the majority of W 12 sections have r_y values in excess of the 2.88 in. minimum value (which would make the L_p value of a section equal to 12 ft.). If the chosen W12 section has an r_y above this value, then it falls into category #1 and the following is true:

$$M_n = Z_x \, F_y \text{ (if category \#1)}$$

$$M_u = 400 \text{ kip-ft.}$$

$$M_u / \phi_b \le M_n$$

$$M_n = 400 \text{ kip-ft} / .90 = 444.4 \text{ kip-ft}$$

$$Z_{xreq'd} = 444.4 \text{ kip-ft. (12 in./ft.)} / 36 \text{ ksi} = 148.1 \text{ in.}^3 \text{ (Eq. 15-7)}$$

From the section tables in Appendix A, select W 12×106 ($Z_x = 164$ in.^{3})

Since a W 12×106 has $r_y = 3.11$ inches it falls into category #1 because $l_b < L_p$ as shown below by Eq. 15-1 (and since it is also compact);

$$L_p = 300 r_y / \sqrt{F_y} \qquad \text{(Eq. 15-1)}$$

$$L_p = 300(3.11) / \sqrt{36} = 155.5 \text{ in.} / 12 = 12.9 \text{ ft.}$$

Therefore the nominal moment capacity can be calculated as follows;

$$M_n = Z_x \, F_y \text{ (if category \#1)}$$

$$M_n = 164 \text{ in.}^3 \times 36 \text{ ksi} = 5904 \text{ kip-in. or 492 kip-ft.}$$

The design moment capacity is;

$$\phi_b \, M_n = .90(492 \text{ kip-ft.}) = 442.8 \text{ kip-ft.}$$

Since the design moment capacity is greater than the required capacity of 400 kip-ft, the beam is adequate.

EXAMPLE 15.4

Design a W 12 section to hold a factored moment of 220 kip-ft . Steel is A36 and the beam is 12 feet long and unbraced ($C_b = 1.0$)

Using the logic stated above, the unbraced length can be set equal to the L_p value and a minimum value of r_y can be found that would place a section into category #1. This can be performed as follows:

$$l_b = 144 \text{ in.}$$

Setting $l_b = L_p$, we can solve the following:

$$r_{y\ min} = 144 \times \sqrt{36} / 300 = 2.88 \text{ in.}$$

If the chosen section has an r_y above this value, then it falls into category #1 and the following is true:

$$M_n = Z_x F_y$$

$$M_u = 220 \text{ kip-ft.}$$

$$M_n = 220 \text{ kip-ft.} / .90 = 244.4 \text{ kip-ft.}$$

If the assumption of case #1 is correct, then $M_n = Z_x F_y$ and the process continues by solving Equation 15-7 as follows:

$$Z_{x\ req'd} = M_n/F_y \qquad \qquad \text{(Eq. 15-7)}$$

$$Z_{x\ req'd} = 244.4 \text{ kip-ft. (12 in./ft.) / 36 ksi} = 81.4 \text{ in.}^3$$

From the property tables in Appendix A, choose W 12 × 58 ($Z_x = 86.4$ in.^{3})

Realizing that a W 12 × 58 has $r_y = 2.51$ in., which is less than $r_{y\ min}$, this section does not fall into category #1.

Using Eq. 15-1 and 15-2 for a W 12 × 58 section, this can be verified:

$$L_p = 300(r_y) / \sqrt{F_y} \qquad \qquad \text{(Eq. 15-1)}$$

$$L_p = 300(2.51)/ \sqrt{36} = 125.5 \text{ in.} = 10.4 \text{ ft.}$$

$$L_r = (2.51(3070 \text{ ksi}) / (36 \text{ ksi} - 10 \text{ ksi})) \left[\sqrt{1 + \sqrt{1 + 0.001470 \left(36 \text{ ksi} - 10 \text{ ksi}\right)^2}} \right]$$

$$= 460.3 \text{ in. or } 38.4 \text{ ft.} \qquad \qquad \text{(Eq. 15-2)}$$

$$L_r = 38.4 \text{ ft.}$$

Since this W 12 × 58's unbraced length falls between L_p and L_r (10.4 ft. < 12 ft. < 38.4 ft.) the beam falls into category #2 and our initial assumption is incorrect. However, since the W 12 × 58 just barely falls into category #2, this section may work because its plastic section modulus was a bit larger than necessary. Check to see if it does work anyway, using Equation 15-4.

$$M_n = C_b [M_p - (M_p - M_r)(l_b - L_p/L_r - L_p)] \qquad \qquad \text{(Eq. 15-4)}$$

$$C_b = 1.0$$

$$M_p = Z_x F_y = 86.4 \text{ in}^3(36 \text{ ksi}) / 12 = 259.2 \text{ kip-ft.}$$

$$M_r = (F_{yf} - F_r)S_x = (36 \text{ ksi} - 10 \text{ ksi}) \ 78 \text{ in.}^3 /12 = 169 \text{ kip-ft.}$$

$$l_b = 12 \text{ ft.}$$

$$L_p = 10.4 \text{ ft.}$$

$$L_r = 38.4 \text{ ft.}$$

Calculating Equation 15-4:

$$M_n = 1.0 \ [259.2 \text{ k-ft.} - (259.2 \text{ k-ft.} - 169 \text{ k-ft.})$$
$$[(12 \text{ ft.} - 10.4 \text{ ft.} / 38.4 \text{ ft.} - 10.4 \text{ ft.})]] = 254.1 \text{ kip-ft.}$$

Since this W 12×58 has nominal moment capacity (254.1 kip-ft.) greater than that required (244.4 kip-ft), the section is adequate. (The section is also compact, therefore flange buckling is not a consideration).

15.5   AISC Rolled-Beam Philosophy Using ASD

The most basic beam design requirement in the ASD method is that actual bending stress, f_b, must be less than or equal to the allowable bending stress, F_b. This most basic requirement can be simply stated as follows:

$$f_b \le F_b$$

As discussed earlier, bending strength is most affected by unbraced length of the compression flange. Since rolled shapes are primarily bent about the strong axis and compact, we will restrict our evaluation of the ASD requirements to those that focus on the unbraced length of the compression flange.

When considering the unbraced length of the beam's compression flange, l_b, there is a primary limiting length that we must compare this against. This value, referred to as L_c, is the maximum unsupported length of the compression flange of a section that still allows the section to reach its plastic moment.

The AISC specification sets the criteria for determining allowable bending stress, F_b. The allowable bending stress determination for steel rolled-beams and channels can be thought of as falling into one of the three following categories:

- Category #1. Members with compact sections and $l_b \le L_c$
- Category #2. Members with non-compact sections and $l_b \le L_c$
- Category #3. Compact or non-compact sections with $l_b > L_c$

These categories will be discussed in the upcoming pages and should be thought of as a hierarchy of allowable bending stress levels that range from the best (Category #1) to the worst (Category #3). The AISC philosophy that governs rolled-beams is graphically represented in Figure 15-8. Realize that the vast majority of steel rolled-beams are compact, therefore the upcoming discussion will focus primarily with a compact rolled-beam.

Category #1 Compact Sections and $l_b \le L_c$

Beams that fall into this category are able to achieve their plastic capacity by preventing local buckling (being compact) and lateral-torsional-buckling (having $l_b \le L_c$). Since failure by these two behaviors is prevented, this category represents

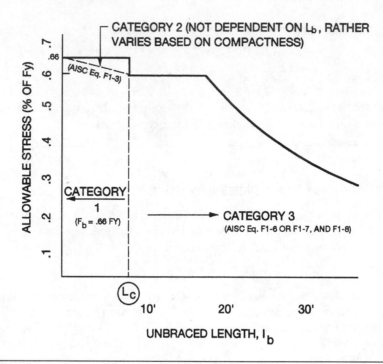

Figure 15-8   Effect of Laterally Unbraced Length on Allowable Bending Stress-ASD

the best possible type of beam behavior. The AISC rewards beams falling in this category by setting the allowable bending stress level at $0.66 \, F_y$. This is the highest allowable bending stress level awarded to beams bent about their strong axis.

L_c is given as the smaller of the two following values:

$$76 \, b_f / \sqrt{F_y}$$
$$20000 / (d/A_f) \, F_y$$

where

$$b_f = \text{flange width}$$

$$d/A_f = \text{ratio of section depth to area of the compression flange}$$

The compactness criteria are checked for the flanges and the web of the rolled-beam section. These criteria are acceptable if the following are met:

$$b_f / 2t_f \leq 65 / \sqrt{F_y}$$
$$d / t_w \leq 640 / \sqrt{F_y}$$

where

$$t_f = \text{flange thickness}$$

$$t_w = \text{web thickness}$$

$$d = \text{depth of section}$$

Category #2 Noncompact Sections and $l_b \le L_c$

Beams that meet the unbraced length requirement $(l_b \le L_c)$, but fail to meet flange compactness criteria, will fall into this category. Recognizing that local buckling will not let the section reach its full plastic capacity, the AISC provides an equation setting the allowable stress at a transitional value between 0.66 F_y and 0.60 F_y. This equation is as follows:

$$F_b = F_y \, [0.79 - .002(b_f / 2t_f)\sqrt{F_y}\,] \qquad \text{(Eq. 15-8)}$$

If a beam should fail both the web and flange criteria regarding compactness, but still meet the unbraced length criteria, the AISC sets the allowable bending stress to 0.60 F_y.

Category #3 Any Beam (Compact or Noncompact) with $l_b > L_c$

The primary mode of failure for beams in this category is by lateral-torsional-buckling, since the unbraced length criteria $(l_b > L_c)$ is exceeded. The compactness of the rolled-beam section is irrelevant because the lateral torsional behavior is dominant. The allowable bending stress that the AISC specifies is the larger calculated from two equations, *but in no case can the allowable stress be higher than 0.60 F_y for beams in this category.*

The two equations that will be used, consist of a torsional equation and a lateral buckling equation of the compression flange. However, as with column buckling, when the lateral buckling behavior is considered, we will have to decide whether elastic or inelastic behavior controls. As the unbraced length of the compression flange gets longer, elastic behavior will control, and as the unbraced length gets smaller, inelastic behavior will control. The slenderness ratio of the compression flange, termed l/r_T, is compared to certain AISC dividing lines to determine if the lateral behavior is elastic or inelastic.(In the slenderness ratio of the compression flange, l is equal to the unbraced length and r_T is the radius of gyration of the compression flange, which can easily be found in the section tables. It should be noted that the "compression flange" actually is considered to also encompass a small part of the web). If elastic behavior controls, Equation 15-10 is calculated, but if inelastic behavior controls, Equation 15-9 is calculated. The torsional equation is Equation 15-11 and is always calculated. After calculating *either* Equation 15-9 *or* 15-10 *and* Equation 15-11, the larger value calculated is used for the allowable stress. However, it is important to remember, that in no case can the allowable bending stress exceed 0.60 F_y in Category #3. The AISC equations for allowable stress are as follows:

$$\text{When } \sqrt{\frac{102,000 C_b}{F_y}} \leq l/r_T \leq \sqrt{\frac{510,000 C_b}{F_y}}$$

$$F_b = \left[\frac{2}{3} - \frac{F_y \left(\dfrac{l}{r_t} \right)^2}{1,530,000 C_b} \right] F_y \qquad \text{(Eq. 15-9)}$$

$$\text{When } l/r_T \geq \sqrt{\frac{510,000 C_b}{F_y}}$$

$$F_b = \frac{170,000 C_b}{\left(l/r_T \right)^2} \qquad \text{(Eq. 15-10)}$$

In every case, calculate

$$F_b = \frac{12,000 C_b}{ld/A_f} \qquad \text{(Eq. 15-11)}$$

The C_b term is a modifier used to define the effect of moment gradient along the beam. It can conservatively be taken as 1.0 and will be, for our purposes, in this text. This modifier is discussed further in the AISC specification and the readers are urged to refer to it if they are interested in further exploration of the subject.

The following examples will demonstrate the use of these allowable bending stress formulas in some typical problems.

EXAMPLE 15.5

Determine the adequacy of a W 24 × 117 beam that has full lateral support across a 10 ft. span, loaded with a uniform load of 12 kips/ft. Steel is A36.

For an evaluation of an existing beam, determine the allowable bending stress per AISC specification criteria and compare that to the actual bending stress. Since the beam is said to have full lateral support, the unbraced length, l_b, is zero and less than any calculated value of L_c. Therefore, if the section meets compactness criteria it will fall into Category #1.

Check section for compactness.

$$b_f / 2t_f \leq 65 / \sqrt{F_y}$$

$$d / t_w \leq 640 / \sqrt{F_y}$$

$$b_f/2_{tf} = 7.5 \leq 65/ \sqrt{36} = 10.8 \text{ - flanges are compact}$$

$$d/t_w = 44.1 \leq 640/ \sqrt{36} = 106.6 \text{ - web is compact}$$

Since the section is compact and fully supported against lateral buckling, the allowable bending stress falls into Category #1 and the allowable bending stress is:

$$F_b = 0.66F_y$$

$$F_b = 0.66(36\text{ksi}) = 23.76 \text{ ksi}$$

(Note: Many designers prefer to round 23.76 ksi to 24 ksi.) Now we must find the actual stress caused by the uniform load of 12 kips/ft. over this beam. We can find this moment by drawing shear and moment diagrams, or by the following equation for maximum moment on a uniformly loaded beam.

$$M = wl^2 / 8 = (12 \text{ kip} / \text{ft.})(10 \text{ ft.})^2 / 8 = 150 \text{ kip-ft.}$$

From Appendix A a W 24 × 117 is found to have a S_x of 291 in³ and calculating the actual bending stress yields:

$$f_b = M / S_x = 150 \text{ ft.-kips (12 in. / ft.) / 291 in.}^3 = 6.19 \text{ ksi}$$

Since the actual bending stress (6.19 ksi) is less than the allowable bending stress (23.76 ksi), the beam is adequate in bending. (It should be noted that the compactness values for $b_f/2t_f$ and d/t_w are already precalculated in the section tables in Appendix A.)

EXAMPLE 15.6

Determine the adequacy of a W 12 × 87 that is 15 ft. long, if the compression flange is unbraced. The beam is loaded with a uniform load of 3 kips/ft. and the steel is A36. $C_b = 1.0$.

Begin by determining the value of allowable bending stress for this section under the stated conditions. Being unsupported means that the unbraced length of the compression flange (l_b) is equal to 15 ft., or 180 in. Before checking the compactness values listed for a W 12 × 87 (since we won't need them if this falls into category #3), calculate L_c as the smaller of:

$$L_c = 76b_f/ \sqrt{F_y} = 76(12.125 \text{ in.}) / \sqrt{36} = 153.6 \text{ in. (smaller)}$$

or

$$L_c = 20000 / (d/A_f)F_y = 20000 / (1.28)36 = 434 \text{ in.}$$

Therefore, $L_c = 153.6$ in., or 12.7 ft., and $l_b > L_c$. This means the section falls into Category #3 for determination of allowable bending stress. In Category #3 the allowable stress is the larger of the calculated stresses from Equation 15-9 *or* Equation 15-10 *and* Equation 15-11; however, in no case can it ever be greater than $0.60\ F_y$.

First, determine whether to use Equation 15-9 or 15-10. This is based on the slenderness ratio of the compression flange, l/r_T. Calculate l/r_T:

$$l/r_T = 180 \text{ in. } / 3.32 \text{ in.} = 54.2$$

where

l = unbraced length

r_T = radius of gyration of the compression flange (taken from the section tables in Appendix A)

With $C_b = 1.0$, our l/r_T value falls between the two AISC dividing lines for inelastic lateral buckling shown below.

$$\sqrt{102,000\,(1.0)/36} = 53.2 \leq l/r_T \leq \sqrt{510,000\,(1.0)/36} = 119.0$$

Therefore, calculate Equation 15-9 and 15-11, and use the larger.
For Equation 15-9

$$F_b = (2/3 - [36(54.2)^2 / (1,530,000(1.0)]) \times 36\text{ksi} = 21.52 \text{ ksi} \leq 21.6 \text{ ksi}$$

For Equation 15-11

$$F_b = 12,000(1.0) / (180(1.28)) = 52.08 \text{ ksi (not less than } 0.60\ F_y\text{). Therefore, use}$$
$0.60\ F_y = 21.6$ ksi.

The larger of the two equations is 21.6ksi.
Now calculate the actual maximum moment placed on this beam, as we did in the preceding example:

$$M = (3 \text{ kips/ft.})(15 \text{ ft.})^2 / 8 = 84.38 \text{ kip-ft.}$$

From Appendix A, a W 12×87 is found to have a S_x of 118 in.[3] and calculating the actual bending stress yields:

$$f_b = M / S_x = 84.38 \text{ kip-ft.}(12 \text{ in/ft.}) / 118 \text{ in}^3 = 8.58 \text{ ksi}$$

Since the actual stress (8.58 ksi) is less than the allowable stress (21.6 ksi), the beam is adequate in bending.

The design of steel rolled-beams is a bit more difficult than evaluation of existing beams, since the value of allowable bending stress changes as a beam moves from one of the three aforementioned categories. There are many charts and aids that are available to a steel designer. One of the most commonly used charts to help in the design of steel rolled-beams, is published in the ASD manual. A sample of these charts is provided below in Figure 15-9.

These charts have plotted the allowable bending moment that a beam can carry at various values of unbraced length, l_b. The charts given in the manual can be used for the aforementioned shapes made from steels with $F_y = 36$ ksi, and $F_y = 50$ ksi. These charts are graphical representations utilizing Equations 15-9, 15-10, and 15-11, mentioned earlier in this chapter. They are to be used for cases when the bending coefficient, C_b, is equal to 1.0.

To use the AISC charts, a designer can simply enter the unbraced length, l_b, on the bottom scale and intersect that with the moment needed to be resisted on the vertical scale. Any beam listed above and to the right of this intersection will satisfy the necessary requirements (considering that $C_b = 1.0$). Solid lines on the graph indicate the most economical section by weight in a given region, whereas the dashed line indicates that a lighter section will satisfy the strength requirement. The author would encourage the readers to use this as a preliminary design step because the fundamentals of beam behavior must still be fully understood. Items such as compactness and the bending coefficient must still be checked, in order to ensure proper usage of these charts.

If such charts are not available, the following is a basic process that can be used to help the beginner in ASD design of rolled-beams. As previously mentioned, compact rolled-beams fall into one of three categories, based on the unbraced length of the compression flange, l_b. Since l_b will be given or assumed in any problem, the task begins by looking at the dividing line between a beam falling into Category #1 or Category #3, which is referred to as L_c. If the beam sections under consideration fall into Category #1 (in which beams can reach their full plastic moment), then the section's unbraced length would have to be less than or equal to L_c. Knowing that lateral buckling controls the majority of sections, we can use this strategy and set $l_b = L_c$ and solve for a minimum value of b_f by rearranging the L_c equation as follows:

$$L_c = 76\, b_f / \sqrt{F_y}$$

Setting $L_c = l_b$;

$$l_b = 76\, b_f / \sqrt{F_y}$$

or

$$b_{f\,min} = l_b\left(\sqrt{F_y}\right) / 76$$

If a majority of beam sections under consideration meet the minimum requirement for b_f, there is a very good chance of the beam selected being in Category #1. This being the case, the allowable bending stress assumed at the beginning of the

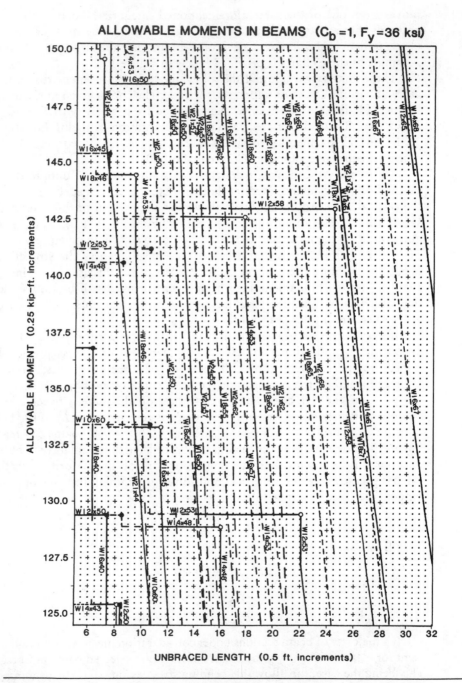

Figure 15-9 Standard ASD Steel Rolled-Beam Design Chart. (Courtesy American Institute of Steel Construction)

problem should be 0.66 F_y. Therefore, the first trial section should be based on the following:

$$S_{x \text{ req'd}} = M / 0.66\ F_y$$

Should the majority of beam sections not meet this requirement, it would seem likely that the beam chosen would stand a good chance of being in Category #2 or #3. The allowable stress decreases as $l_b > L_c$ we can use logic to help us in our assumptions when our beam falls into category #2 and #3. Therefore, the first trial section should be based on the following:

$$S_{x \text{ req'd}} = M / 0.60\ F_y$$

The following example will illustrate this technique in the design of a rolled-beam section, using the aforementioned logic.

EXAMPLE 15.7

Choose the most economical W 12 section to hold a uniform load of 3 kips/ft. over a 16 ft. simple span. The steel is A36 and the beam is braced at midspan. Neglect self weight.

Calculating the moment which is applied to the beam, we find the beam must hold a design moment as follows:

$$M = wl^2 / 8 = (3 \text{ kips / ft.}) (16 \text{ ft.})^2 / 8 = 96 \text{ kip-ft.}$$

Setting the beam's unbraced length, l_b, equal to the L_c equations, we can solve for $b_{f\text{min}}$

$$l_b = 8 \text{ ft. or } 96 \text{ in.}$$

Setting $L_c = l_b$;

$$96 \text{ in} = 76\ b_f / \sqrt{F_y}$$

or

$$b_{f \text{ min}} = 96 \text{ in.} (\sqrt{36 \text{ ksi}}) / 76 = 7.6 \text{ in.}$$

Looking at the range of possibilities for W12 sections in Appendix A, almost all W12 sections have a flange width over the minimum. In fact, any section over a W 12 × 35 will fall into Category #1. Therefore, because there is a good probability that any beam selected will fall into Category #1, assume $F_b = 0.66$ (36 ksi) = 23.76 ksi.

Calculate a required section modulus based on this assumption.

$S_{x \text{ req'd}} = M / 0.66 F_y$

$S_{x \text{ req'd}} = 96 \text{ ft.-kip}(12 \text{ in. /ft.}) / 23.76 \text{ ksi} = 48.5 \text{ in.}^3$

From AISC section tables in Appendix A, select a W 12 × 40 (S_x = 51.9 in.³) and check applicable unbraced length and compactness requirements to ensure that the assumed allowable stress of 0.66 F_y is correct.

$L_c = 8.5 \text{ ft} > 8 \text{ ft. (falls in Category \#1)}$

$b_f / 2\, t_f = 7.8 < 65 / \sqrt{F_y}$ —flanges are compact

$d / t_w = 40.5 < 640 / \sqrt{F_y}$ —web is compact

Calculate the allowable moment capacity to prove that a W 12 × 40 does supply the necessary resistance.

$M_{\text{allow}} = (F_b)\, S_x$

$M_{\text{allow}} = 23.76 \text{ ksi} \times 51.9 \text{ in.}^3 = 1233.1 \text{ in.-kip or } 102.8 \text{ ft.-kip}$

Since 102.8 ft.-kip > 96 ft.-kip, the W 12 × 40 works and is economical.

Should the initial assumption of allowable bending stress prove to be wrong, the first selection becomes the first iteration in a trial and error procedure. If the beam is undersized the next step would be to size the section up and vice versa.

15.6   Timber Beam Design

Timber beams can be either solid-sawn beams or glue-laminated beams (Figure 15-10). Solid-sawn beams are beams that are cut whole from a single piece of wood, while glue-laminated beams are made from various plies that are typically glued and pressed together. Solid-sawn beams are used extensively in construction today. Glue-laminated beams have many advantages; such as superior strength, decreased shrinkage, and the ability to use the more beautiful hardwoods as veneer on structural members. This section will explore the fundamental concepts of timber beam design as they relate to solid-sawn beams. The design information contained herein will closely follow the AITC reference.[3]

Timber beam design is based upon the allowable stress design philosophy, which states that the actual stress of some behavior must be less than or equal to the allowable stress for that behavior. The allowable stresses are given in the AITC *Timber Construction Manual.*

Although timber beam design can be involved, the two primary items of concern are the following:

- bending stresses
- horizontal shear stresses

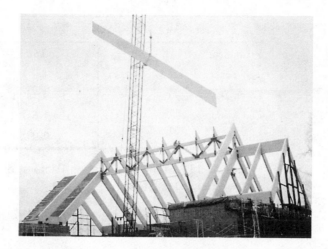

Figure 15-10   Timber Beam Construction (Courtesy of Southern Forest Products Association)

Deflection and bearing stress at the supports can also be considerations with timber beams, although they are typically not the primary concerns. Should the students be interested in these and other in-depth topics, they are referred to in the AITC reference.

Bending stress

Timber beams subjected to moment will resist those moments by developing an internal bending stress. This bending stress cannot exceed the design bending stress, F_b'. This design bending stress can be calculated as follows, in general terms:

$$F_b' = F_b \times \text{modifiers}$$

where

F_b = tabular allowable bending stress value listed per species in the AITC manual (psi)

modifiers = separate modifers as explained below

The primary modifying factors for bending stress may include the following:

C_D - duration factor

Due to the creep-related behavior of wood, the amount of time that loads are applied to a timber beam can directly affect its behavior. This modifier is meant to reduce the tabular value of allowable bending stress, F_b, if the loads are long term, and increase the tabular value for allowable bending stress if loads are short term. Values are listed in Figure 15-11.

Duration of Load	C_D
Permanent	0.90
Ten Years	1.0
Seven Days	1.25
Ten Minutes	1.6
Impact	2.0

Figure 15-11 Standard Duration Factors for Timber Beam Design

C_F - size factor

Solid-sawn beams with depths greater than 12 in. will have a reduction of tabular bending stress values. This is done to account for the effectiveness due to the depth of the beam and is based also on species. It is typically based on the following formula:

$$C_F = (12 / d)^{1/9}$$

where

$$d = \text{depth of the beam}$$

C_L- beam stability factor

Is used to account for lateral instability before the tabular value of allowable bending stress can be reached. If the beam is braced continually along its compression flange, C_L is 1.0. Also, if the depth of the beam is less than its width, C_L is 1.0. If these conditions do not exist, then C_L is calculated based on equations found in Chapter 5 of the AITC reference. For our purposes, C_L will be taken equal to 1.0.

Other modifying factors may also be used to account for moisture (C_M), temperature(C_T), and form (C_f). These modifiers are considered beyond the scope of this text, but can be explored further in any text on timber design. The design bending stress value for timber beams would then take the following form:

$$F_b' = F_b\,(C_D)(C_F)(C_L)(C_M)(C_T)(C_f) \qquad \text{(Eq. 15-13)}$$

where

$$F_b' = \text{design allowable stress}$$

$$F_b = \text{tabular allowable stress value}$$

$$C_D,\ C_F,\ \text{etc.} = \text{modifiers}$$

Horizontal Shear Stresses

Timber beams will also be subjected to shear force as the beam is loaded. This shear force will have to be resisted by the beam developing an internal shearing stresses. Unlike steel beams, timber beams can be very susceptible to excessive shearing stresses. Such stresses tend to cause the beam to split horizontally along the grain. To have a safe design relative to this failure mode, these shear stresses cannot exceed the design shearing stress value, F_v'. This design shear stress value is the tabular allowable shear stress value, F_v, as listed in the AITC reference, multiplied by several modifying factors that may also be involved. This can be shown as follows:

$$F_v' = F_v \times \text{modifiers}$$

where

$$F_v = \text{tabular allowable shear stress value listed per species}$$
$$\text{in the AITC manual (psi)}$$

$$\text{modifiers} = \text{separate modifers as explained below}$$

The modifiers which may be used are the duration factor (C_D), the moisture factor (C_M), the temperature factore (C_t), and the shear stress factor (C_H). Besides the shear stress factor, all others have been previously mentioned.

The shear stress factor may be used with sawn lumber to account for a change of the tabular allowable shear stress due to the presense of a known defect. Such a defect may include a split, check, or shake. This factor should be used cautiously by the designer, making sure each piece of lumber is visually inspected. In this text, the shear stress factor will be assumed to be 1.0.

The following example will illustrate the use of these criteria in evaluating a timber beam.

EXAMPLE 15.8

A 2 × 8 joist made of Select Structural Spruce-Pine-Fir is to span a length of 12 ft. It is carrying a total uniform load of 100 lbs/ft. that is assumed to be a permanent load. If the beam has full lateral support, will it be adequate?

This is an evaluation problem of an existing beam that will be handled by determining actual shear and bending stresses and comparing those with design values. To begin with, the maximum shear and bending moment can be calculated as follows (formulas can be found in Appendix C):

$$V_{max} = w\,l/2 = (100 \text{ lbs / ft.})(12 \text{ ft.}) / 2 = 600 \text{ lbs}$$

$$M_{max} = w\,l^2 / 8 = (100 \text{ lbs / ft.})(12 \text{ ft.})^2 / 8 = 1800 \text{ ft.-lbs or } 21,600 \text{ in.-lbs}$$

Bending stress

The properties (S_x and area) of a 2 × 8 are known values and are found in Appendix D. Actual bending stress can be calculated as follows:

$$f_b = M/S_x = 21600 \text{ in.-lbs} / 13.1 \text{ in.}^3 = 1648 \text{ psi}$$

The tabular value of allowable bending stress taken from Appendix D is as follows:

$$F_b = 1250 \text{ psi}$$

Since the joist is not deeper than 12 in. and is laterally braced, only the duration modifier is effective. Since this load is indefinite (permanent) the duration modifier (C_D) is 0.90. Therefore, the design value of bending stress is as follows:

$$F_b' = 1250 \text{ psi } (0.90) = 1125 \text{ psi}$$

Comparing actual bending stress to the design value, we find that the actual is greater than the design value (1648 psi > 1125 psi), therefore this will not be acceptable. Check shear stresses anyway.

Horizontal shear stress

Actual shear stress on a rectangular section can be calculated as follows:

$$f_v = VQ/Ib = 3V / 2A = 3(600 \text{ lb}) / 2(10.9 \text{ in.}^2) = 82.8 \text{ psi}$$

The tabular value of allowable shear stress taken from the AITC reference is as follows:

$$F_v = 70 \text{ psi}$$

Again the only applicable modifier in this case is due to the duration, and since this load is indefinite (permanent), the duration modifier (C_D) is 0.90. Therefore, the design value of bending stress is as follows:

$$F_v' = 70 \text{ psi } (0.90) = 63 \text{ psi}$$

Comparing actual shear stress to the design value, we find that the actual is greater than the design value (82.8 psi > 63 psi), therefore this will also not be acceptable.

The beam must be larger to accomodate both bending and shear considerations. As a final note, it is typically necessary to check a beam relative to deflection and bearing stress requirements as well, although these are usually not controlling criteria.

15.7   Reinforced Concrete Beam Design

Reinforced concrete beams are composite beams made from two distinctly different materials—concrete and steel reinforcing bars. Concrete is a material that is basically made from a mixture of portland cement, sand, gravel, and water. The properties of concrete can be highly variable, although designers are usually most interested in its compressive strength, $f'c$. Concrete is very good at resisting the compressive stresses generated when a beam is subjected to bending. The steel reinforcing bars are rolled bars that come in various sizes, as listed in Figure 15-12. These bars (sometimes referred to as rebar) have high tensile strength and are ideal for resisting tensile stresses in the beam. Reinforced concrete beams can be designed and constructed in almost any arrangement and size (Figure 15-13). This section will explore the fundamental concepts of standard reinforced concrete beam design, as specified in the ACI reference.[4]

Reinforced concrete beams need steel in the tension regions because concrete is assumed to have no tensile strength. Longitudinal steel bars are placed in the beam to withstand tensile stresses. Shear stresses are resisted by the use of U-shaped bars, known as stirrups, which are spaced at various intervals along the beam. The remainder of this section focuses on the design of reinforced concrete beams to resist bending and shear effects through a method known as **strength design**.

The strength design method ensures safety by designing the capacity of a member to be greater than the load effects produced by factored loads. The calculated or nominal strength of a member is multiplied by a strength reduction factor to lower its anticipated capacity while the loads effects are increased by the load factors. The strength design philosophy is stated as follows:

Bar no.	Diameter (in.)	Area (in.^{2})
3	.375	.11
4	.50	.20
5	.625	.31
6	.75	.44
7	.875	.60
8	1.0	.79
9	1.128	1.0
10	1.27	1.27
11	1.41	1.56

Figure 15-12 Sizes of Standard Reinforcing

Figure 15-13 Steel Placement in Reinforced Concrete Beam Construction (Courtesy Concrete Reinforcing Steel Institute)

$$\text{Design strength} \geq \text{Required strength}$$

where

Design strength = (nominal strength of a member)(reduction factor)

Required strength = factored load effects

To determine the required strength for a given member, the ACI Code specifies that load factors for dead load, live load, and environmental loads be applied as shown in Table 15-1. A complete listing of these load effects can be found in Chapter 9 of the ACI Code.

Table 15-1 Provisions of Required Strength for Reinforced Concrete

Loading Condition	Required Stength Combination
Basic	1.4(Dead Load) + 1.7(Live Load)
Wind	0.75[Basic + 1.7(Wind Load)]
	or
	0.90(Dead Load) + 1.3(Live Load)
Earth Pressure	Basic + 1.7(Earth Pressure)

Moment Capacity of Reinforced Concrete Beams

The strength design philosophy for the behavior of reinforced concrete beams in bending is as follows:

$$\text{Design strength} \geq \text{Required strength}$$

or

$$\phi M_n \geq M_u$$

where

ϕ = strength reduction factor for bending, 0.90

M_n = nominal moment capcity of the beam

M_u = factored moment applied on the beam

The nominal moment capacity of a beam is based on the assumption that all the concrete below the neutral axis at the point of immenent failure is cracked, and therefore ineffective in resisting moment. The tension below the neutral axis is carried completely by the steel, and the tension resultant at the level of the steel centroid is as follows:

$$T = A_s f_y$$

where

A_s = area of steel reinforcing, in.^{2}

f_y = yield stress of steel reinforcing, psi or ksi

The compressive stress above the neutral axis is approximated by a rectangular block having a magnitude of $.85 f'c$. The block stretches over an area that is the product of the width of the beam, b, and the depth of the block, a. The compression resultant is as follows:

$$C = .85 f'c (a)(b)$$

where

$f'c$ = compressive strength of concrete, psi or ksi

a = depth of the rectangular stress block, in

b = width of the beam, in

The depth of the rectangular stress block, a, can be found by rearranging the previous equation as:

$a = C / (.85 f'c)(b)$

The tension (T) and compression (C) resultants form an internal moment couple that counteracts the external moment applied to the section (Figure 15-14).

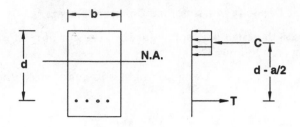

Figure 15-14 Reinforced Concrete Beam Mechanics

The tension and compression resultants must always be equal to each other so that internal equilibrium prevails. Notice that the distance between the resultant is:

$$d - (a/2)$$

where

d = distance from the exterior compressive surface to tension resultant, sometimes referred to as the "design depth"

Therefore, the nominal moment capacity of a beam is given by Equation 15-14 as follows:

$$M_n = T(d - a/2) \qquad \text{(Eq. 15-14)}$$

This equation is also expressed in the following form:

$$M_n = \rho f_y b d^2 (1 - 0.59\rho(f_y/f'c) \qquad \text{(Eq. 15-15)}$$

where

ρ = reinforcement ratio (A_s / bd)

b = width of beam

d = design depth of beam

f_y = yield strength of the steel

$f'c$ = compressive strength of the concrete

The design capacity of a reinforced concrete beam is found by multiplying either Equation 15-14 or Equation 15-15 by the strength reduction factor of 0.90. The following example will illustrate the calculation of a beam's design capacity.

EXAMPLE 15.9

The beam shown below is reinforced with #9 bars that are Grade 60 (f_y = 60 ksi). Determine the design capacity, ϕM_n, of the beam if the $f'c$ is 4000 psi.

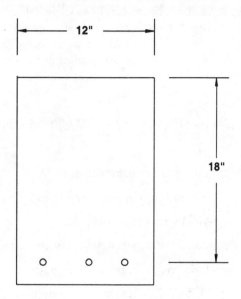

The area of steel using three- #9 bars can be found from Figure 15-12 as 3 in.² Therefore the reinforcement ratio, ρ, can be calculated as follows:

$$\rho = A_s \, / \, bd$$

$$\rho = 3 \text{ in}^2 / (12 \text{ in.})(18 \text{ in.}) = .01388$$

Utilizing Equation 15-15, the nominal moment capacity can be found as follows:

$$M_n = \rho f_y bd^2 \, (1 - 0.59\rho(f_y/f'c)) \qquad (\text{Eq. 15-15})$$

$$M_n = (.01388)(60 \text{ ksi})(12 \text{ in.})(18 \text{ in.})^2 \, (1 - 0.59(.01388)$$
$$(60/4)) = 2840 \text{ kip-in.}$$

or 236.7 kip-ft.

The design moment capacity of this beam is:

$$\phi M_n = 0.90(236.7 \text{ kip-ft.}) = 213 \text{ kip-ft.}$$

Therefore, the maximum required moment that can be applied to this beam is 213 kip-ft.

Upon considering Equation 15-15, the design of reinforced concrete beams is a function of three variables, if the strength of steel and concrete are known. The three variables are: reinforcement ratio (ρ), width of the beam (b), and design depth (d). It is possible for two beams to have the same design moment capacity and have very different overall dimensions of width and design depth. To attain a certain design capacity, a beam can have large values of width and design depth, using a smaller value for reinforcement ratio, or a beam can use smaller dimensions of width and design depth and a larger value for reinforcement ratio. The ACI sets the following limits for the minimum and maximum values that can be used for reinforcement ratio. These are as follows:

$$\rho_{min} = 200 \,/\, f_y \qquad\qquad\qquad\qquad \text{(Eq. 15-16)}$$

$$\rho_{max} = 0.75[(.85)(f'c \,/\, f_y)\beta_1(87,000 \,/\, (87,000 + f_y))] \qquad \text{(Eq. 15-17)}$$

where

ρ_{min} = minimum reinforcement ratio

ρ_{max} = maximum reinforcement ratio

f_y = yield strength of steel, psi

$f'c$ = compressive strength of concrete, psi

β_1 = 0.85 for concrete with strength of 4000 psi or less

0.80 for concrete with strength of 5000 psi

It is typical for the reinforcement ratio to have a value near the middle of the range between ρ_{min} and $\rho_{max}.$ The following example will illustrate the design of a reinforced concrete beam.

EXAMPLE 15.10

Design the steel required in a beam which is 12 inches wide with a design depth of 20 inches required to carry a factored moment of 275 kip-ft. The steel is to be Grade 60 (f_y = 60 ksi) and the $f'c$ is 4000 psi.

From Equation 15-15, the nominal moment capacity is:

$$M_n = \rho f_y b d^2 \,(1 - 0.59\rho(f_y/f'c)) \qquad (\text{ Eq. 15-15})$$

and the design capacity is then:

$$\phi M_n = \phi\rho f_y b d^2 \,(1 - 0.59\rho(f_y/f'c))$$

Since the design moment capacity must be at least equal to the required moment of 275 kip-ft., the previous equation becomes:

$$275 \text{ kip-ft.} = \phi\rho f_y bd^2 (1 - 0.59\rho(f_y/f'c))$$

or

$$3300 \text{ kip-in.} = \phi\rho f_y bd^2 (1 - 0.59\rho(f_y/f'c))$$

with

$$\phi = 0.90$$

$$b = 12 \text{ in.}$$

$$d = 20 \text{ in.}$$

$$fy = 60 \text{ ksi}$$

$$f'c = 4 \text{ ksi}$$

The equation can be rearranged as follows:

$$3300 \text{ kip-in.} = 259{,}200\rho (1 - 8.85\rho)$$

$$3300 \text{ kip-in.} = 259{,}200\rho - 2{,}293{,}920\rho^2$$

$$2{,}293{,}920\rho^2 - 259{,}200\rho + 3300 = 0$$

This is a quadratic equation that can be solved directly for ρ:

$$\rho = .0146$$

Upon checking the calculated ρ with the minimum and maximum reinforcement ratios set forth in Equations 15-16 and 15-17, it is found to lie in between these limits. Therefore, the area of steel needed can be found as:

$$A_s = \rho(b)(d)$$

$$A_s = .0146(12 \text{ in.})(20 \text{ in.}) = 3.5 \text{ in.}^2$$

Shear Capacity of Reinforced Concrete Beams

The strength design philosophy for the shear behavior of reinforced concrete beams is as follows:

$$\text{Design shear strength} \geq \text{Required shear strength}$$

or

$$\phi V_n \geq V_u$$

where

ϕ = strength reduction factor for shear, 0.85

V_n = nominal shear capcity of the beam

V_u = factored shear applied on the beam

Shear cracking can potentially occur in areas of high shear and manifests itself by a pattern of diagonal cracking. The nominal shear capacity of a beam is assumed to come from two sources—the shear strength of the concrete, V_c, and the shear strength from the shear reinforcing bars referred to as stirrups, V_s. Stirrups are located at specified locations along the length of the beam. They are generally made from #3 or #4 bars and are closely spaced near regions of high applied shear and more widely spaced where the shear forces are lower (Figure 15-15).

The nominal shear strength of a beam is calculated as follows, per ACI requirements:

$$V_n = (V_c + V_s) \qquad \text{(Eq. 15-18)}$$

where

V_n = nominal shear capacity of the beam

V_c = concrete shear capacity

V_s = stirrup shear capacity

The concrete shear capacity is approximated by the following formula:

$$V_c = 2\sqrt{f'c}\,(b)(d) \qquad \text{(Eq. 15-19)}$$

where

$f'c$ = concrete compressive strength

b = width of the beam

d = design depth of the beam

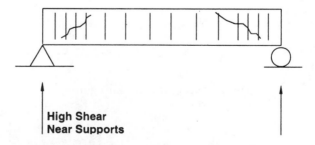

Figure 15-15   Potential Shear Crack Development in a Reinforced Concrete Beam

The shear capacity from the stirrups is given by the following:

$$V_s = (A_v \, f_y \, d) \, / \, s \qquad \text{(Eq. 15-20)}$$

where

A_v = area of the shear reinforcing, generally two times the bar area

f_y = yield strength of the steel, psi or ksi

d = design depth

s = spacing of the stirrups, inches

The area of the shear reinforcing is generally two times the area of the bar because a normal stirrup has two "legs" that are spanning across a potential shear crack. The design shear capacity in large part is a function of stirrup spacing because by the time shear design is performed, the beam dimensions and materials are usually determined. The design shear capacity is calculated as follows:

$$\phi V_n = 0.85(V_c + V_s)$$

The following example will illustrate the evaluation of the shear capacity of a beam.

EXAMPLE 15.11

Determine the adequacy of a section of a beam which is carrying a factored shear of 80 kips if the beam is 12 inches wide with a design depth of 20 inches. Stirrups are #3 bars (Grade 60) spaced at 8 inches and the concrete has an $f'c$ of 4000 psi.

The beam will be adequate in shear capacity if the design capacity, ϕV_n, is at least equal to 80 kips. The design capacity can be expressed as follows:

$$\phi V_n = 0.85(V_c + V_s)$$

The concrete shear capacity is calculated from Eq. 15-19 as shown:

$$V_c = 2 \sqrt{f'c} \ (b)(d) \qquad \text{(Eq. 15-19)}$$

$$V_c = 2 \sqrt{4000 \text{ psi}} \ (12 \text{ in.})(20 \text{ in.}) = 30{,}350 \text{ lb or } 30.35 \text{ kips}$$

The shear capacity of the #3 stirrups (A_v = .22 in²) is calculated from Equation 15-20 as follows:

$$V_s = (A_v \, f_y \, d) \, / \, s \qquad \text{(Eq. 15-20)}$$

$$V_s = (.22 \text{ in.}^2)(60 \text{ ksi})(20 \text{ in.}) \, / \, 8 \text{ in} = 33 \text{ kips}$$

The design shear capacity is as shown below:

$$\phi V_n = 0.85(30.35 \text{ kips} + 33 \text{ kips})$$

$$\phi V_n = 53.8 \text{ kips}$$

Since the design capacity is less than the factored shear on this section, the beam is inadequate and would have to be redesigned.

15.8   Summary

Rolled-beams are one of the basic building blocks for steel construction. Steel beam failure mechanisms may initiate in several different manners, depending on the stability or buckling of the compression flange. A common instability is referred to as lateral-torsional-buckling and is associated with the unbraced length of the compression flange. If a beam is prevented (usually by lateral bracing) from this instability, it will typically be able to reach its maximum or plastic moment.

Design of steel beams is typically accomplished using either the Load and Resistance Factor Design (LRFD) method or the Allowable Stress Design (ASD) method. Bending behavior of steel beams is typically the controlling failure mechanism. Shear failure on steel beams is typically not a controlling feature of steel design; unless the beams are very short, heavily loaded, or have coped sections.

Timber beams are designed using an allowable stress philosophy and such designs focus primarily on preventing bending stress failure and horizontal shear stress failure. The design values used for bending and shear stresses are based on the species of wood and various modifiers. The tabular value for such behaviors is taken out of the AITC manual and modified appropriately to get the final design value.

Reinforced concrete beams are typically reinforced with steel bars, which are ideal in resisting tension forces. The design of these beams is complicated, although the two main design features are bending capacity and shear capacity. Reinforced concrete beams are designed by a method known as strength design, which ensures safety by designing the capacity of a member to be greater than the load effects produced by factored loads.

EXERCISES

1. (LRFD). A W 12×120 is used as a beam that is 20 ft. long and braced along its compression flange at 5 ft. intervals. If the maximum factored moment is 400 kip-ft. (occurring at the beam's midpoint) and the steel is A36, will the beam be adequate per AISC criteria? Assume $C_b = 1.0$.

2. (LRFD). A W 310×86 is used as a beam to span 5 m and is unbraced. If the beam holds a factored load of 10 Kn / m, is it adequate per LRFD criteria? The steel is A36. Assume $C_b = 1.0$.

3. (LRFD). If the beam in Exercise 1 is braced at its midpoint, how does this affect its design capacity?

4. (LRFD). Using the same information as in problem 1, calculate the maximum uniform load that the beam can safely carry.

5. (LRFD). Calculate the adequacy per LRFD specifications of a W 12 × 79 beam that carries a factored uniform load of 5 kips/ft. over a span length of 25 ft. The beam is made from A36 steel and is braced at points located 5 ft. from each end. Check bending requirements.

6. (LRFD). Calculate the adequacy per LRFD specifications of a W 310 × 143 beam that carries a factored uniform load of 25 KN / m over a span length of 9 m. The beam is made from A36 steel and is braced at points located 1.5 m from each end. Check bending requirements.

7. (LRFD). Design an economical W section to carry a service dead load of 2 kips/ft. and a service live load of 10 kips located at midspan. The beam is 20 ft.long and is unbraced. The steel is A36, also check shear requirements. Depth of the beam is limited to 18 in. $C_b = 1.0$.

8. (LRFD). A W 12 × 58 spans 8 ft. and is fully braced. Determine the magnitude of factored uniform load that would produce a shear equal to the beam's design shear capacity. The steel is A36.

9. (LRFD). For the beam shown below, design an economical W - section made from A36 steel. The beam is braced only at the supports and the loads shown are factored. Consider bending moment requirements and assume $C_b = 1.0$.

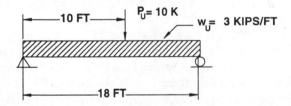

10. (ASD). A W 12 × 120 is used as a beam that is 20 ft. long and unbraced along its compression flange. If the maximum moment is 300 kip-ft. (occurring at the beam's midpoint) and the steel is A36, will the beam be adequate per AISC bending stress criteria?

11. (ASD).Using the same information as in Exercise #10, calculate the maximum uniform load that the beam can safely carry.

12. (ASD). Check the adequacy of a W 24 × 84 to hold 350 kip-ft. that is 20 ft. long and unbraced along its compression flange.The beam is made from A36 steel. Assume $C_b = 1.0$.

13. (ASD). Recheck the beam in Exercise 12 if the beam is braced at its midpoint.

14. (ASD). Determine the adequacy of a W 12×50 that is 16 ft long and braced at midspan. The beam is under a uniform load of 3 kips/ft. and the steel is A36.

15. (ASD). Design the most economical W 12 section to carry a maximum moment of 250 kip-ft. located at midspan .The beam is 20 ft. long and is braced at midspan.The steel is A36 and assume $C_b = 1.0$, if necessary.

16. (ASD) Design the lightest W 10 section to carry a 20 kip concentrated loads at its third points. The beam is 15 ft. long and unbraced. Steel is A36.

17. (ASD). For the beam shown below, design the lightest W18 section, if it is:

    a. unbraced
    b. braced at mid-span
    c. braced at quarter points
       Steel is A36 and C_b is assumed to equal to 1.0.

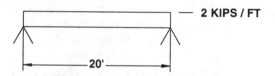

2 KIPS / FT

20'

18. A timber beam is made from two 2×10's that are nailed and made from Eastern Softwood No.1. Calculate the maximum uniform load that the beam can carry if the beam is fully braced and the full load is assumed to be permanent. The beam is 15 ft. long.

19. If the uniform load in Exercise 18 was assumed to be 50% permanent load and 50% impact load, how would this change the answer?

20. A 2×6 made from Spruce Pine Fir No.3 is to carry a uniform load of 50 lbs/ft. over 12 ft. If the load is assumed to be a two month load, will the beam be adequate? Consider selfweight of the beam.

21. If the beam in exercise #20 does not work, what value would the load have to be reduced in order to be adequate?

22. A timber beam made from two 2×8's spans a distance of 16 ft. and carries a uniform load of 100 lbs/ft. If the lumber is Spruce-Pine-Fir No. 1 and the load is assumed to be 20% permanent and 80% 7-day load, will the beam be adequate?

23. What would be the maximum length of the beam be in Exercise 22?

24. A simply supported reinforced concrete beam spanning 20 ft. carries a uniform factored load of 2.5 kips/ft. (Includes all loads). If the beam has a width of 14 in., a design depth of 22 in., and is reinforced with four #9 bars, will the beam be adequate? Steel has $f_y = 60$ ksi and concrete has $f'c = 4$ ksi.

25. What would be the maximum factored uniform load that the beam in Exercise 24 could support?
26. Design the steel in a beam that is 10 in. wide with a design depth of 18 in. and is required to carry a factored moment of 175 kip-ft. The steel is to be Grade 60 (f_y = 60 ksi) and the $f'c$ is 5000 psi.
27. Calculate the design depth required in a beam that needs to carry 125 kip-ft. The beam is 14 in. wide and has a ρ = .01. The steel is to be Grade 60 (f_y = 60 ksi) and the $f'c$ is 3000 psi
28. Determine the shear capacity adequacy of a section of a beam that is carrying a factored shear of 55 kips, if the beam is 12 in. wide with a design depth of 18 in. Stirrups are #4 bars (Grade 60) spaced at 7 in. and the concrete has an $f'c$ of 4000 psi.

REFERENCES

1. AISC. *Manual of Steel Construction, Load and Resistance Factor Design*, First Edition, Chicago, American Institute of Steel Construction, 1986.
2. AISC. *Manual of Steel Construction, Allowable Stress Design*, Ninth Edition, Chicago, American Institute of Steel Construction, 1989.
3. AITC. *Timber Construction Manual*, Fourth Edition, Englewood, Colorado, American Institute of Timber Construction, 1994.
4. ACI. Building Code Requirements for Reinforced Concrete, ACI 318-89, Detroit, American Concrete Institute, 1989.
5. Salmon, Charles G. and Johnson, John E., *Steel Structures, Design and Behavior*, Harper-Row Publishers, 2nd edition, New York, 1980, p. 450-517.
6. Salmon, Charles G. and Johnson, John E., *Steel Structures, Design and Behavior*, Harper-Collins Publishers, 3rd Edition, New York, 1990, p.516-543.

16

STEEL COLUMN DESIGN

16.1 Introduction

Compression members are found in all types of construction—from the skeleton framing of a building, to the massive pier towers of the Golden Gate Bridge (Figure 16.1). If compression members are set vertically, they are commonly referred to as columns. The student should note that the terms "compression members" and "columns" will be used interchangeably throughout this chapter.

The importance of compression members cannot be underestimated, for they are extremely important in the overall design in buildings or bridges. The student must remember that if a column fails, everything supported above that column will most probably collapse.

This chapter will present the AISC technique for concentric compression member design using the Load and Resistance Factor Design (LRFD) and the Allowable Stress Design (ASD) methods. Discussion will include column instability and the factors relating this behavior to the AISC equations.

It should be stated that columns are rarely, if ever, loaded concentrically in the real world. They are more typically subjected to some bending moment; either through eccentric axial loads, beam reactions, lateral loadings, or a combination thereof. When such situations induce bending stresses into the axially loaded compression member, these members are typically referred to as beam-columns. This topic is covered in a conceptual manner in Chapter 14, however, the design of beam-columns is beyond the scope of this text.

Figure 16-1 The Golden Gate Bridge, San Francisco. (Courtesy Bethleham Steel Corporation)

16.2 Potential Modes of Column Failure

Column behavior is notably different from that of tension members. The tension member under stress tends to "straighten out" in the direction of the load, whereas a column tends to "move out" of the plane of loading. This tendency to "move out" is referred to as **buckling** and is a serious concern in column design. Buckling constitutes failure because the column is unstable and cannot accept any additional load.

Another method of column failure is the yielding (crushing) of the material, due to the actual stress on the member being greater than the yield. It is evident that as columns get longer, the failure mode that dominates is buckling. Only very short columns will fail due to yielding, and the intermediate length column will fail by a combination of these two types of behavior (Figure 16-2).

This intermediate behavior occurs because these columns do not buckle until stresses have reached a sufficiently high level, thereby initiating some yielding of the fibers. The vast majority of columns fall into the category of intermediate behavior.

16.3 Column Behavior

The understanding of columns and the use of today's design equations can be traced back to a Swiss mathematician, Leonard Euler. In the mid-eighteenth cen-

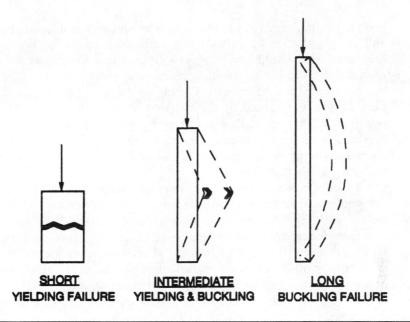

SHORT
YIELDING FAILURE

INTERMEDIATE
YIELDING & BUCKLING

LONG
BUCKLING FAILURE

Figure 16-2 Various Failure Modes Relative to Column Length

tury, Euler studied column buckling and derived a formula that would predict the load that would cause a column with rounded ends to buckle [1]. This formula is the basis of all modern day column equations and is shown below.

$$P = (\pi^2 \, E \, I)/l^2$$

Where

P = Buckling load

E = Material's modulus of elasticity

I = Moment of inertia about the bending axis

l = Length of column between points of zero moment

Usually this formula is written in terms of buckling stress (P/A) by substituting Ar^2 for the moment of inertia, I. We can then express Euler's equation in terms of buckling stress as follows:

$$P/A = (\pi^2 \, E)/(l/r)^2$$

Where

r = radius of gyration about the buckling axis

From Euler's equation, the student can see that buckling is dependent on basically two factors—the l/r term (this is typically referred to as slenderness ratio) and the column's modulus of elasticity. The student should study Euler's equation and realize that as the slenderness ratio (l/r) increases, the buckling stress decreases. This would mean that as a column gets longer (therefore, more slender), the stress to initiate buckling becomes smaller.

Because of the incorporation of the modulus of elasticity in the Euler equation, its use was limited by the fact that it would describe only buckling that occurred in the material's elastic region. When buckling in the elastic region occurs, it is referred to as **elastic buckling**. This phenomena of elastic buckling is limited to long, slender columns that buckle under low stress levels, typically under the proportional limit.

This limitation makes the equation accurate in predicting buckling behavior for long, slender columns. Columns that are of intermediate length do not exhibit elastic buckling, and therefore Euler's equation tended to overestimate their capacity. (Figure 16-3) Buckling for intermediate length columns occurs at higher stress levels and this type of buckling is referred to as **inelastic buckling**, since some fibers will have yielded as buckling commences. Because many typical columns are of intermediate length, the Euler equation was ignored for many years[2] while engineer's searched for a single equation to describe buckling over all column lengths.

A single column equation defining column behavior over all ranges of slenderness ratios has never been found. In lieu of this absence, modern day column design equations are based on the two types of buckling that occur—elastic and inelastic. Elastic buckling occurs as stresses on a column are rather low and still

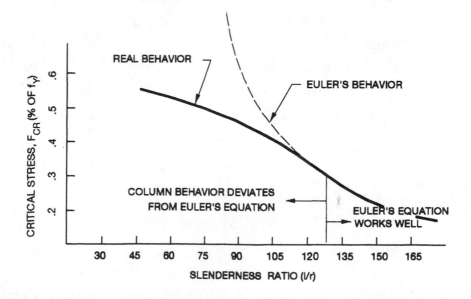

Figure 16-3   Actual Column Behavior versus Euler Behavior

under the material's proportional limit. Inelastic buckling occurs at higher stress levels and cannot be well defined by the original Euler equation. All design specifications recognize this fact, and therefore will introduce multiple column equations (at least one for elastic and one for inelastic behavior) for use in design. We will specifically discuss the American Institute of Steel Construction's (AISC) equations further in Section 16.5.

16.4   Effective Length

The amount of fixity that exists at the ends of a column is known to dramatically affect a column's buckling resistance. Because of this, there has been some modification of Euler's original equation, for his testing dealt only with one type of end-restraint condition. Actually, columns may have many types of end-restraint conditions, so a modification factor that estimates effective length of the column has been introduced. This modification factor is referred to as the **effective length factor, K**.

This effective length factor approximates the length over which a column actually buckles. This buckling length can be longer or shorter than the actual length of the column, based on end restraints of the column or whether the column is subjected to lateral movement. The student should realize that as the length of a column is reduced, the column actually increases its resistance to buckling; conversely, as the column length is increased, the resistance to buckling decreases. End-restraint conditions on a column will dramatically influence the buckling length or **effective length, Kl,** of this member. The effective length factor, *K,* is an attempt to accurately describe the true buckling length of the column.

Much research has been devoted to the investigation of K factors, in both braced and unbraced frames. The following chart (Figure 16.4) contains *K* values that the American Institute of Steel Construction recognizes for ideal end-restraint conditions (pin, fixed, free, etc.). The AISC recognizes that in real practice there are no ideal conditions, therefore, recommended *K* values are also shown. To be conservative, the recommended *K* values are always greater than or equal to the theoretical values.

The following example illustrates the power of the effective length factors as they are taken from this chart. Keep in mind that a column's effective length and its critical stress will be shown to have an inversely proportional relationship. That is, as a column becomes "shorter", the calculated critical stress will increase.

EXAMPLE 16.1

Calculate the effective length, Kl, of a 20-foot-long column under the following three end-restraint conditions. Use the recommended values from the chart.

1. pin–pin
2. pin–fixed
3. fixed–fixed

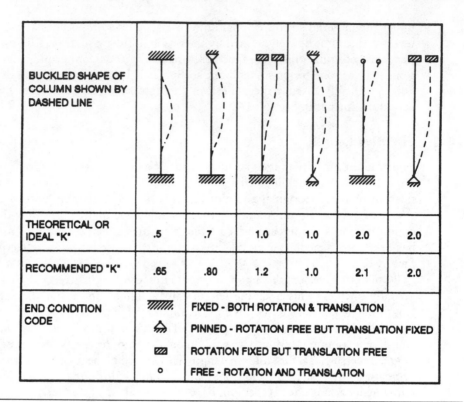

BUCKLED SHAPE OF COLUMN SHOWN BY DASHED LINE						
THEORETICAL OR IDEAL "K"	.5	.7	1.0	1.0	2.0	2.0
RECOMMENDED "K"	.65	.80	1.2	1.0	2.1	2.0
END CONDITION CODE	🖤 FIXED - BOTH ROTATION & TRANSLATION					
	🔺 PINNED - ROTATION FREE BUT TRANSLATION FIXED					
	▨ ROTATION FIXED BUT TRANSLATION FREE					
	○ FREE - ROTATION AND TRANSLATION					

Figure 16-4 Typical Values of "K" for Idealized End Conditions

The effective length should be calculated as follows:

1. Since the recommended k value for a pin - pin end condition is 1.0, the effective length, Kl, of this value is 1.0×20 ft. = 20 ft.
2. The recommended K value for a pin - fixed end condition is .80, therefore the effective length is $.80 \times 20$ ft. = 16 ft. (Essentially because of its end conditions this column behaves similarly to a 16 ft. long column with pin–pin end conditions).
3. With the recommended K value equal to .65, the effective length is equal to $.65 \times 20$ ft = 13 ft (Again, the column has essentially the same behavior as a 13 ft. long column with pinned ends).

16.5   AISC Column Design Philosophy Using LRFD

The LRFD method for column design follows a format that was discussed in the previous chapter on beam design. Its basic criterion is as follows:

Design Strength ≥ Required Resistance

For columns in particular, this general format will take the specific form as shown below:

$$\phi_c \, P_n \geq P_u$$

where

$\phi_c = .85$, column reduction factor

$P_n = A_g(F_{cr})$, nominal column strength, kips

$A_g =$ area of the column, in.^{2}

$F_{cr} =$ critical column stress, ksi

$P_u =$ maximum factored column load, kips

The AISC formulas for column design are the result of research committed to finding a formula that accurately predicts column behavior in all slenderness ranges. There have been many formulas proposed; including the Gordon-Rankine, the parabolic, and the secant formulas. The AISC uses a parabolic formula in the inelastic range and a hyperbolic Euler formula in the elastic range. The AISC philosophy recognizes that the behavior of a column changes from the elastic to the inelastic range. Because of this, the AISC sets a value of a slenderness parameter, called λ_c, as the dividing line between the two aforementioned ranges. The value of this slenderness parameter is actually a ratio of yield stress to Euler buckling stress, and can be written as shown below:

$$\lambda_c = (Kl/r\pi) \, \sqrt{F_y/E} \qquad \text{(Eq. 16-1)}$$

The Load and Resistance Factor Design (LRFD) specification uses the following two formulas for steel column design. For elastic buckling (when $\lambda_c > 1.5$), the critical stress is given by the following equation:

$$F_{cr} = (.877/\lambda_c^2)F_y \qquad \text{(Eq. 16-2)}$$

where

$F_{cr} =$ critical column stress

$F_y =$ specified steel yield stress

For inelastic buckling stress (when $\lambda_c \leq 1.5$), the LRFD specification presents the equation that is shown below:

$$F_{cr} = (.658^{\lambda_c^2})F_y \qquad \text{(Eq. 16-3)}$$

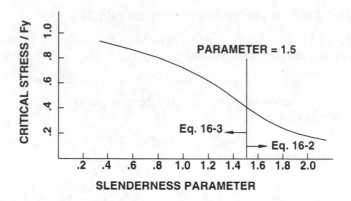

Figure 16-5   Use of AISC Column Equations-LRFD

The reader should notice that the aforementioned Equations 16-2 and 16-3 become equal when the slenderness parameter, λ_c, equals 1.5. At this point, both equations become equal to $.39F_y$. Therefore, if the stress in the column (P_u/A) exceeds $.39F_y$ the column is assumed to behave inelasticly. This AISC philosophy for column design using the LRFD method is outlined in Figure 16-5.

As stated before, there are two general types of problems encountered in steel design. In the evaluation problem, a designer is given a member and corresponding load, and asked to determine its adequacy. The design problem typically gives the loads and subsequently asks to design the member size. The evaluation problem is very easy to perform and the following examples will attest to this. All that is necessary, is to compare design strength ($\emptyset_c P_n$) to factored load (P_u). The following examples will provide some insight to these problems.

EXAMPLE 16.2

Calculate the maximum factored load (P_u) that a W 14 × 53 column can hold if it is 20 ft. long with a $K = 1.0$. Steel is A36.

From Appendix A, it can be found that a W 14 × 53 has $A = 15.6$ in.^{2} and $r_x = 5.89$ in. and $r_y = 1.92$ in.

Because the column is not braced, the weak axis (r_y) will control.

$$Kl/r_y = 1.0 \times 20 \times 12/1.92 = 125$$

Calculating the controlling slenderness parameter from Equation 16-1 we find:

$$\lambda_c = (Kl/r\pi) \sqrt{F_y/E} \qquad \text{(Eq. 16-1)}$$

$$\lambda_c = (125/\pi) \sqrt{36000/29 \times 10^6} = 1.401$$

Since $\lambda_c = 1.401 < 1.5$, use Equation 16-3 to calculate the critical buckling stress, F_{cr}, as follows:

$$F_{cr} = (.658^{\lambda_c^2})F_y \qquad \text{(Eq. 16-3)}$$

$$F_{cr} = (.658^{1.401^2})\ 36\ \text{ksi} = 15.81\ \text{ksi}$$

Calculating the design capacity of the section, $\phi_c\ P_n$, we find:

$$\phi_c\ P_n = (\phi_c\ A_g)\ F_{cr}$$

$(0.85 \times 15.6\ \text{in.}^2)(15.81\ \text{ksi}) = 209.6\ \text{kips}$

Since the design capacity of the section represents the maximum factored load on the column, P_u is equal to 209.6 kips.

16.6   AISC Column Design Using ASD

The allowable stress philosophy is based on keeping the actual stress in a member below some prescribed value, which is set forth in a code or specification. This prescribed value is referred to as the allowable stress. Therefore, the ASD philosophy for columns can simply be summarized as follows:

Actual axial stress ≤ Allowable axial stress

The AISC formulas for column design calculate the allowable axial stress permitted on a column, in all slenderness ranges. There have been many formulas proposed, all realizing that the slenderness ratio of the column dictates whether elastic or inelastic buckling will occur. The AISC uses a parabolic formula in the inelastic range and a modified Euler's formula in the elastic range.

The AISC philosophy recognizes that the behavior of a column changes from the elastic to the inelastic range. Because of this, the ASD sets a value of a slenderness parameter, called C_c, as the dividing line between the two aforementioned ranges. The value of this slenderness parameter (C_c) is actually the slenderness ratio, which would cause the buckling stress to be 50% of a steel's yield stress by using Euler's equation. The reason for using a value of only 50% of F_y to delineate elastic from inelastic buckling is due to the effect of residual stress in rolled members. Residual stresses are those stresses that are "built-in" rolled members as a result of uneven cooling during the manufacturing process.

Residual stresses can be as high as 20 to 30% of F_y and exist in steel members before they are ever loaded! These stresses can lead to intermediate columns buckling at stresses below their theoretical critical load[3], therefore the AISC specification uses the 50% of F_y stress level as the boundary between elastic and inelastic behavior.

The value of C_c can be calculated as follows:

$$C_c = \sqrt{\frac{2\pi^2 E}{F_y}} \qquad \text{(Eq. 16-4)}$$

The AISC uses C_c to distinguish whether elastic or inelastic behavior will occur in a steel column. When a column's individual slenderness ratio, Kl/r, is less than C_c, inelastic buckling predominates. In a similar manner, when a column's slenderness ratio, Kl/r, is greater than C_c, elastic buckling controls.

When a column is controlled by inelastic buckling, the AISC uses the parabolic formula (Equation 16-5) to accurately predict its allowable stress. This formula is as follows:

$$F_a = \frac{\left[1 - \dfrac{(Kl/r)^2}{2C_c^2}\right]F_y}{\dfrac{5}{3} + \dfrac{3}{8}\left(\dfrac{Kl}{r}/C_c\right) - \dfrac{1}{8}\left(\dfrac{Kl}{r}/C_c\right)^3} \qquad \text{(Eq. 16-5)}$$

Although this equation looks rather imposing, the denominator is actually just its safety factor, which can theoretically range from 5/3 (when our column's $Kl/r = 0$) to 23/12 (when our column's $Kl/r = C_c$). This variable safety factor increases as the column becomes more slender to account for the detrimental effects caused by possible column crookedness.

When a column is controlled by elastic behavior, the AISC recognizes the validity of Euler's equation, and therefore uses the same formula, modified only by a safety factor and the effective length factor, K. The safety factor for columns in the elastic range is given as 23/12 (~1.92). When the slenderness ratio, Kl/r, of an individual column exceeds C_c, the AISC uses the following equation (Equation 16-6) to describe the allowable buckling stress:

$$F_a = \frac{12\pi^2 E}{23\left((Kl)/r\right)^2} \qquad \text{(Eq. 16-6)}$$

The student should note the striking similarity between this equation for allowable stress and the 250-year-old Euler equation for actual buckling stress. The AISC philosophy is easily seen in Figure 16-6.

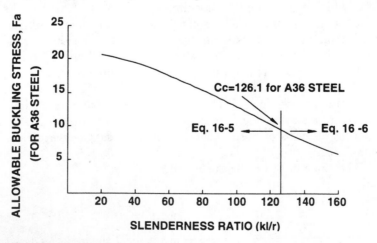

Figure 16-6   Use of AISC Column Equations -ASD

It should be noted that helpful aids exist in the calculation of Equation 16-5 that can be used to quickly solve for the allowable axial stress. The AISC provides a table in its ASD specification, which is shown below, in Table 16-1. This table has precalculated ordinates (referred to as C_a) along the curve for allowable stress in the inelastic region (when $Kl/r < C_c$).The student simply enters the table using the ratio of slenderness ratio to C_c ($Kl/r/C_c$) and selects the appropriate value of C_a. The value of C_a is then multiplied by the steel's yield stress (F_y) to attain the correct value of allowable stress.The student is urged to rework the problem in Example 2, using this method, in order to be familiar with this design aid.

Table 16-1 C_a Values for the Determination of Allowable Stress-ASD

$\frac{Kl/r}{C_c}$	C_a	$\frac{Kl/r}{C_c}$	C_a	$\frac{Kl/r}{C_c}$	C_a	$\frac{Kl/r}{C_c}$	C_a
.01	.599	.26	.548	.51	.472	.76	.375
.02	.597	.27	.546	.52	.469	.77	.371
.03	.596	.28	.543	.53	.465	.78	.366
.04	.594	.29	.540	.54	.462	.79	.362
.05	.593	.30	.538	.55	.458	.80	.357
.06	.591	.31	.535	.56	.455	.81	.353
.07	.589	.32	.532	.57	.451	.82	.348
.08	.588	.33	.529	.58	.447	.83	.344
.09	.586	.34	.527	.59	.444	.84	.339
.10	.584	.35	.524	.60	.440	.85	.335
.11	.582	.36	.521	.61	.436	.86	.330
.12	.580	.37	.518	.62	.432	.87	.325
.13	.578	.38	.515	.63	.428	.88	.321
.14	.576	.39	.512	.64	.424	.89	.316
.15	.574	.40	.509	.65	.420	.90	.311
.16	.572	.41	.506	.66	.416	.91	.306
.17	.570	.42	.502	.67	.412	.92	.301
.18	.568	.43	.499	.68	.408	.93	.296
.19	.565	.44	.496	.69	.404	.94	.291
.20	.563	.45	.493	.70	.400	.95	.286
.21	.561	.46	.489	.71	.396	.96	.281
.22	.558	.47	.486	.72	.392	.97	.276
.23	.556	.48	.483	.73	.388	.98	.271
.24	.553	.49	.479	.74	.384	.99	.266
.25	.551	.50	.476	.75	.379	1.00	.261

Source: Courtesy of the American Institute of Steel Construction, Inc.

*When ratios exceed the noncompact section limits of AISC Sect. B5.1, use $\frac{Kl/r}{C'_c}$ in lieu of $\frac{Kl/r}{C_c}$ values and equation $F_a = C_a Q_a Q_s F_y$ (Appendix Sect. B5).

Note: Values are for all grades of steel.

The evaluation problem is easy to perform and the following examples will demonstrate how the typical evaluation problem is performed. Following these evaluation examples, a technique for solving the column design problem will be presented.

EXAMPLE 16.3

Calculate the allowable load, P_{all}, that an unbraced W14 × 53 column can hold if it is 20 ft. long with K = 1.0 (in both directions). Steel is A36.

From Appendix A, a W 14 × 53 has the following properties:

$$A = 15.6 \text{ in.}^2$$

$$r_x = 5.89 \text{ in.}$$

$$r_y = 1.92 \text{ in.}$$

The weak axis (r_y) will control, since column is not braced, and therefore:

$$Kl_y/r_y = 1.0 \times 20 \text{ ft.} \times 12 \text{ in./ft.}/ 1.92 \text{ in.} = 125$$

Since the steel is A36, C_c can be calculated as follows:

$$Cc = \sqrt{2\pi^2 (29,000) / 36} = 126.1$$

Since the controlling Kl/r < 126.1, use Equation 16-5 to calculate the allowable stress, F_a.

Calculating:

$$F_a = \frac{\left[1 - \dfrac{(125)^2}{2(126.1)^2}\right] 36 \text{ Ksi}}{\dfrac{5}{3} + \dfrac{3}{8}\left(\dfrac{125}{126.1}\right) - \dfrac{1}{8}\left(\dfrac{125}{126.1}\right)^3}$$

$$F_a = 9.55 \text{ ksi}$$

Therefore, the allowable load, P_{all}, is as follows:

$$P_{all} = (9.55 \text{ ksi})(15.6 \text{ in.}^2) = 149 \text{ kips}$$

(Remember, when bracing is used, the slenderness ratios about both the strong and weak axis should be checked with the larger controlling the design. Note that Table 16-1 could have also been used to calculate the value for allowable stress. This could have been done by entering the table with the ratio of $Kl/r/C_c$ (125/126.1) and finding the coefficient C_a. The allowable stress, F_a, would simply be the product of C_a and F_y.)

16.7 Braced Columns

We have already discussed how columns will behave as the stress on them becomes increasingly large. Designers must realize, in order to carry more load, a member's area can always be increased. But can the load carrying capacity of these columns be increased without using a larger rolled section?

Yes, with columns, we can accomplish an increase in strength by reducing the effective length, Kl, through the use of bracing. Bracing of columns can take various forms—such as steel X-bracing, the framing in of beams, or the bracing effects of a floor system. All of these will reduce the overall buckling length of column, and as this length is reduced, the stress that the column is allowed to carry will increase.(Figure 16-7)

The location of this bracing is also very important to the load carrying capacity of the column. The student will remember the buckling load formulas all incorporate the λ_c term, known as the slenderness parameter (for LRFD) or the slenderness ratio (for ASD). A closer look at these parameters will reveal that the only variable, once a particular grade of steel has been chosen, is the radius of gyration, r. The radius of gyration is an elastic property of the individual rolled section, based on the axis under consideration. Although difficult to describe in practical terms, the radius of gyration can most easily be thought of as defining the relationship between a shape's moment of inertia and its area. This is expressed by the formula:

$$r = \sqrt{I/A}$$

where

I = moment of inertia about the axis being considered

A = the cross-sectional area of the shape

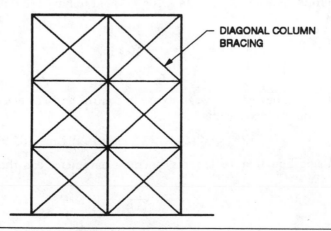

DIAGONAL COLUMN BRACING

Figure 16-7 Bracing of Structural Frames

As the students view the section property tables, they will notice that r_y is always less than r_x, for wide-flange sections. Therefore, for an unbraced wide-flange column, the slenderness ratio Kl/r_y will always be larger than Kl/r_x (assuming that K remains the same about both axes). This means that buckling of an unbraced column will always occur about the weak or y-y axis because the column is more slender in this direction. By bracing the column in the weak direction (y-y), we also reduce the Kl/r_y value to produce a larger critical or allowable stress. By lowering Kl/r_y, we might even reach the point where the weak axis no longer controls ($Kl/r_y < Kl/r_x$).

The design equations for columns with bracing are exactly the same as outlined in the previous sections, although when bracing is used, the student cannot assume that buckling will necessarily occur about the weak axis. *When bracing is used, or if the effective length factor is different between the x and y axis, the slenderness parameter about the strong and weak axis must both be calculated. The higher of these two slenderness parameters will control the design, since the higher of the two will lead to a smaller value of critical stress or allowable stress.*

When bracing is used, the slenderness parameters, λ_c, or slenderness ratio, Kl/r, about both the strong and weak axis should be checked, with the larger controlling the design. The following example will illustrate this type of problem.

EXAMPLE 16.4

Calculate the maximum factored load, P_u, for a W 24 × 104 that is 20 ft. long with a $K = .80$. Then, recalculate if the column is braced at midheight in the weak axis (still assume $K = .80$). Use A36 steel.

Part a:

From Appendix A, a W 24 × 104 is found to have the following properties:

$$A = 30.6 \text{ in.}^2$$

$$r_x = 10.1 \text{ in.}$$

$$r_y = 2.91 \text{ in.}$$

Being unbraced, the weak axis will control and calculating the slenderness ratio about this axis we find:

$$Kl/r_y = 0.8(20 \text{ ft.} \times 12 \text{ in./ft.})/2.91 \text{ in.} = 65.98$$

Using Equation 16-1, the slenderness parameter can then be calculated as follows:

$$\lambda_c = (Kl/r\pi) \sqrt{F_y/E} \qquad \text{(Eq. 16-1)}$$

$$\lambda_c = (65.98/\pi) \sqrt{0.00124} = .7399 < 1.5$$

Since the slenderness parameter is less than 1.5, inelastic behavior controls and Equation 16-3 is used to find the critical buckling stress.

$$F_{cr} = (.658^{\lambda_c^2})F_y \qquad \text{(Eq. 16-3)}$$

$$F_{cr} = (.658^{.7399^2})36 \text{ ksi} = 28.62 \text{ ksi}$$

Calculating the design capacity of the section, $\phi_c P_n$, we find:

$$\phi_c P_n = (\phi_c A_g) F_{cr}$$

$$= (0.85 \times 30.6 \text{ in.}^2)\ 28.62 \text{ ksi} = 744.4 \text{ kips}$$

Since the design capacity of the section represents the maximum factored load on the column, P_u is equal to 744.4 kips in this case.

Part b: Column Braced Midheight in Weak Direction

Calculating the slenderness ratio, we find that the weak axis still controls in this scenario.

$$Kl/r_x = 0.8(20 \text{ ft.} \times 12)/\ 10.1 \text{ in.} = 19.01$$

$$Kl/r_y = 0.8(10\text{ft.} \times 12)/\ 2.91 \text{ in.} = 32.99$$

Using Equation 16-1, the slenderness parameter can then be calculated as follows:

$$\lambda_c = (Kl/r\pi)\ \sqrt{F_y/E} \qquad \text{(Eq. 16-1)}$$

$$\lambda_c = (32.99/\pi)\ \sqrt{0.00124}\ = .369 < 1.5$$

Since the slenderness parameter is again less than 1.5, inelastic behavior controls and Equation 16-3 is used to find the critical buckling stress.

$$F_{cr} = (.658^{\lambda_c^2})F_y \qquad \text{(Eq. 16-3)}$$

$$F_{cr} = (.658^{.369^2})36 = 34 \text{ ksi}$$

Calculating the design capacity of the section, $\phi_c P_n$, we find:

$$\phi_c P_n = (\phi_c A_g) F_{cr}$$

$$\phi_c P_n = (0.85 \times 30.6 \text{ in.}^2)\ 34 \text{ ksi} = 884.4 \text{ kips}$$

Since the design capacity of the section represents the maximum factored load on the column, P_u is equal to 884.4 kips in this case. Therefore, bracing increases capacity 884.4 − 744.4/744.4 = 18.8%.

Should the designer encounter a situation where different unbraced lengths occur on the same column, the unbraced length with the largest slenderness parameter would control the section. The largest slenderness parameter will translate into a lower critical stress, thereby reducing the section's design capacity. The next example will address this occurrence.

EXAMPLE 16.5

Calculate the design capacity of a W 12 × 79 column whose length is 20 feet. The column is braced along its weak axis at five feet from the top and bottom of the column and the steel is A36. K = 1.0 and assume the braces are likewise.

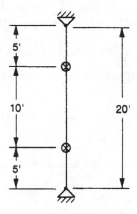

In this case, the unbraced lengths along the weak axis are 5 ft. at the ends and 10 ft. in the middle, while the unbraced length along the strong axis is 20 feet. The effective length along the weak axis, to be used for comparison with the strong axis, is the larger of the following:

$$Kl_y = 1.0(5 \text{ ft.}) = 5 \text{ ft.}$$

$$Kl_y = 1.0(10 \text{ ft.}) = 10 \text{ ft.—controls}$$

From Appendix A, the values for area and radius of gyration can be found. Calculating the slenderness ratio, we find that the strong axis begins to control this scenario.

$$Kl/r_x = 1.0(20 \text{ ft.} \times 12 \text{ in./ft.})/5.34 \text{ in.} = 44.9$$

$$Kl/r_y = 1.0(10 \text{ ft.} \times 12 \text{ in./ft.})/3.05 \text{ in.} = 39.3$$

Using Equation 16-1, the slenderness parameter can then be calculated as follows:

$$\lambda_c = (Kl/r\pi) \sqrt{F_y/E} \qquad \text{(Eq. 16-1)}$$

$$\lambda_c = (44.9/\pi) \sqrt{0.00124} = .503 < 1.5$$

Since the slenderness parameter is again less than 1.5, inelastic behavior controls and Equation 16-3 is used to find the critical buckling stress.

$$F_{cr} = (.658^{\lambda_c^2})F_y \qquad\qquad \text{(Eq. 16-3)}$$

$$F_{cr} = (.658^{.503^2})36 \text{ ksi} = 32.4 \text{ ksi}$$

Calculating the design capacity of the section, $\phi_c\,P_n$, we find:

$$\phi_c\,P_n = (\phi_c\,A_g)\,F_{cr}$$

$$\phi_c\,P_n = (0.85 \times 23.2 \text{ in.}^2)\,32.4 \text{ ksi} = 638.9 \text{ kips}$$

16.8   Column Design Using LRFD

The design problems are a little more difficult than the evaluation problems. The design procedure outlined herein proceeds on the assumption that the student does not have access to the many design tables that are available. This initial design procedure is a trial and error approach, but is found to be very quick using LRFD. If the reader realizes the relationship between critical stress, F_{cr}, and the slenderness parameter, λ_c, is inversely proportional—then the process can begin by assuming a reasonable value for design stress, $\phi_c\,F_{cr}$. Typically, this value may be approximately 25–30 ksi, tending to be larger when the slenderness parameter is small and vice versa. Once a value of design stress is assumed, a trial section can be selected, based on the following formula:

$$A_{g\text{ trial}} = P_u/\phi_c\,F_{cr}$$

This trial section can then be checked for its design capacity, $\phi_c\,P_n$, relative to the applied factored load on the column, P_u. If the design capacity is equal to or just a bit larger than the required (factored) capacity, then the trial section is good. If the trial section's design capacity is smaller than or much greater than the required capacity, then the section should be resized accordingly. The step by step process might be listed as follows:

1. Assume a design stress, $\phi_c\,F_{cr}$, typically around 25–30 ksi
2. Using an assumed $\phi_c\,F_{cr}$, calculated $A_{g\text{ trial}}$
3. Choose trial member based on $A_{g\text{ trial}}$
4. Calculate F_{cr} of trial member & compare $\phi_c\,P_n$ to P_u
5. If $\phi_c\,P_n = P_u$ or just a bit larger, the section is perfect
   If $\phi_c\,F_{cr}\,A_g >>> P_u$, choose a smaller section
   If $\phi_c\,F_{cr}\,A_g < P_u$, choose a larger section

To aid the designer in the efficient utilization of the aforementioned equations (Equations 16-2 and 16-3), the LRFD manual provides many different design charts. One such helpful chart is shown in Table 16-2. This table lists design

Table 16-2 Design Stresses versus Slenderness for A36 Steel-LRFD

$\dfrac{Kl}{r}$	$\phi_c F_{cr}$ (ksi)	$\dfrac{Kl}{r}$	$\phi_c F_{cr}$ (ksi)	$\dfrac{Kl}{r}$	$\phi_c F_{cr}$ (ksi)	$\dfrac{Kl}{r}$	$\phi_c F_{cr}$ (ksi)	$\dfrac{Kl}{r}$	$\phi_c F_{cr}$ (ksi)
1	30.60	41	28.01	81	21.66	121	14.16	161	8.23
2	30.59	42	27.89	82	21.48	122	13.98	162	8.13
3	30.59	43	27.76	83	21.29	123	13.80	163	8.03
4	30.57	44	27.64	84	21.11	124	13.62	164	7.93
5	30.56	45	27.51	85	20.92	125	13.44	165	7.84
6	30.54	46	27.37	86	20.73	126	13.27	166	7.74
7	30.52	47	27.24	87	20.54	127	13.09	167	7.65
8	30.50	48	27.11	88	20.36	128	12.92	168	7.56
9	30.47	49	26.97	89	20.17	129	12.74	169	7.47
10	30.44	50	26.83	90	19.98	130	12.57	170	7.38
11	30.41	51	26.68	91	19.79	131	12.40	171	7.30
12	30.37	52	26.54	92	19.60	132	12.23	172	7.21
13	30.33	53	26.39	93	19.41	133	12.06	173	7.13
14	30.29	54	26.25	94	19.22	134	11.88	174	7.05
15	30.24	55	26.10	95	19.03	135	11.71	175	6.97
16	30.19	56	25.94	96	18.84	136	11.54	176	6.89
17	30.14	57	25.79	97	18.65	137	11.37	177	6.81
18	30.08	58	25.63	98	18.46	138	11.20	178	6.73
19	30.02	59	25.48	99	18.27	139	11.04	179	6.66
20	29.96	60	25.32	100	18.08	140	10.89	180	6.59
21	29.90	61	25.16	101	17.89	141	10.73	181	6.51
22	29.83	62	24.99	102	17.70	142	10.58	182	6.44
23	29.76	63	24.83	103	17.51	143	10.43	183	6.37
24	29.69	64	24.67	104	17.32	144	10.29	184	6.30
25	29.61	65	24.50	105	17.13	145	10.15	185	6.23
26	29.53	66	24.33	106	16.94	146	10.01	186	6.17
27	29.45	67	24.16	107	16.75	147	9.87	187	6.10
28	29.36	68	23.99	108	16.56	148	9.74	188	6.04
29	29.28	69	23.82	109	16.37	149	9.61	189	5.97
30	29.18	70	23.64	110	16.19	150	9.48	190	5.91
31	29.09	71	23.47	111	16.00	151	9.36	191	5.85
32	28.99	72	23.29	112	15.81	152	9.231	192	5.79
33	28.90	73	23.12	113	15.63	153	9.11	193	5.73
34	28.79	74	22.94	114	15.44	154	9.00	194	5.67
35	28.69	75	22.76	115	15.26	155	8.88	195	5.61
36	28.58	76	22.58	116	15.07	156	8.77	196	5.55
37	28.47	77	22.40	117	14.89	157	8.66	197	5.50
38	28.36	78	22.22	118	14.70	158	8.55	198	5.44
39	28.25	79	22.03	119	14.52	159	8.44	199	5.39
40	28.13	80	21.85	120	14.34	160	8.33	200	5.33

stresses ($\phi_c F_{cr}$) for varying slenderness ratios. By simply multiplying the tabular value by the section's area, the designer can quickly calculate the column's design capacity ($\phi_c P_n$). The following examples will illustrate the design of compression members, using LRFD.

EXAMPLE 16.6

Find the most economical W 12 section to hold a compressive factored load of 168 kips. Steel is A36, $K = 1.0$, $l = 16$ feet.

Since the effective length, Kl, is 16 ft., we will start by assuming a design stress of 25 ksi. The trial area can then be calculated as follows:

$$A_{g\,trial} = 168 \text{ kips}/25 \text{ ksi} = 6.72 \text{ in.}^2$$

Therefore, we would tend to select a W 12×26 ($A = 7.65$ in.^{2}) as the trial member. From Appendix A, the properties for the W 12×26 can be found and from here we can calculate the section's slenderness ration as:

$$Kl/r_y = (1)(16 \text{ ft.} \times 12)/1.51 \text{ in.} = 127.1$$

Using Equation 16-1 the slenderness parameter can then be calculated as follows:

$$\lambda_c = (Kl/r\pi) \sqrt{F_y/E} \qquad \text{(Eq. 16-1)}$$

$$\lambda_c = (127.1/\pi) \sqrt{36{,}000/29 \times 10^6} = 1.42 < 1.5$$

Since the slenderness parameter is again less than 1.5, inelastic behavior controls and Equation 16-3 is used to find the critical buckling stress.

$$F_{cr} = (.658^{\lambda_c^2}) F_y \qquad \text{(Eq. 16-3)}$$

$$F_{cr} = (.658^{1.42^2}) \, 36 \text{ ksi} = 15.4 \text{ ksi}$$

Calculating the design capacity of the W 12×26 section, $\phi_c P_n$, we find:

$$\phi_c P_n = \phi_c A_g F_{cr}$$

$$= .85 \times 7.65 \text{ in.}^2 \times 15.4 \text{ ksi} = 100.1 \text{ kips}$$

Since the design capacity of this section is less than the required capacity, we need a bigger section.

From Appendix A, try W 12×40 ($A = 11.8$ in.^{2} and $r_y = 1.93$ in.).

Calculating this section's controlling slenderness ratio as follows:

$$Kl/r_y = (1)(16 \text{ ft.} \times 12)/1.93 \text{ in.} = 99.5$$

Using Equation 16-1 the slenderness parameter can then be calculated as follows:

$$\lambda_c = (Kl/r\pi)\sqrt{F_y/E}$$

$$\lambda_c = (99.5/\pi)\sqrt{36{,}000/29 \times 10^6} = 1.116 < 1.5$$

Since the slenderness parameter is again less than 1.5, inelastic behavior controls and Equation 16-3 is used to find the critical buckling stress.

$$F_{cr} = (.658^{\lambda_c^2})F_y \qquad\qquad\qquad (\text{Eq. 16-3})$$

$$F_{cr} = (.658^{1.116^2})\ 36\ \text{ksi} = 21.37\ \text{ksi}$$

Calculating the design capacity of the W 12×40 section, $\phi_c P_n$, we find:

$$\phi_c P_n = \phi_c A_g F_{cr}$$

$$= .85 \times 11.8\ \text{in.}^2 \times 21.37\ \text{ksi} = 214.3\ \text{kips}$$

Since the W 12×40 has a design capacity greater than the required capacity it works. At this point the reader might decide that a section smaller than a W 12×40 might work since there seems to be a sizeable gap between the required strength (168 kips) and the design strength. Try a W 12×35—you'll find this will not work, and therefore the W 12×40 is most economical.

16.9   Column Design Using ASD

As discussed in previous chapters, there are two general types of problems encountered in the design arena—evaluation and design. In the evaluation problem, you will be given a member and corresponding load—and then asked to determine its adequacy, or to determine how much load it can safely withstand per the specification. For the design problem, you will be given loads and asked to design the member size, based on your knowledge of the applicable specification criteria using Allowable Stress Design philosophy.

The realization that the allowable stress, F_a, is dependent only on the slenderness ratio of a column, is the key point that must be grasped by the student in column design. As the student becomes more experienced in column design and behavior, he will be able to estimate values of allowable stress, based only on familiarity with a general range of slenderness ratio. A step-by-step procedure for column design may be outlined as follows:

1. Estimate a value of allowable stress, $F_{a\ est}$, based on the controlling slenderness ratio. (The accuracy of this step depends greatly on the designer's experience).
2. Using $F_{a\ est}$, calculate the required area as $A_{req'd} = P/F_{a\ est}$.
3. Choose a trial member based on $A_{req'd}$. Realize that $A_{req'd}$ is only a "ballpark" area, which is only as good as your estimate of allowable stress.
4. Calculate the allowable stress, F_a, of the trial member, based on its controlling slenderness ratio. Use Equation 16-5 or 16-6 (use Table 16-1 when $Kl/r < C_c$, which is much faster).

5. Calculate the allowable load, P_{all}, of the trial member as follows:

$$P_{all} = A_{\text{trial member}} \times F_a$$

6. If the allowable load of the trial member is equal to or just a little greater than the actual load, then the member is good. If the allowable load of the trial member is less than the actual load, then the member is no good. If the allowable load of the trial member is much greater than the actual load, the member works but there is probably a smaller and more economical member available.

The following example will introduce the student to this design procedure.

EXAMPLE 16.7

Find the most economical W 12 section to hold a compressive load of 120 kips. Steel is A36, $K = 1.0$, $l = 20$ ft. (for both axes). The column is unbraced.

Since the column is unbraced, we know that the weak axis slenderness ratio controls the design. With an effective length, Kl, equal to 1.0 × (20 ft. × 12 in./ft.) = 240 in., let's take a look at the r_y values for W 12 sections in Appendix A. Notice, except for the very small sections, most r_y values range from ~ 2 to ~ 3.5. If we assumed an average r_y value of 2.75, that would make our slenderness ratio ~ 240 in./2.75 in. ~ 87. This value shows that this column will most probably be controlled by inelastic buckling, since 87 < C_c (126.1 for A36 steel, by Equation 16-4). With this assumed slenderness ratio in the inelastic region, start the design process by estimating a value of allowable stress of 15 ksi.

1. Estimate $F_a = 15$ ksi
2. $A_{req'd} = 120$ kips/15 ksi = 8 in.^{2}
3. From tables in Appendix A, choose a trial member of W 12 × 30 (A = 8.79 in.^{2}).
4. Calculate the allowable stress, F_a, for the trial section of W 12 × 30. Since a W 12 × 30 has an r_y of 1.52 in., the slenderness ratio is calculated as:

$$Kl/r_y = 1.0(20\text{ft.})(12 \text{ in./ft.})/1.52 \text{ in.} = 157.9$$

157.9 > C_c so use Equation 16-6 to calculate F_a

Find $F_a = 5.99$ ksi

5. The allowable load, P_{all}, for this W 12 × 30 is:

$$P_{all} = (5.99 \text{ ksi})(8.79 \text{ in.}^2) = 52.6 \text{ kips}$$

6. Since allowable load of 52.6 kips < 120 kips this section does not work. Our original estimate of F_a was probably too high. Start again with $F_{a\,est} = 10$ ksi

1. $F_{a\ est} = 10$ ksi
2. $A_{req'd} = 120$ kips$/10$ ksi $= 12$ in.^{2}
3. Choose W 12×45 (A $=13.2$ in.^{2}) as the new trial member
4. Calculate the allowable stress, F_a, for this W 12×45. The controlling slenderness ratio, Kl/r_y, equals $= 123.7$.which is less than C_c. Therefore, by using Equation 16-5 or Table 16-1, we find $F_a = 9.75$ ksi.
5. The allowable load for this trial member is 9.75 ksi $\times 13.2$ in.$^2 = 128.7$ kips.
6. Since the allowable load (128.7 kips) is greater than the actual load (120 kips) the member works. Since the allowable load is only a small percentage over the actual, it practically guarantees that this is the smallest section (most economical) that works!!

(If there should be a doubt as to whether the trial section is indeed the most economical, the student should calculate the allowable load for the next smallest section in the table and check its capacity as well.)

16.10 Summary

Columns, and compression members in general, are some of the most important elements in the scope of structural design. Failure of such members primarily occurs due to an out-of-plane bending, which is referred to as buckling. Buckling of a column is very dependent on the length of the member, which is quantified in design by the concept of the slenderness parameter, λ_c or the slenderness ratio. Another important element contained in the slenderness parameter is the effective length factor, K. This factor takes into account the column's end-restraint affect and is used to estimate the actual buckling length.

The AISC specification considers two potential types of buckling behavior for steel columns—elastic and inelastic. When columns are shorter (and less slender) they will tend to fail due to inelastic buckling, and therefore the AISC specification will use a larger critical or allowable stress. Conversely, as a column becomes longer it will be more likely to fail due to elastic buckling, and therefore proper use of the specifications will lead to a smaller critical stress.

EXERCISES

1. Given a column with a rectangular cross-section of 6 in. $\times$ 10 in., calculate the critical buckling load by Euler's equation. The column is 20 ft. long with a modulus of elasticity of 3×10^6 psi.
2. Repeat problem #1 using a column with a circular cross section measuring 8 in. in diameter.
3. For a given column, Euler's equation is considered accurate up to a stress level of 15000 psi. If the material has a modulus of elasticity of 10×10^6

psi, calculate the smallest value of slenderness ratio for which the equation is still applicable.

4. A W 12 × 50 steel column has a length of 38 feet. Determine the critical Euler buckling stress if $E = 29{,}000{,}000$ psi. The ends are considered pinned. How would this compare to Equation 16-6 if the steel were A36. What is the difference?

5. Given an unbraced W 10 × 49 column 20 ft. long, pinned at both ends, calculate the ultimate factored load, P_u. Steel is A242 ($F_y = 50{,}000$ psi).

6. An unbraced W 12 × 79 is 15 ft. long and carries a factored load of 208 kips. If the $K = 1.0$, and the column is made from A36 steel, determine if it is adequate using the ASD and LRFD methods.

7. Design the most economical W 12 section to carry a concentric factored axial load of 375 Kips. Column is pin-fix at its ends and is unbraced. Use the LRFD method. Steel is A36 and the column is 20 ft. long.

8. Design the most economical W10 section that is to hold a factored load of 240 kips, if it is 13 ft. long and unbraced with a $K = 1.0$. Use the LRFD method. Steel is A242 ($F_y = 50$ ksi)

9. Rework problem #8 using a column that is braced at mid height in the weak axis.

10. Calculate the utimate load, P_u, on a W 12 × 40 column that is 10 ft. long with pinned end conditions, if it is braced at midheight in the strong direction. Steel is A36.

11. Design an unbraced column from a rectangular steel tube to hold a factored axial load of 400 kips. The column is 15 ft. long with pinned end conditions.Use the LRFD method. Steel has an $F_y = 46$ ksi.

12. Given an unbraced W 360 × 196 which is 8 m long, calculate its design capacity, $\phi_c\, P_n$. The column ends are assumed to be fixed and the steel is A36.

13. A W 310 × 179 is 10 m long and is braced along its strong axis at the third points. If all the column ends are assumed to be pinned, what is the ultimate factored load (P_u) which this column can carry? Steel is A36.

14. Design an economical column that is 7.5 m long and unbraced using the LRFD method. The column has pinned ends and carries a factored loads of 1250 KN. Steel is A36.

15. An unbraced W 12 × 79 is 15 ft. long and carries a load of 149 kips. If the $K = 1.0$, and the column is made from A36 steel, determine if it is adequate, using the ASD method.

16. An unbraced W 14 × 90 is 19 ft. long and carries a load of 280 kips. If the $K = 1.0$, and the column is made from A36 steel, determine if it is adequate using the ASD method.

17. Design the most economical W 12 section to carry a concentric axial load of 375 Kips using the ASD method. Column is pin-fix at its ends and is unbraced. Steel is A36. Column is 20 ft. long.

18. Design the most economical W 10 section to hold a load of 240 kips, if it is 13 ft. long using the ASD method. The column is unbraced, with a $K = 1.0$. Steel is A242 ($F_y = 50$ ksi)

19. Rework problem 17 if the column is braced at mid height in the weak axis.

20. Calculate the allowable load, P_{all}, on a W 12 × 40 column that is 10 ft. long with pinned end conditions, if it is braced at midheight in the strong direction.Steel is A36.

21. Calculate the allowable load, P_{all}, on a W 310 × 179 column which is 7 m long with pinned end conditions, if it is braced at midheight in the weak direction. Steel is A36.

22. If the column in the previous example is braced at midheight in the strong direction, how does the allowable load change?

23. Design the most economical W 360 section to hold a load of 2000 KN, if it is 7.5 m long using the ASD method. The column is unbraced, with a $K = 1.0$. Steel is A36.

REFERENCES

1. Euler, L., with English translation by J. A. Van den Broek, "Euler's Classic Paper 'On the Strength of Columns'", *American Journal of Physics*, January 1947, p.309-18.

2. McCormac, Jack, *Structural Steel Design*, Harper and Row, New York, 1981, p.84.

3. Dowling, Patrick, Knowles, Peter, and Owens, Graham, *Structural Steel Design*, The Steel Construction Institute, London, 1988, p.156.

CHAPTER

17

CONNECTIONS

17.1 Introduction to Bolted Connections

The need to fasten structural members together exists with all types of materials. Timber and steel members are the primary focus of structural fasteners and connection design. This chapter will focus solely on the requirements dictating the connection of structural steel members, although the general philosophy of design can be extrapolated to timber connections as well. The requirements shown in this chapter regarding steel design are based upon the American Institute of Steel Construction's (AISC) *Manual of Steel Construction*.[1,2] For requirements concerning timber connections, the student is referred to the American Institute of Timber Construction's (AITC) *Timber Construction Manual*.[3]

The need to join steel members together has existed since the introduction of steel as a building material, in the latter half of the nineteenth century. Steel structures of every type require the fastening of individual members to achieve their ultimate structural form (Figure 17-1). Two of the most common methods of connecting steel structures—high strength bolting and welding—will be discussed in this chapter.

The fastening of steel members by bolting is very common in steel construction today. However, prior to the 1960's the primary method of connecting steel members was riveting. Riveting consisted of heating a "slug" of steel to approximately 1800°F and inserting this slug into holes that joined the members together. Once in the hole, the ends of the soft slug were formed by using a pneumatic hammer which would shape the ends of the rivet while also filling up more of the original hole. Many end shapes for rivets were utilized, as shown in Figure 17-2, but by far the most common is the "buttonhead" shape.

Figure 17-1   Use of Fasteners in Structural Steel Building Frames. Sheraton Hotel, Philadelphia. Courtesy Bethlehem Steel Corporation.

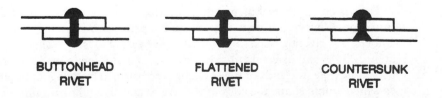

| BUTTONHEAD RIVET | FLATTENED RIVET | COUNTERSUNK RIVET |

Figure 17-2   Types of Rivets

Rivets have virtually been replaced in today's world with the advent of high strength bolting, due to economic considerations. Riveting is a labor intensive operation requiring a crew of approximately four to five skilled people and is a slower process in terms of installation. High strength bolting is much quicker, requiring a much smaller two-person crew that does not have to be as skilled.

High strength bolting is one of the most common procedures used today in the connection of structural steel members. High strength bolts are made from medium carbon or alloy steel and have very large tensile strengths. Besides the economic advantages that bolting enjoys, it also has other important advantages—such as higher fatigue strength and easier retrofitting ability.

The two most common high strength bolts are an A325 bolt and an A490 bolt. The A325 bolt is the most common high strength bolt used today and is made from heat-treated medium carbon steel. The A490 bolt is a higher strength bolt, manufactured from an alloy steel and is used in situations requiring its improved properties.

The proper installation of bolts in some connections requires that the bolts be "fully tensioned," while in other connections, the bolts need only be "snug-tight". The snug-tight condition will bring the plates being joined into firm contact and is achieved by a few impacts of an impact wrench or the complete effort of a worker using a spud wrench. A fully tensioned connection is based on achieving adequate clamping forces between the members being joined. Minimum bolt tension for fully tensioned A325 and A490 bolts is accomplished by using one of the four accepted methods that are outlined by the AISC's Specification for Structural Joints Using ASTM A325 or A490 Bolts.[4] These methods are briefly outlined below.

Turn-of-the-Nut Method

This method installs bolts by first placing them in a snug-tight condition and then turning the nut by the rotation listed in Table 17-1. During this operation there must be no rotation of the part that is not being turned by the wrench.

Calibrated Wrench Method

This method utilizes an automatic wrench that is calibrated to "stall" at a torque which produces the required tension. Proper calibration must occur on a daily basis and maintenance of the wrench is critical in achieving the desired results. Tightening should proceed from the most rigid part of the connection to the free edges in a systematic fashion. The initial bolts in the tightening sequence maybe "touched up" with the wrench later, since some relaxation may occur as adjacent bolts are tighened.

Direct Tension Indicators (DTI's)

This method installs bolts using a hardened washer with arched protrusions placed on its bottom surface. This washer is typically placed between the bolt

Table 17-1 Requirements for "Snug-tight" Condition. Courtesy of the American Institute of Steel Construction, Inc.

Disposition of Outer Fact of Bolted Parts			
Bolt length (Under side of head to end of bolt)	Both faces normal to bolt axis	One face normal to bolt axis and other sloped not more than 1:20 (beveled washer not used)	Both faces sloped not more than 1:20 from normal to the bolt axis (beveled washer not used)
Up to and including 4 diameters.	$^1/_3$ turn	$^1/_2$ turn	$^2/_3$ turn
Over 4 diameters but not exceeding 8 dia.	½ turn	$^2/_3$ turn	$^5/_6$ turn
Over 8 diameters but not exceeding 12 dia.[c]	$^2/_3$ turn	$^5/_6$ turn	1 turn

[a]Nut rotation is relative to bolt regardless of the element (nut or bolt) being turned. For bolts installed by 1/2 turn and less, the tolerance should be plus or minus 30 degrees; for bolts installed by 2/3 turn and more, the tolerance should be plus or minus 45 degrees.

[b]Applicable only to connections in which all material within the grip of the bolt is steel.

[c]No research has been performed by the Council to establish the turn-of-nut procedure for bolt lengths exceeding 12 diameters. Therefore, the required rotation must be determined by actual test in a suitable tension measuring device which simulates conditions of solidly fitted steel.

head and the edge of a connection plate. These protrusions form a gap, and as the bolt is tightened, these protrusions become flattened. By measuring the gap between the bolt head and the washer (or between the plate and the washer), one can assess the correct tension placed on the bolt (Figure 17-3). Research[5] has shown that DTI's are very reliable in achieving proper bolt tension and effectively inhibit the loss of pretension which may occur over extended periods of time.[6]

Calibrated Bolt Assemblies

This method utilizes a specially calibrated assembly of bolts, nuts, and washers used to estimate the proper tension in the bolt. In this method, bolts having splined ends which extend beyond the threaded portion are tightened with a special wrench. This tightening causes the tip of the bolt to shear off, once the proper tension has been achieved (Figure 17-4).

A) Direct Tension Indicator (DTI).

B) DTI before tensioning.

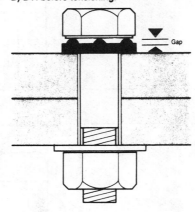

C) DTI after tensioning.

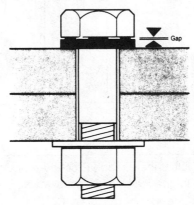

Figure 17-3 Direct Tension Indicators. Courtesy of J & M Turner, Inc.

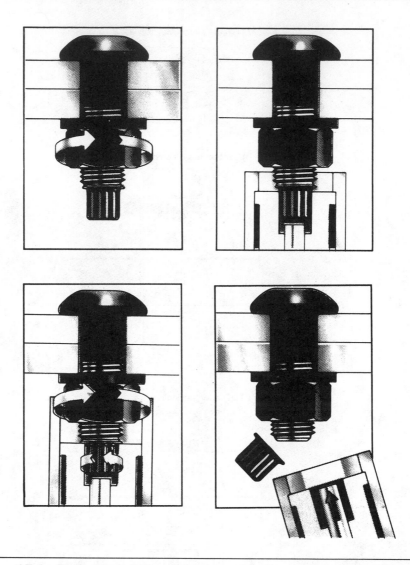

Figure 17-4   Typical Installation Procedure of Calibrated Bolt Assemblage, Rapid Tension Bolting System. Courtesy of NSS Industries.

17.2   Types of Bolted Connections

No matter what types of steel shapes are utilized, the connection of steel members almost always boils down to a "plate-to-plate" configuration. Therefore, the two most common configurations for connections are the "two-plate" and "three-plate"—sometimes referred to as the lap joint and butt joint (See Figure 17-5).

Although two-plate and three-plate systems might seem simplistic, their use is wide-spread throughout steel construction. If the student ponders the typical beam-to-column connection shown in Figure 17-6, they will notice that the angle-to-column flange connection is a two-plate system and the angle-to-beam web connection is a three-plate system.

One of the most important distinctions in any connection system is the number of shear planes that pass through a single bolt. In the two-plate system, the student will notice that as the plates are pulled, movement of the connection plates will attempt to rip the bolts apart along one shear plane, should slippage occur. This phenomena is referred to as **single shear.** Consequently, in the three-plate system, the bolts are trying to be ripped apart along two shear planes. This phenomena is referred to as **double shear.** The most important point of difference between single and double shear is the number of areas per bolt which are resisting stress. This will be a major focus when analyzing and designing bolted connections later in this chapter.

Another category of connection types revolves around the performance of the connection under loading. A connection can be designed as a **bearing type connection** or a **slip-critical type connection.** The basic difference between the two types is an assumption of slippage under service load that results in the use of different design strength values.

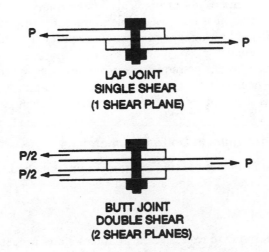

Figure 17-5   Common Types of Bolted and Riveted Connections.

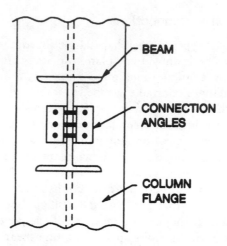

Figure 17-6 Typical Beam-Column Connection

The slip-critical type connection is assumed to be slip-free under all service load conditions and is always supposed to transfer load through the joint by the clamping forces generated between the connected plates. Therefore, the AISC specification requires that slip-critical connection be fully tensioned using one of the aforementioned procedures. This type of connection is primarily used for structures that have high impact load cases or where slippage in the joint is considered undesirable by the designer.

The bearing-type connection is assumed to slip only under very high load conditions and if this slippage occurred, the joint would transfer load through shear on the bolts and plate bearing. This type of connection is used when the designer feels the structure is less susceptible to impact, stress-reversals, or vibration. The installation of bearing type connections requires only that they be tightened to the snug-tight condition.

As the AISC design requirements are discussed in Section 17.4 and 17.5, it will become evident that the type of connection—two-plate versus three-plate and bearing versus slip-critical—is very important in the evaluation of bolted connections.

17.3 Modes of Failure in Bolted Connections

The possible modes of failure in the vicinity of bolted or riveted connections are shown in Figure 17-7 and are listed as follows:

Shearing of Bolts

Bolts break due to excessive shear forces along predetermined shearing planes. The typical number of shearing planes is either one (single shear) or two (double shear), although in special configurations there could be more.

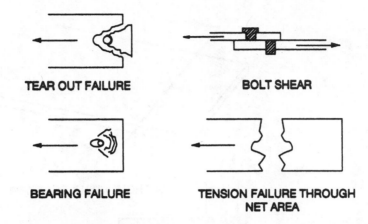

TEAR OUT FAILURE   **BOLT SHEAR**

BEARING FAILURE   **TENSION FAILURE THROUGH NET AREA**

Figure 17-7   Common Potential Failures in Bolted Connections

Plate Crushing by Bolt

Commonly referred to as bearing failure, where the bolt comes into contact with the edge of the bolt hole and causes a crushing type of failure of the plate material, this failure becomes more common as the connecting plates are thin (relative to the bolt size). Typically this does not contol the connection design.

Tear Out Failure of Plate

Ripping failure in the plate, due to excessive shear forces near the edge of a member, can be common if the plate is thin or bolts are close to the edge of the plate. This failure can easily be avoided by following minimum edge and center-to-center spacing requirements, which are given in the AISC specification. These will be discussed later.

Member Failure in Tension—Gross Area or Net Area

Failure in a tension may initiate in the member when the member is in tension. Such failure is not truly a "connection" failure, but should be considered along with connections, since it is closely related. Such failure may occur in two potential areas—the gross area and the net area.

Gross area refers to the full area of the member. Failure of a tension member over the gross area initiates by **yielding** of the plate material because stresses exceed the yield strength of the steel. Such a failure would be illustrated by excessive deformation of the member.

Net area refers to the section of the member through the bolt holes (Figure 17-8). This section has the bolt hole area subtracted from the full (gross) area. The net area can be calculated from the following equation:

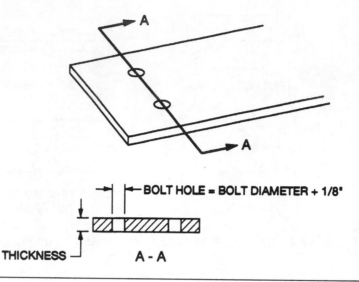

Figure 17-8 Illustration of Net Area in a Flat Plate

$$\text{Net Area} = \text{Gross Area} - [(\text{number of bolts}) \times (\text{diameter of bolt} + 1/8") \\ \times (\text{plate thickness})]$$

In this formula, the area of a bolt hole is considered to be its rectangular projection on the proposed failure plane. The AISC specification considers the width of the actual bolt hole to be 1/16 in. larger than the *nominal* diameter of the hole, which is 1/16 in. larger than the bolt. This makes the width of the bolt hole simply the bolt diameter plus 1/8 in. The failure at this location initiates by stresses around the bolt holes reaching very high levels, and therefore causing a fracture across the member through the bolt holes.

Of the potential failure modes listed previously, the primary mode of failure in bolted connections (and therefore the usual controlling feature of connection design) is either bolt shear or tension on the member. The shear resistance in most cases is assumed to be split evenly among all bolts in the connection, because we assume the plate to be rigid and nondeforming. This is the standard assumption by the average designer, although the approach is oversimplified, because the typical connection is flexible enough to allow unequal deformations resulting in lower stresses on the bolts in the center of the connection area.[7] The designer must keep in mind all possible failure mechanisms and design the members such that it is safe for all.

17.4   AISC Requirements for Bolted Connections - ASD

The allowable stress design (ASD) philosophy is a design method that strives to keep the actual stress in a member below the allowable (safe) stress level, as dictated in a relevant code or specification. This section will outline the general requirements for bolted steel connections outlined in the AISC specifications.[1]

The primary modes of failure—as outlined in section 17.3—are shear on the bolts, bearing on the plates, and tension across the gross and the net areas. The actual stress must be less than or equal to the particular value of allowable stress in each given case, in order for the connection to be adequate. The AISC requirements for allowable stress design will be outlined in the following paragraphs.

Shear on the Bolts

Shear on the bolts is assumed to be evenly distributed to each bolt in a group, with the actual stress on any bolt simply being calculated as:

$$f_v = V / (n \times A) \qquad \text{(Eq. 17-1)}$$

where

f_v = actual shear stress

V = total shearing force or load

n = number of bolts

A = individual area per bolt resisting the shear

It should be noted that the value of A will be dependent on the number of shear planes present within the particular bolt (single shear or double shear). The values of allowable shear stress are given in Figure 17-9.

Bearing on the Plates

The next potential mode of failure is the crushing of the plate by the bolt bearing on it. This failure mechanism is relatively rare unless very thin plates are used. Still, it must be checked as part of any analysis or design. The bearing stress of a bolt on a plate is given again by the direct stress formula as follows:

$$f_b = P / (n \times A) \qquad \text{(Eq. 17-2)}$$

where

f_b = actual bearing stress

P = total load

n = number of bolt areas making contact with the plate

A = bearing area = (diameter of bolt) $\times$ (thickness of plate)

Description of Fasteners	Allowable Tension[g] (F_t)	Allowable Shear[g] (F_v)					
		Slip-critical Connections[e,j]					Bearing-type Connections[i]
		Standard size Holes	Oversized and Short-slotted Holes	Long-slotted holes			
				Transverse[j] Load	Parallel[j] Load		
A502, Gr. 1, hot-driven rivets	23.0[a]						17.5[f]
A502, Gr. 2 and 3, hot-driven rivets	29.0[a]						22.0[f]
A307 bolts	20.0[a]						10.0[b,f]
Threaded parts meeting the requirements of Sects. A3.1 and A3.4 and A449 bolts meeting the requirements of Sect. A3.4, when threads are not excluded from shear planes	0.33F_u[a,c,h]						0.17F_u[h]
Threaded parts meeting the requirements of Sects. A3.1 and A3.4, and A449 bolts meeting the requirements of Sect. A3.4, when threads are excluded from shear planes	0.33F_u[a,h]						0.22F_u[h]
A325 bolts, when threads are not excluded from shear planes	44.0[d]	17.0	15.0	12.0	10.0		21.0[f]
A325 bolts, when threads are excluded from shear planes	44.0[d]	17.0	15.0	12.0	10.0		30.0[f]
A490 bolts, when threads are not excluded from shear planes	54.0[d]	21.0	18.0	15.0	13.0		28.0[f]
A490 bolts, when threads are excluded from shear planes	54.0[d]	21.0	18.0	15.0	13.0		40.0[f]

[a]Static loading only.

[b]Threads permitted in shear planes.

[c]The tensile capacity of the threaded portion of an upset rod, based upon the cross-sectional area at its major thread diameter A_b shall be larger than the nominal body area of the rod before upsetting times 0.60F_y.

[d]For A325 and A490 bolts subject to tensile fatigue loading, see Appendix K4.3.

[e]Class A (slip coefficient 0.33). Clean mill scale and blast-cleaned surfaces with Class A coatings. When specified by the designer, the allowable shear stress, F_v, for slip-critical connections having special faying surface conditions may be increased to the applicable value given in the RCSC Specification.

[f]When bearing-type connections used to splice tension members have a fastener pattern whose length, measured parallel to the line of force, exceeds 50 in., tabulated values shall be reduced by 20%.

[g]See Sect. A5.2

[h]See Table 2, Numerical Values Section for values for specific ASTM steel specifications.

[i]For limitations on use of oversized and slotted holes, see Sect. J3.2.

[j]Direction of load application relative to long axis of slot.

Figure 17-9 Allowable Stresses on Fasteners-ASD (Courtesy American Institue of Steel Construction)

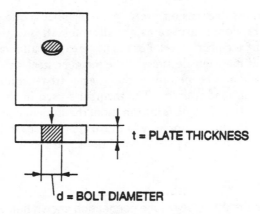

Figure 17-10   Bearing Area of Bolt in Bolt Hole

The resisting area is the projection of the contact area of the bolt on the plate. This area, as noted, is the thickness of the plate multiplied by the diameter of the bolt. (See Figure 17-10)

The allowable bearing stress requirements from the AISC specification are based on proper distances between bolts and edge spacings. These can fall into the following categories. Where deformation is not a consideration, adequate bolt spacing exists (usually $3 \times$ bolt diameter) and edge distance (usually $1.5 \times$ bolt diameter) requirements are met; the allowable bearing stress is $1.5 \, F_u$, where F_u is the specified minimum ultimate tensile strength of the connected parts. For standard holes, where there are more than two bolts in the line of stress (and where deformation is a consideration), the allowable bearing stress is given as $1.2 \, F_u$. When proper bolt and edge spacings are not met, refer to the allowable stress requirements found in Section J3.7 of the AISC specification[2].

Tension on the Member

Tension stresses on the member must be checked relative to both the gross and the net areas. As discussed previously, the gross area is the full cross-sectional area of the member—while the net area is typically a perpendicular cross-section of the member through the bolt holes. The actual tension stress is calculated as follows:

$$f_t = P \, / \, A \qquad\qquad \text{(Eq. 17-3)}$$

where

f_t = actual tension stress

P = load or force on the member

A = gross area or net area

The allowable tension stress on the member is different for the gross area and the net area. For the gross area, the allowable stress is $0.60F_y$, where F_y is the yield stress of the member's steel. For the net area, the allowable stress is $0.50F_u$, where F_u is the ultimate tensile stress of the member steel.

It should also be mentioned that the net area is sometimes reduced due to ineffectiveness in load transfer. The member's used in this chapter will not include such a reduction. For more information, the student can refer to many references.[9]

The following examples will illustrate the use of these requirements in both analyzing and designing steel connections, per the ASD method.

EXAMPLE 17.1

Determine the adequacy of the connection shown below. The bolts are A325 with threads not excluded from the shear plane (A325-N) and the steel plates are made from A36 steel ($F_y = 36$ ksi, $F_u = 58$ ksi). The applied load is 75 kips and the connection is bearing-type.

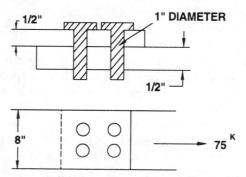

The items necessary to check—as far as the connection is concerned—are shear, bearing, and tension. For this problem, the bolt spacing and edge spacing layout are adequate and deformation is not a consideration. This is an evaluation type of problem and will evaluate by comparing actual stress to the allowable stress.

Shear

In the two-plate, single shear condition we have one shear plane in each bolt, therefore:

$$n \times A = 4 \text{ bolts} \times \pi/4(1\text{in.})^2 = 3.14 \text{ in.}^2$$

Allowable stress = 21 ksi (per Figure 17-9 for A325 bolts with the threads in the shear plane, bearing-type connection)

Calculate the actual stress:

$$f_v = 75 \text{ kips} / 3.14 \text{ in.}^2 = 23.88 \text{ ksi}$$

Compare this to the allowable shear stress:

$$23.88 \text{ ksi} > 21 \text{ ksi}$$

Since the actual shear stress exceeds the allowable, the connection is inadequate relative to shear. However, we will continue to check bearing and tension (always) even though shear failed to detect further inadequacies.

Bearing

Area of bearing;

$$n \times A = 4 \text{ bolts} \times (1/2 \text{ in.} \times 1 \text{ in.}) = 2 \text{ in.}^2$$

The allowable stress in bearing (considering no deformation and spacings are adequate) is:

$$\text{Allowable bearing stress, } F_b = 1.5 \ (58\text{ksi}) = 87 \text{ ksi}$$

Calculate actual bearing stress as follows:

$$f_b = 75 \text{ kips} / 2 \text{ in.}^2 = 37.5 \text{ ksi}$$

Compare actual stress to allowable stress:

$$37.5 \text{ ksi} < 87 \text{ ksi}$$

Since the actual bearing stress is less than the allowable, the connection is adequate, relative to plate bearing.

Tension

$$\text{Gross area} = 8 \text{ in.} \times 1/2 \text{ in.} = 4 \text{ in.}^2$$

Calculate actual tension stress:

$$f_t = 75 \text{ kips} / 4 \text{ in.}^2 = 18.75 \text{ ksi}$$

Allowable tension stress over gross area $= 0.60 F_y = 0.60(36 \text{ ksi}) = 21.6 \text{ ksi}$

Compare actual stress to allowable over the gross area:

$$18.75 \text{ ksi} < 21.6 \text{ ksi}$$

Since the actual tensile stress over the gross area is less than the allowable, the connection is adequate with respect to the gross area. Check the net area as follows:

$$\text{Net area} = \text{Gross} - [(\text{number of bolts}) \times (\text{diameter of bolt} + 1/8") \\ \times (\text{plate thickness})]$$

$$= 4 \text{ in.}^2 - [2 \text{ bolts}(1 \ 1/8 \text{ in.})(1/2 \text{ in.})] = 2.875 \text{ in.}^2$$

Calculate actual tension stress over net area:

$$f_t = 75 \text{ kips} / 2.875 \text{ in.}^2 = 26.1 \text{ ksi}$$

Allowable net area tension stress = 0.50 F_u = 0.50(58 ksi) = 29 ksi
Compare actual stress to allowable over the net area

26.1 ksi < 29 ksi — therefore the connection is adequate over the net area

The shear on the bolts is the only mechanism that failed in this connection. One solution would be to use more bolts. However, this may affect the other pieces of the design, most notably tension in the net area.

EXAMPLE 17.2

The connection to be made in the lap-joint shown below is to withstand a tensile force of 100 kips. If the connection is a slip-critical type and the steel is A36 (F_y = 36000 psi and F_u = 58000 psi) and bolts are A325-X (threads not in the shear plane), design the bolt size and number of bolts to be used.

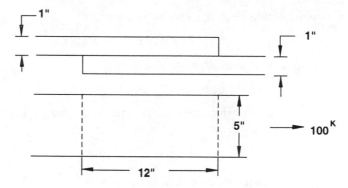

MAXIMUM CONNECTION LENGTH

Calculate bolt area required, based on the shear and bearing requirements. Afterwards, check the adequacy of the tension stresses over the gross and net areas.

Shear

$$P = 100 \text{ kips}$$

Allowable shear stress = 17.0 ksi (per Figure 17-9)

$$A_{req'd} = 100 \text{ kips}/17.0 \text{ ksi} = 5.88 \text{ in.}^2 \text{ (Try 1 in. diameter bolts)}$$

$$A_{\text{1" diamter bolt}} = \pi(d)^2 / 4 = .785 \text{ in.}^2/\text{bolt}$$

$$A_{req'd} = n \times A_{\text{1" diamter bolt}}$$

$$5.88 \text{ in.} = n \times (.785 \text{ in.} / \text{bolt})$$

$$n = 7.49 \text{ bolts}$$

Therefore, use 8 bolts. This is probably a wise choice because you could lay out the bolts as shown below, which maintains adequate bolt spacing. A smaller diameter of bolt would lead to a larger number of bolts and a minimum spacing of bolts might be in jeopardy.

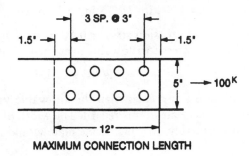

MAXIMUM CONNECTION LENGTH

Bearing

Check bearing area using 1 in. diameter bolts to ensure its adequacy, assuming all spacings are adequate and that deformation is not considered.

$$P = 100 \text{ kips}$$

Allowable bearing stress, $F_b = 1.5 \ F_u = 1.5(58 \text{ ksi}) = 87 \text{ ksi}$

$$A_{req'd} = 100 \text{ kips } /87 \text{ ksi} = 1.15 \text{in.}^2$$

$$A_{\text{bearing 1" diamter bolt}} = 1 \text{ in.} \times 1 \text{ in.} = 1 \text{ in.}^2 \text{ per bolt}$$

$$A_{req'd} = n \times A_{\text{bearing 1" diamter bolt}}$$

$$1.15 \text{ in.}^2 = n \times 1 \text{ in.}^2$$

$$n = 1.15 \text{ bolts needed for bearing}$$

Since 8 bolts are used for shear, this is adequate

Tension

$$\text{Gross area} = 5 \text{ in.} \times 1 \text{ in.} = 5 \text{ in.}^2$$

Calculate actual tension stress:

$$f_t = 100 \text{ kips} / 5 \text{ in.}^2 = 20 \text{ ksi}$$

Allowable gross area tension stress = $0.60F_y = 0.60(36 \text{ ksi}) = 21.6 \text{ ksi}$

Compare actual stress to allowable over the gross area:

20 ksi < 21.6 ksi—therefore the stresses are adequate over the gross area

Net area = Gross − [(number of bolts) × (diameter of bolt + 1/8") × (plate thickness)]

$$= 5 \text{ in.}^2 - [2 \text{ bolts}(1 \ 1/8 \text{ in.})(1 \text{ in.})] = 2.75 \text{ in.}^2$$

Calculate actual tension stress in net area:

$$f_t = 100 \text{ kips} / 2.75 \text{ in.}^2 = 36.4 \text{ ksi}$$

Allowable net area tension stress = $0.50 \ F_u = 0.50(58 \text{ ksi}) = 29 \text{ ksi}$
Compare actual stress to allowable over the net area:

$$36.4 \text{ ksi} > 29 \text{ ksi}$$

Because the stress is greater than the allowable value over the net area, the connection is not adequate in this regard. The width of the plate would have to be increased, most probably. A-6-inch-wide plate would make the connection work adequately.

17.5   AISC Requirements for Bolted Connections - LRFD

The Load and Resistance Factor Design (LRFD) philosophy of steel design strives to keep the design capacity (ϕR_n) of a connection greater than the required (factored) strength (R_u). This can be shown as follows:

$$\phi R_n \geq R_u$$

The design capacity is typically the design strength (stress) multiplied by the product of resisting area and the appropriate reduction factor. The design strength for different connection behaviors will be discussed below. This section will outline the general requirements for bolted steel connections per the AISC specifications.

The primary modes of failure (as outlined in Section 17.3) are shear on the bolts, bearing on the plates, and tension across the gross and the net areas. The design capacity of each of these behaviors must equal, or exceed, the required or factored capacity. The required capacity (factored capacity , R_u) is the effect caused by using load factors on various loads or combinations of loads. These load factors are based on probabilistic studies of loads on structures, and therefore give a more rational and reliable design. The AISC requirements for LRFD connection design will be outlined in the following paragraphs.

Shear on the Bolts

Shear on the bolts is assumed to be evenly distributed to each bolt in a group with the design capacity (ϕR_n) of the bolt group in shear being calculated as:

$$\phi R_n = \phi \text{ (Nominal strength} \times A \times n) \qquad \text{(Eq.17-4)}$$

where

Nominal strength = value from Figure 17-11

ø = .75 for all bolts in bearing type connections

ø = 1.0 for all slip-critical connections (except when long-slotted holes are used, where ø = .85)

n = number of bolts

A = individual area per bolt resisting the shear

It should be noted that the value of A will be dependent on the number of shear planes present within the particular bolt, as discussed earlier in section 17.3 (single shear or double shear).

Bearing on the Plates

The next potential mode of failure is the crushing of the plate by the bolt bearing on it. This failure mechanism is relatively rare, unless very thin plates are used. Still, it must be checked as part of any analysis or design.The design bearing capacity (ø × nominal strength) of a bolt on a plate is given again by the direct stress formula, or as follows:

If deformation is not a consideration and adequate bolt spacing and edge distance requirements are met, the design capacity is as follows:

$$ø \, R_n = ø \, (3.0)(d)(t)F_u \qquad \text{(Eq. 17-5)}$$

where

F_u = minimum tensile strength of the plate

d = diameter of the bolt

t = thickness of the plate

ø = .75 (reduction factor for plate, not bolt)

If deformation is to be considered, adequate spacing criteria are met, and where there are more than two bolts in the line of stress the design capacity is given as follows:

For standard or short-slotted holes:

$$ø \, R_n = ø \, (2.4)(d)(t)F_u \qquad \text{(Eq. 17-6)}$$

For long-slotted holes perpendicular to the load:

$$ø \, R_n = ø \, (2.0)(d)(t)F_u \qquad \text{(Eq. 17-7)}$$

Description of Fasteners	Tensile Strength		Shear Strength in Bearing-type Connection	
	Resistance Factor φ	Nominal Strength, ksi	Resistance Factor φ	Nominal Strength, ksi
A307 bolts		45[a]		24 [b,e]
A325 bolts, when threads are not excluded from shear planes		90 [d]		48 [e]
A325 bolts, when threads are excluded from shear planes		90 [d]		60 [e]
A490 bolts, when threads are not excluded from shear planes		113[d]		60 [e]
A490 bolts, when threads are excluded from the shear planes	0.75	113[d]	0.75	75[e]
Threaded parts meeting the requirements of Sect. A3, when threads are not excluded from the shear planes		$0.75 F_\mu.$ [a,c]		$0.40 F_\mu$
Threaded parts meeting therequirements of Sect. A3, when threads are excluded from the shear planes		$0.75 F_\mu$ [a,c]		$0.50 F_\mu$ [a,c]
A502, Gr. 1, hot-driven rivets		45 [a]		25 [e]
A502, Gr. 2 & 3, hot-driven rivets.		60 [a]		33 [e]

[a] Static loading only.
[b] Threads permitted in shear planes.
[c] The nominal tensile strength of the threaded portion of an upset rod, based upon the cross-sectional area at its major thread diameter, A_D, shall be larger than the nominal body area of the rod before upsetting times F_y,
[d] For A325 and A490 bolts subject to tensile fatigue loading, see Appendix K3.
[e] When bearing-type connections used to splice tension members have a fastener pattern whose length, measured parallel to the line of force, exceeds 50 in., tabulated values shall be reduced by 20 Percent.

Figure 17-11 Design Strengths of Bolts and Rivets-LRFD. Courtesy of the American Institute of Steel Construction

where

$$F_u = \text{minimum tensile strength of the plate}$$

$$d = \text{diameter of the bolt}$$

$$t = \text{thickness of the plate}$$

$$\emptyset = .75 \text{ (reduction factor for plate, not bolt)}$$

For other cases regarding bearing, the nominal strength requirements can be found in Section J6 of the LRFD specification[1].

Tension on the Member

The design tension capacity $(\emptyset P_n)$ on the member must be checked relative to both the gross and the net areas. The design tension capacity must equal or exceed the required or factored tension (P_u) placed on the member. As discussed previously, the gross area is the full cross-sectional area of the member, while the net area is typically a perpendicular cross-section of the member through the bolt holes. The design tension capacity $(\emptyset_t P_n)$ on the gross area is calculated as follows:

$$\emptyset_t P_n = \emptyset_t (F_y A_g) \qquad \text{(Eq.17-8)}$$

where

$$\emptyset_t = .90$$

$$F_y = \text{steel yield stress}$$

$$A_g = \text{gross area}$$

The design capacity of Equation 17-8 considers the limit state for this behavior as yield stress of the particular steel being used. This is rather logical, since the mode of failure at this location is a yielding mechanism.

Net area is the cross-sectional area through the bolt holes and is given by the following equation:

Net area = Gross − [(number of bolts) × (diameter of bolt + 1/8 ") × (plate thickness)]

The design tesion capacity $(\emptyset_t P_n)$ on the net area is calculated as follows:

$$\emptyset_t P_n = \emptyset_t (F_u A_n) \qquad \text{(Eq. 17-9)}$$

where

$$\emptyset_t = .75$$

$$F_u = \text{steel's ultimate tensile stress}$$

$$A_n = \text{net area}$$

Notice that because the limit state is now a fracture type of behavior, the critical stress level is set to the steel's ultimate tensile strength. This is logical, since the limit state involves cracking or fracture-type mechanism.

The following examples will illustrate the use of these requirements in both analyzing and designing steel connections per the LRFD method.

EXAMPLE 17.3

Determine the adequacy of the connection shown below. The bolts are A325 with threads not excluded from the shear plane, and the steel plates are A36 (F_y = 36000 psi and F_u = 58000 psi). The factored load is 75 kips and the connection is bearing-type.

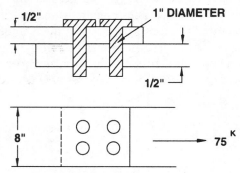

Remember the items necessary to check as far as the connection is concerned are shear, bearing, and tension. The bolt spacing and edge layout are adequate and deformation is not a concern. This is an evaluation type of problem and will evaluate by comparing required strength (factored load) to the connection's design capacity in shear, bearing and tension.

Shear

Since this is a two-plate system, single shear exists and therefore there is only one shear plane in each bolt. Therefore:

$$\text{Area} = (Area \text{ per bolt})(n)$$

$$= \pi/4(1 \text{ in.})^2 \times 4 \text{ bolts} = 3.14 \text{ in.}^2$$

From Figure 17-11, for A325 bolts in a bearing-type connection (with the threads not excluded from the shear plane), we find that the nominal strength of such a bolt is 48.0 ksi. This makes the design capacity in shear, using Equation 17-4 for this connection, as follows:

$$\text{ø } R_n = \text{ø (Design strength} \times A) \tag{Eq. 17-4}$$

$$\text{ø } R_n = (0.75)(48.0 \text{ ksi})(3.14 \text{in.}^2) = 113 \text{ kips}$$

Compare this to the factored load:

$$113 \text{ kips} > 75 \text{ kips}$$

Since the design capacity is greater than the required capacity, the connection is adequate in shear.

Bearing

Area of bearing surface:

$$\text{Area} = (n)(\text{Area per bolt}) = n \times (d \times t)$$

$$A = 4 \text{ bolts} \times (1/2 \text{in.} \times 1 \text{in.}) = 2 \text{ in.}^2$$

Calculate the design capacity in bearing (considering no deformation and spacings are adequate), by using Equation 17-5 as follows:

$$\emptyset \, R_n = \emptyset \, (3.0)(d)(t)F_u \qquad \text{(Eq. 17-5)}$$

$$\emptyset \, R_n = (.75)(3.0)(2 \text{in.}^2)(58 \text{ ksi}) = 261 \text{ kips}$$

Compare this to the factored load:

$$261 \text{ kips} > 75 \text{ kips}$$

Since the design capacity is greater than the required capacity, the connection is adequate in bearing.

Tension

The design capacity in tension for the member over the gross area can be calculated as follows:

$$A_g = 8 \text{ in.} \times 1/2 \text{ in.} = 4 \text{ in.}^2$$

$$\emptyset_t \, P_n = \emptyset_t \, (F_y \times A_g) \qquad \text{(Eq. 17-8)}$$

$$= .90 \, (36 \text{ ksi} \times 4 \text{ in.}^2) = 129.6 \text{ kips} > 75 \text{ kips—OK}$$

Since this capacity is greater than the factored load, the connection is adequate across the gross area.

The design capacity for the member across the net area can be found as follows:

$$A_n = 4 \text{ in.}^2 - [2(1 \text{ in.} + 1/8 \text{ in.})(1/2 \text{ in.})] = 2.875 \text{ in.}^2$$

$$\emptyset_t \, P_n = \emptyset_t \, (F_u \times A_n) \qquad \text{(Eq. 17-9)}$$

$$= .75 \, (58 \text{ ksi} \times 2.875 \text{ in.}^2) = 125.1 \text{ kips} > 75 \text{ kips}$$

Since the design capacity over the net area again exceeds the required capacity, the member is adequate across the net area. This connection is therefore adequate in shear, bearing, and tension.

EXAMPLE 17-4

The connection to be made in the lap-joint shown below is to withstand a factored tensile force of 200 kips. The connection is a bearing-type, the steel is A36 (F_y = 36000 psi and F_u = 58000 psi), and the bolts are A325-X (threads not in the shear plane). Design the bolt size and the number of bolts to be used.

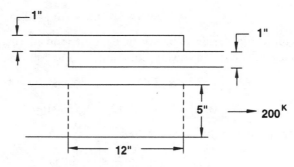

MAXIMUM CONNECTION LENGTH

Calculate the bolt area required, based on the shear and bearing requirements. Afterwards, check the adequacy of the tension stresses over the gross and net areas.

Shear

$$P = 200 \text{ kips}$$

$$\text{Nominal strength} = 60.0 \text{ ksi (from Figure 17-11)}$$

$$\phi = 0.75$$

$$A_{req'd} = 200 \text{ kips}/(0.75)(60.0 \text{ ksi}) = 4.44 \text{ in.}^2 \text{ (Try 1 in.}$$
$$\text{diameter bolts)}$$

$$A_{1\text{" diameter bolt}} = .785 \text{ in.}^2/\text{bolt}$$

$$A_{req'd} = n \times A$$

$$4.44 \text{ in.}^2 = n \times (.785 \text{ in.}^2/\text{ bolt})$$

$$n = 5.66 \text{ bolts}$$

Therefore, use 6 bolts. This is probably a wise choice because you could lay out the bolts in three rows of two, which maintains adequate bolt spacing. A smaller diameter of bolt would lead to a larger number of bolts and the minimum spacing of bolts might be in jeopardy.

Bearing

Check bearing area using 1" diameter bolts to ensure its adequacy assuming all spacings are adequate and that deformation is not considered. Calculate the design capacity in bearing (considering no deformation and spacings are adequate) by using Equation 17-5, as follows:

$$d = 6 \text{ bolts} \times 1 \text{ in. diameter} = 6 \text{ in.}$$

$$\emptyset \, R_n = \emptyset \, (3.0)(d)(t)F_u \tag{Eq. 17-5}$$

$$\emptyset \, R_n = (.75)(3.0)(6\text{in.}^2)(58 \text{ ksi}) = 783 \text{ kips}$$

Since 783 kips > 200 kips, the connection is adequate in bearing.

Tension

The design capacity across the gross area is calculated as:

$$A_g = 5 \text{ in.} \times 1\text{in.} = 5 \text{ in.}^2$$

$$\emptyset_t \, P_n = \emptyset_t \, (F_y \times A_g) \tag{Eq. 17-8}$$

$$= .90 \, (36 \text{ ksi} \times 5 \text{ in.}^2) = 162 \text{ kips} < 200 \text{ kips}$$

Because the design capacity is only 162 kips, which is less than the required capacity of 200 kips, the member fails across the gross area.

The design capacity across the net area is calculated as follows:

$$A_n = 5 \text{ in.}^2 - [2(1 \text{ in.} + 1/8 \text{ in.})(1 \text{ in.})] = 2.75 \text{ in.}^2$$

$$\emptyset_t \, P_n = \emptyset_t \, (F_u \times A_n) \tag{Eq. 17-9}$$

$$= .75 \, (58 \text{ ksi} \times 2.75 \text{ in.}^2) = 119.6 \text{ kips} < 200 \text{ kips}$$

Likewise, the member also fails across the net area. To remedy this situation, the plate width would most likely be increased.

17.6 Introduction to Welded Connections

Welding can be thought of as the fusing of two pieces of metal together to form a continuous, rigid plate. The earliest welding was probably accomplished by craftsmen and artists in ancient times. Although there are many modern day welding techniques, the two basic categories are gas and arc welding. These modern techniques can trace their roots back to the late nineteenth century, when arc welding was first patented and used on a limited scale.

Generally, all welding processes will have the following common elements:

- Base metals
- Heat source
- Electrode or Welding Rod
- Shielding Mechanism

The base metals are simply the pieces to be joined together. They are joined together using some type of heat source and in structural steel welding this heat source is most frequently generated through an electric arc that creates a confined temperature in excess of 6000°F. In gas welding, the heat source is generated through the burning of gas (typically acetylene and oxygen) at the end of a welder's "torch". In both cases, the heat source melts not only the base metal, but also an electrode or welding rod. As the electrode or welding rod is melted, it is deposited as additional steel into the area of the weld. The electrode or welding rod may be thought of as the "weld metal".

The weld must also be protected from coming in contact with the surrounding air during its cooling period. This is usually accomplished through some type of shielding mechanism. This shielding can be accomplished by a gaseous cloud or by immersion of the electrode into a material generally referred to as flux. Flux is a material that will help prevent the intrusion of undesirable contaminants into the pool of molten weld. A common contaminant, which hopefully is minimized by good welding techniques, is air or other gases. If air is allowed to penetrate the weld during its cooling period, it can greatly reduce the strength and quality of the weld, due to pitting or high porosity. Flux can be either a loose, granular material through which the electrode is moved; or it can be contained on a coating that shrouds the electrode. Such a coating would then melt as the electrode is consumed, thereby creating a gaseous cloud.

The use of welding is very popular in steel construction because it has a number of advantages. Among these advantages are a savings produced by the reduction of splicing plates, the ease of welding odd shapes (i.e. pipes), and the ease of implementing field changes. Disadvantages of welding include their fatigue behavior and quality assurance. The former of these disadvantages has led to the development of fatigue criteria and special details in practically all welded structures.

As mentioned earlier, the most common method of welding structural steel is arc welding. This chapter will consider the most basic type of arc welding—the shielded metal arc welding (SMAW) process. (Figure 17-12).

The shielded metal arc welding (SMAW) process is the traditional type of welding that is manually produced. The generation of heat is from an electric arc and the electrode, usually designated by a term such as E70XX, is melted into the weld area to fuse together with the base metal. In the aforementioned electrode designation, E70XX, the 70 represents the ultimate tensile strength (ksi) of the electrode. The subscript symbols may reflect a variety of things—such as coating, positions, and other characteristics. In general, A36 steel can be used successfully with either E60XX or E70XX electrodes. Further designations can be found in the American Welding Society's *Structural Welding Code*[8].

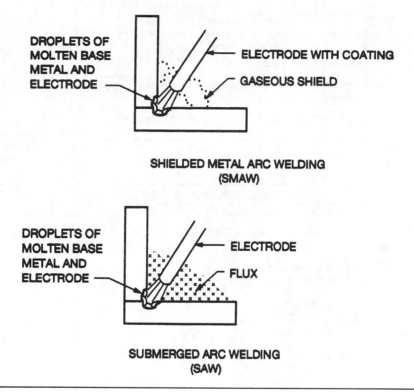

Figure 17-12   Illustration of Common Welding Techniques

17.7   Common Types of Welds

There are many different ways to classify welds—based on the type of weld, the welding position, and the type of end treatment in which a plate may be fabricated to better adapt to the welding procedure. The standard welding symbols, as given by the AISC, are shown in Figure 17-13. In this section, we will be most concerned with identifying the characteristics and terminology for the common weld types.

The most common type of weld is the fillet weld, comprising a large majority of all welds that are produced. These welds are used to join two pieces of steel which form a perpendicular corner where the weld is placed, and can be used in many different types of connections (Figure 17-14). The abundance of fillet welds is due in part to their ease of production, because of the "pocket" formed by the perpendicular edges. This pocket serves as a holder of the molten steel and eliminates the requirement for additional backup plates. Fillet welds also require less precise alignment of the members to be connected, in contrast to the stringent alignment required for groove-welded members.

The most common fillet weld is the equal-leg fillet weld, where the leg dimensions are the same length. The different parts of the fillet weld are shown in Figure

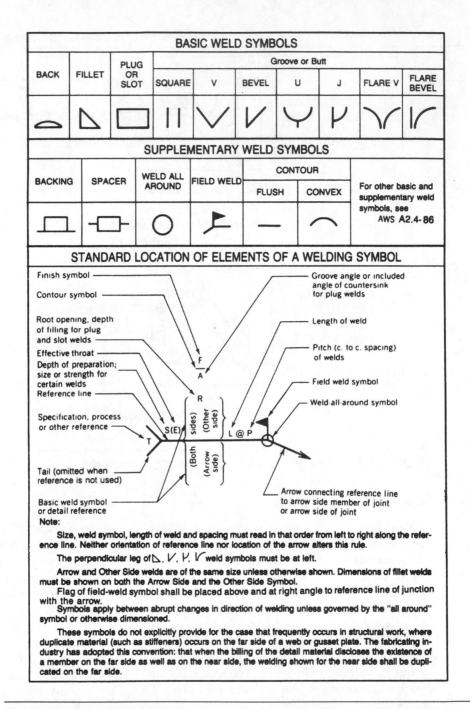

Figure 17-13 Standard Welding Symbols. Courtesy of the American Institute of Steel Construction, Inc.

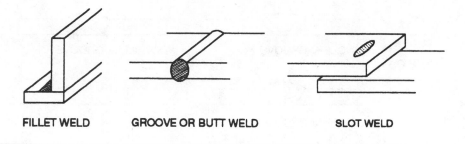

FILLET WELD   GROOVE OR BUTT WELD   SLOT WELD

Figure 17-14   Schematic Illustration of Filet, Groove, and Slot Welds

17-15. The most important part of the fillet weld, as far as design is concerned, is the throat dimension. The throat dimension is the shortest length from the root of the weld to its face. This distance is critical because, although fillet welds have a generally rounded face, the throat distance is the probable line of failure through the weld. In an equal-leg fillet weld this throat distance will be *.707 × the leg*. If a fillet weld should be asymmetrical (having unequal legs), the throat distance should be calculated as the perpendicular distance from the face of the weld to its root, using trigonometric principles. Fillet welds are called out by their leg size in equal leg welds (i.e. a 3/8 in. fillet weld has a leg 3/8 in. long).

The other common type of weld is referred to as a groove or butt weld. This weld is used when connecting two plates that lie in the same plane, and is meant to transfer the load in that plane from one member to the other (Figure 17-14). A groove weld is referred to as a **full penetration groove weld** if the weld extends the full thickness of the plate being joined; and it is referred to as a **partial penetration groove weld** if the weld does not extend the full thickness. The partial-penetration groove welds may occur in a situation where access to the welding area is restricted and can only be accomplished from one side. A groove weld has many other variations, based on the depth of the weld, the edge fabrication of the plates, and the configuration of the plates to be welded. Such configurations may include single and double U, V, or J, welds, which are illustrated in Figure 17-16. In any case, groove welds are more difficult to produce and should be avoided if possible,

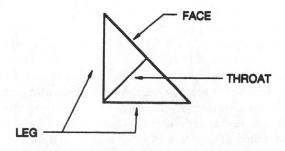

Figure 17-15   Standard Fillet Weld Terminology

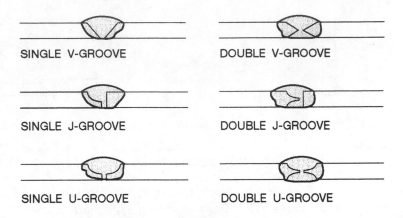

Figure 17-16 Types of Common Groove Welds

because of the cost associated with additional plate fabrication and weld production. Such additional cost is largely because the weld is not confined in a pre-made "pocket," such as the fillet weld.

The critical throat distance in a full penetration groove weld is taken by the AISC to be the thickness of the thinnest plate joined. That is to say, if a 1/4-inch-thick plate were to be groove welded to a 1/2-inch-thick plate, the throat distance of the groove weld would be 1/4 in.

The last type of weld that was shown in Figure 17-14, is known as a slot, or plug, weld. These welds are typically used in lap connections where weld material is placed in standard or slotted holes that have been prepunched in the steel members. These welds increase the shear resistance of the connected parts and inhibit potential buckling of the members along their interface.

17.8 Potential Modes of Failure in Welded Connections

There are basically two manners in which a welded connection can fail—either the weld itself cracks under some type of stress or the base material being joined actually cracks with the weld intact. Both of these mechanisms are induced due to the stresses that act along some plane—either a plane in the weld itself or a plane in the base metal. This plane of failure in the weld is referred to as the **effective weld area**.

The effective weld area is the area that is most susceptible to cracking, assuming the weld to be of uniform quality. In general, this area would be the smallest area resisting the loads placed upon the weld, and therefore the most likely to fail. Whether these loads are placing the weld in tension, compression, or shear; the weld will tend towards resisting a stress with this effective weld area. The student may best picture this area as the surface area that would be remaining *after* the weld had failed. The effective weld area is simply the critical throat distance (sometimes referred to as the effective throat) multiplied by the length of the weld (Figure 17-17).

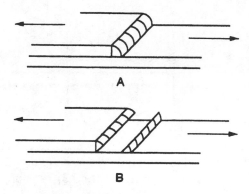

Figure 17-17 Typical Falure Plane in a Welded Connection Exposing the Effective Area. A) Weld Intact B) Weld Broken

With fillet welds, the weld can be loaded along the length of the weld (parallel to the weld axis) or across the length of the weld (perpendicular to the weld axis). Such loadings would cause tension or compression, on a weld stressed parallel to its length, and shear on a weld stressed perpendicular to its length. Although testing has indicated a substantial increase in strength for fillet welds stressed in shear, the AISC considers both cases as equivalent.

The effective weld area for a fillet weld is the area that resists the stresses applied to the weld. As stated earlier, the critical throat distance is theoretically the shortest distance from the root of the weld to its face, which would be .707 × leg for an equal leg fillet weld. This makes the effective weld area for an equal leg fillet weld simply the inclined projection of a rectangular area, *.707 × leg × length of weld*.

To gain a further insight to the requirements of the AISC specification regarding welding, the students should familiarize themselves with the weld requirements in Chapter J of the AISC specification[1,2]. These requirements are especially important with regard to maximum and minimum weld sizes to be used in proper connection design, which will be discussed further in the upcoming section.

17.9   AISC Requirements in Welded Connections - ASD

The allowable stress method in welded connections has the same basic philosophy as it had with bolted connections—actual stresses must be less than or equal to allowable stresses. Remember, both the weld itself and the plates they are connecting, should be checked in order to completely determine the connection's capacity. The planes of failure in a typical welded connection can be summarized in Figure 17-18.

The transfer of stress through fillet welds is assumed to be shear over the effective weld area, and failure is assumed to occur in plane 1-1, as shown in Figure 17-18. When subjected to a shearing force perpendicular to the axis of the

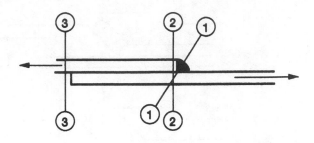

Figure 17-18 Potential Failure Zones in a Welded Connection. 1-1) Shear Through Weld 2-2) Shear Through Plate 3-3) Tension in Plate

weld, the allowable shear stress for a fillet weld is given in the AISC specification as $0.30\ F_u$, where F_u is the ultimate tensile strength of the electrode being used. When a fillet weld is subjected to tension or compression parallel to the weld's longitudinal axis, the allowable stress in the base metal must not exceed previously defined limits (which would be $0.60\ F_y$ in tension and $0.40\ F_y$ in shear).

The stress in full penetration groove welds is transferred exactly in the same manner as it is carried in the plates. The allowable shear stress on the effective area of the groove weld is again $0.30\ F_u$. The allowable stress for tension and compression on the welds, either parallel or perpendicular to the weld's axis, is the same as in the base metal (previously stated). However, when tension stress is applied normal to the weld's effective area, the AISC specification recommends that "matching" weld metal should be used. Matching weld metal refers to using electrode material that has mechanical properties which meet or exceed that of the weakest base metal. A complete table outlining this can be found in the American Welding Society's *Welding Handbook*.[8]

Welds are used to transfer forces from one piece of steel to another. This transfer of force produces stress on the welds and this stress can be quantified in terms of the direct stress equation that we have repeatedly used, $f = P/A$. The area is now the effective weld area that was discussed in a previous section.

The following two examples illustrate solution techniques to these common problems.

EXAMPLE 17.5

Determine the allowable load on the connection shown below. The welds are 5/16" fillet with E70XX electrodes ($F_u = 70$ ksi) using the SMAW process. Plates are made of A36 steel ($F_y = 36$ ksi and $F_u = 58$ ksi).

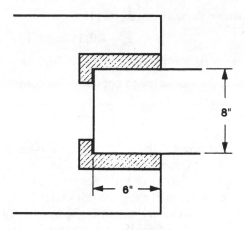

NOTE:
BOTH PLATES ARE 1/2" THICK

Since the problem is to determine the allowable load which the connection can carry, this is an evaluation type of problem. Determine the effective weld area and calculate the AISC allowable stress.

Weld Capacity

Effective weld area = (throat)(length)

total weld length = 6 in. + 6 in. = 12 in.

Effective weld area = $(5/16)(.707) \times (12 \text{ in.}) = 2.65 \text{ in.}^2$

Allowable stress = $0.30\ F_u = .30(70 \text{ ksi}) = 21 \text{ ksi}$

Therefore the allowable load for the weld can be calculated as:

Allowable load = $2.65 \text{ in.}^2 \times 21 \text{ ksi} = 55.6 \text{ kips}$

Also check tension and shear on base plates to make sure they are not weaker than the weld itself.

Plate Tension

$$A_{gross} = 4 \text{ in.}^2$$

Allowable stress, $F_t = 0.60\ F_y = .60(36 \text{ ksi}) = 21.6 \text{ ksi}$

Allowable load = $4 \text{ in.}^2 \times 21.6 \text{ ksi} = 86.4 \text{ kips}$

Plate Shear (around Weld Area)

Area = (length of weld)(plate thickness)

$12 \text{ in.} \times 1/2 \text{ in.} = 6 \text{ in.}^2$

Allowable shear stress, $F_v = 0.40\ F_y$

$$F_v = .40(36\ \text{ksi}) = 14.4\ \text{ksi}$$

Allowable load = 6 in.^{2} × 14.4 ksi = 86.4 kips

Therefore, the weld capacity of 55.6 kips controls the connection's strength.

In actual weld design, the weld size that was chosen should be checked against the minimum and maximum weld sizes allowed. For fillet welds, the maximum weld size is 1/16 in. less than the thickness of the plate for plate thickness over 1/4 in.; and the thickness of the plate, for plates 1/4 in. or smaller. In this example, the maximum fillet weld size would be 7/16 in. (1/2 in. – 1/16 in.).The minimum fillet weld size is given in the AISC specification as 3/16 in. for a 1/2 in. plate. Since our weld is 5/16 in., it fits between the minimum and maximum specified weld sizes.

EXAMPLE 17.6

Design the welds for the connection shown below, using two equal length side fillet welds. Use E70XX electrodes ($F_u = 70$ ksi) with SMAW process and A36 steel ($F_y = 36$ ksi and $F_u = 58$ ksi).

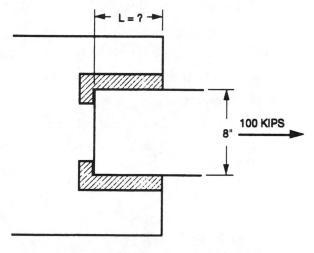

NOTE:
BOTH PLATES ARE 1/2" THICK

This is a design type problem where we are left to design the size and the length of the weld to be used. Again, maximum and minimum fillet weld sizes may be used as a starting point, when deciding on a course of action. Since the plates are each 1/2 in. thick, the maximum size weld per AISC specification is 7/16 in., but let's use 3/8 in. because it will be easier to manually produce.(As in the previous example, when using 1/2 in. plates the minimum fillet weld size would be 3/16 in.).

Weld Capacity

$$\text{Effective weld area} = (.707 \times 3/8) \times \text{length} = .265 \times \text{length}$$

Since the load is given at 100 kips and the allowable weld stress from the specification is 0.30 F_u, the length can be found by rearranging the direct stress equation as follows:

$$A_{req'd} = P \,/\, .30(F_u)$$

$$\text{Area} = .265 \times \text{length}$$

$$(.265)\ \text{Length} = 100 \text{ kips} \,/\, .30(70 \text{ ksi})$$

$$\text{Length} = 17.96 \text{ in.}$$

Therefore, use 2 welds, 9 in. each.
Remember to check tension and shear on plates.

Plate Tension

$$A_g = \text{plate width(thickness)}$$

$$A_g = 8 \text{ in.} \times 1/2 \text{ in.} = 4 \text{ in.}^2$$

Allowable tension stress on gross area, $F_t = 0.60\ F_y$

$$F_t = 21.6 \text{ ksi}$$

$$\text{Allowable load} = 4 \text{ in.}^2 \times 21.6 \text{ ksi} = 86.4 \text{ kips.}$$

Since the allowable tension load of 86.4 kips is less than the actual load, the member would fail. Therefore, a larger member would need to be used.

Plate Shear

$$A = 1/2 \text{ in.} \times 18 \text{ in.} = 9 \text{ in.}^2$$

Allowable shear stress, $F_v = 0.40\ F_y$

$$F_v = 14.4 \text{ ksi}$$

Therefore the shear capacity of the plates would be:

$$\text{Allowable load} = 9 \text{ in}^2 \times 14.4 \text{ ksi} = 129.6 \text{ kips}$$

This is more than the 100 kips which the connection is to carry. Therefore shear capacity of the palte is adequate, however member still needs to be redesigned with regard to tensile capacity.

17.10 AISC Requirements for Welded Connections - LRFD

The LRFD philosophy for welded connections can be summarized as follows:

$$\emptyset R_n \geq R_u$$

where

$\emptyset R_n$ = design capacity of the welded component

R_u = required capacity (factored load) of that component

Fillet Welds

In fillet welds, the transfer of stress through the welds is assumed to be shear over the effective weld area, and failure is assumed to occur in plane 1-1, as shown previously in Figure 17-18. When subjected to a force perpendicular to the axis of the weld, the design strength for a fillet weld is given in the LRFD specification as $.75(0.60F_{Exx})$ for the weld where $.75$ is the reduction factor and $0.60F_{Exx}$ is the nominal strength of the weld. This design capacity can then be illustrated as follows:

$$\emptyset R_n = .75(A \times (0.60F_{Exx})) \qquad \text{(Eq. 17-10)}$$

where

$$A = \text{effective weld area}$$

The term $0.60F_{Exx}$ represents 60% of the ultimate tensile strength of the electrode, since this is approximately the relationship between ultimate shear capacity and ultimate tensile capacity in steel. This design capacity must not be less than the shear capacity of the base material, which can be seen as follows:

$$\emptyset R_n = .75(A \times (0.60 F_u)) \qquad \text{(Eq. 17-11)}$$

where

$$A = \text{shear area of the base material}$$

Also, when a fillet weld is subjected to tension or compression parallel to the weld's longitudinal axis, the design capacity must be checked against a yielding mechanism over the base material; with the reduction factor being 0.90 and the nominal capacity being the yield strength of the base material, F_y.

Groove Welds

When dealing with complete penetration groove welds, the designer should recognize that the stress is transferred exactly in the same manner as it is carried in the plates. Therefore, when a groove welded member is subjected to tension or compressive stresses, either parallel or perpendicular to the weld axis, the design capacities of the base material and weld are as follows:

For the base material:

$$\emptyset R_n = .90(A \times F_y)$$ (Eq. 17-12)

For the weld material:

$$\emptyset R_n = .90(A_w \times F_{yw})$$ (Eq. 17-13)

where

A and A_w = area of the base material resisting stress and the effective weld area respectively

F_y and F_{yw} = tensile yield strength of the base metal and weld metal respectively

However, when tension stress is applied normal to the weld's effective area, the AISC specification recommends that "matching" weld metal shall be used. Matching weld metal refers to using the yield strength of the weakest base material in lieu of the electrode yield strength. A complete table outlining the matching weld metal can be found in the American Welding Society's *Welding Handbook*.[8]

When a groove weld is subjected to shear on the effective area, the design capacity for the base material and the weld are as follows:

For the base material:

$$\emptyset R_n = .90(A \times .60F_y)$$ (Eq. 17-14)

For the weld material:

$$\emptyset R_n = .80(A \times 0.60F_{EXX})$$ (Eq. 17-15)

Weld analysis or design can be viewed in terms of one inch segments of the weld under consideration. This is because a one inch segment of a weld has a particular load capacity based on the strength of the electrode used and the size of the weld. To double this load capacity, the designer merely needs to double the weld length. When designing side or end fillet welds, it should be noted that these welds do not typically terminate at the corner of a connected part, but are "wrapped around" such a corner. The "wrap-around" is referred to as an **end return** and this return should be at least two times the nominal weld size. The end return does not considerably increase the strength of the weld but does delay the initial tearing of the weld.

The following examples illustrate the use of the LRFD requirements in the solutions of these common problems.

EXAMPLE 17.7

Determine the design capacity of the connection shown below. The welds are 5/16 in. fillet with E70XX electrodes ($F_u = 70$ ksi) using the SMAW process. Plates are made of A36 steel ($F_y = 36$ ksi and $F_u = 58$ ksi).

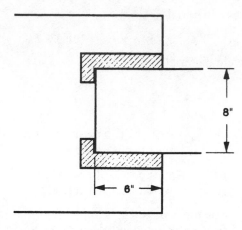

NOTE:
BOTH PLATES ARE 1/2" THICK

Since the problem is to determine the design capacity of the connection, this is an evaluation type of problem. Therefore, we should determine the effective weld area and calculate capacity from the appropriate equation.

Weld Capacity

$$\text{Length of weld} = 6 \text{ in.} + 6 \text{ in.} = 12 \text{ in.}$$

$$\text{Effective weld area} = (5/16)(.707) \times 12 \text{ in.} = 2.65 \text{ in.}^2$$

$$\text{Weld Nominal Strength} = 0.60 F_{EXX} = 0.60(70 \text{ ksi}) = 42 \text{ ksi}$$

$$\phi = 0.75$$

Design capacity based on weld from Equation 17-10 as follows:

$$\phi \, R_n = .75(A \times (0.60 F_{Exx})) \tag{Eq. 17-10}$$

$$= .75(2.65 \text{ in.}^2 \times 42 \text{ ksi}) = 83.5 \text{ kips}$$

Also check tension and/or shear on base material to make sure they are not weaker than the weld itself.

Plate Tension

$$A_{\text{gross}} = \text{plate width} \times \text{thickness}$$

$$A_{\text{gross}} = 8 \text{ in.} \times 1/2 \text{ in.} = 4 \text{ in.}^2$$

Nominal tensile strength of plate $= F_y = 36$ ksi

Calculating design capacity as previously shown:

$$\phi_t \, P_n = \phi_t \, (F_y \, A_g) \tag{Eq. 17-8}$$

$$\text{Design capacity} = 0.90(4 \text{ in.}^2 \times 36 \text{ ksi}) = 129.6 \text{ kips}$$

Plate Shear (around Weld Area)

$$Area = 12 \text{ in.} \times 1/2 \text{ in.} = 6 \text{ in.}^2$$

$$\text{Nominal shear strength of plate} = .60(F_u) = 34.8 \text{ ksi}$$

Calculating the shear design capacity from Eq. 17-11;

$$\emptyset \, R_n = .75(A \times (0.60 \, F_u)) \qquad \text{(Eq. 17-11)}$$

$$\text{Design capacity} = .75(6 \text{ in.}^2 \times 34.8 \text{ ksi}) = 156.6 \text{ kips}$$

Therefore, the connection's design capacity, $\emptyset \, R_n$, is 83.5 kips. The weld strength controls the connection's strength.

EXAMPLE 17.8

Design the welds for the connection shown below using two equal length side fillets welds. Use E70XX electrodes (F_u = 70 ksi) with SMAW process and A36 steel (F_y = 36 ksi and F_u = 58 ksi). The factored load is 200 kips.

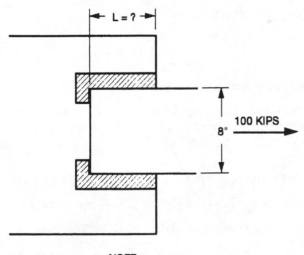

NOTE:
BOTH PLATES ARE 1/2" THICK

This is the design type problem where we are left to design the size and the length of weld to be used. Because the student may not be familiar with maximum weld sizes, please refer to Section J of the LRFD specification for the logic used below.

Since the plates are each 1/2 in. thick, the maximum size weld per the LRFD specification is 7/16 in., but use 3/8 in. because it may be easier in some cases to produce.

Weld Capacity

$$\text{Effective weld area} = (.707 \times 3/8) \times \text{Length} = .265 \times \text{Length}$$

Since the factored load is given at 200 kips, the design capacity ($\emptyset R_n$) must equal or exceed 200 kips and the length can be found as follows:

$$\text{Area} = 200 \text{ kips} / 0.75(.60 \times 70 \text{ ksi}) = 6.35 \text{ in.}^2$$

$$6.35 \text{ in.}^2 = .265 \times \text{Length}$$

$$\text{Length} = 23.96 \text{ in.}$$

Therefore, use 2 welds that are each 12 in. long, with an end return of at least two times the nominal weld size. In this case the end returns would be a minimum of 3/4 in.

Remember to check tension and shear on plates.

Plate Tension

$$A_g = \text{plate width} \times \text{thickness}$$

$$A_g = 8 \text{ in.} \times 1/2 \text{ in.} = 4 \text{ in.}^2$$

Nominal tensile strength = F_y = 36 ksi

Calculating design capacity:

$$\emptyset_t P_n = \emptyset_t (F_y A_g)$$

Design capacity = $0.90(36 \text{ ksi} \times 4 \text{ in.}^2) = 129.6$ kips

Since the design tensile capacity is less than the required capacity of 200 kips, the member would have to be redesigned appropriately. This redesign will be accomplished in the following example. However, check shear for adequacy.

Plate Shear (using 12 in. side welds, but neglecting the end returns)

$$A = 1/2 \text{ in.} \times (24 \text{ in.}) = 12 \text{ in.}^2 \text{ (plate area susceptible to shear)}$$

Nominal shear strength of plate = $.60(F_u)$ = 34.8 ksi

Calculating the shear design capacity from Equation 10-2:

$$\emptyset R_n = .75(A \times (0.60 F_u)) \qquad \text{(Eq. 17-11)}$$

Design capacity = $.75(12 \text{ in.}^2 \times 34.8 \text{ ksi}) = 313$ kips

Since the design shear capacity is greater, the required strength of 200 kips, the plate shear is adequate. However, the connection still has to be redesigned for tension across the plate tension.

17.11 Summary

Bolted and welded connections are the most common methods available to join structural steel members together. The two predominant methods used for designing these connections in steel are the allowable stress design (ASD) method and the load and resistance factor design (LRFD) method. Bolted connections can be categorized as bearing-type or slip-critical. This distinction is made by the designer on the basis of slippage under maximum load conditions. Bolted connections should be designed for shear on the bolts, bearing on the plates, and tension over the gross and net areas.

Welded connections are made by fusing the base metals together, using some heat source, typically an electric arc. The electric arc melts the base metal and the electrode while depositing the molten metal in the weld area. As this molten pool of steel is cooled, the weld is formed. The most common type of weld is the fillet weld and these connections should be designed for capacity of the weld as well as tension and shear on the plate.

EXERCISES

1. (ASD). Determine the allowable load on the bearing-type connection shown below, if the bolts are A325-N, 7/8 in. diameter. Steel is A36, check plate tension also.

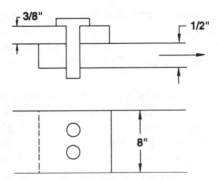

2. (ASD). Check the adequacy of the slip-critical connection shown below if the bolts are A490-X, 3/4 in. diameter. Steel is A36, check plate tension also.

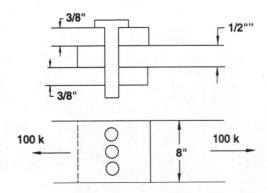

3. (ASD). Design the size and number of 3/4 in. diameter A325-X bolts required in the bearing connection shown below. Check shear, bearing, and tension. Assume required edge distance is 1 1/2 in. to bolt center and minimum spacing between centers of bolts is 3d. Steel is A36

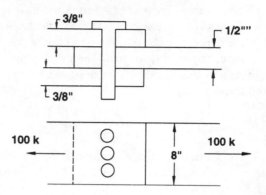

4. (ASD). Redesign the connection in problem #3 if the two outside plates are 1/4 in. thick and consider deformation around the holes.
5. (ASD). Recalculate the adequacy of the connection shown in problem #2 if the connection is now a bearing type. How does this change your answer?
6. (ASD). Check the adequacy of the bearing-type connection shown below in shear, bearing, and tension. The bolts are A325-N, 3/4 in. diameter and deformation is to be considered. Steel is A36. Check by comparing actual stresses to allowable stresses.

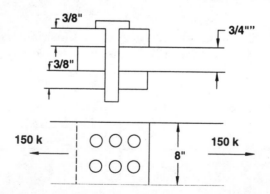

7. (LRFD). Determine the maximum factored load on the bearing-type connection shown below if the bolts are A325-N, 19.1 mm diameter. Steel is A36, check plate tension also. Deformation is not a consideration.

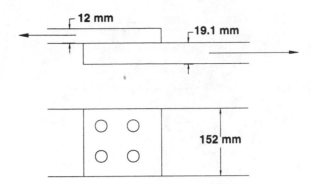

8. (LRFD). Check the adequacy of the slip-critical connection shown below if the bolts are A490-X, 3/4 in. diameter. The load shown is a factored load with the service portion being 60 kips. Steel is A36, check plate tension also.

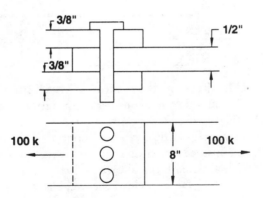

9. (LRFD). Determine the maximum factored load on a bearing-type lap joint that is connected by six 25.4 mm diameter, A490-X bolts arranged in three rows of two bolts each. The plates are 200 mm wide by 20 mm thick and the steel is A36.

10. (LRFD). Design the size and number of A325-X bolts required in the bearing-type connection shown below under the factored load of 150 kips. Check shear , bearing, and tension. Also assume required edge distance is 1 1/2 in. to bolt center and minimum spacing between centers of bolts is 3d. Steel is A36.

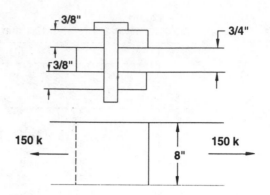

11. Design the size and number of 25.4 mm diameter A325-N bolts required in the bearing-type connection shown below. Check shear and bearing, and assume that deformation is not a consideration. Steel is A36.

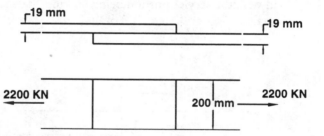

12. (LRFD). Redesign the connection in exercise #10 if the two outside plates are 1/4 in. thick and consider deformation around the holes.

13. (LRFD). Redesign the connection shown in exercise #11 if the connection is now a slip-critical type of connection. The service load is 1500 KN. Also consider the tension requirements in the member as well.

14. (ASD). Calculate the allowable load which can be held on the welded connection shown below. The weld is a 3/8 in. fillet weld made from E70XX electrodes. The base steel is A36 and the weld is made by the SMAW process.

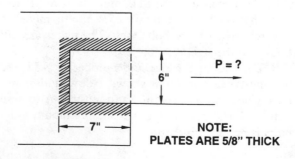

NOTE:
PLATES ARE 5/8" THICK

15. (ASD). Calculate the length of the full penetration groove weld shown below, if the connection is to hold 80 kips. The electrode is an E60XX and the base steel is A36. The SMAW process is used.

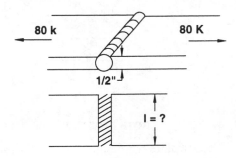

16. (ASD). If the welded connection shown below is made from a 1/4 in. fillet weld around the total plate perimeter using the SMAW process, will the connection work? Steel is A242 ($F_y = 50$ ksi, $F_u = 70$ ksi) and electrodes are E70XX.

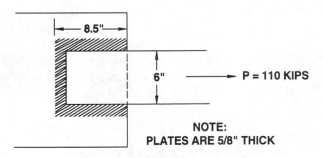

17. (ASD). If the connection in problem #16 used E60XX electrodes, would the connection's capacity be decreased? Calculate and explain.

18. (ASD). Design a fillet weld (size and length) to hold a 200 kip tensile force in the connection shown below. Use side welds initially and an end weld if needed. The electrodes are E80XX and the steel is A242. The SMAW process is used.

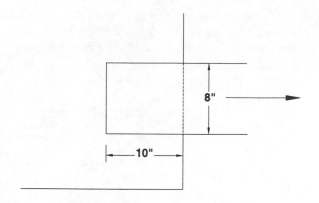

BOTH PLATE ARE 3/4" THICK

19. (LRFD). Calculate the design capacity on the welded connection shown below. The weld is a 3/8 in. fillet weld made from E70XX electrodes. The base steel is A36 and the weld is made using the SMAW process.

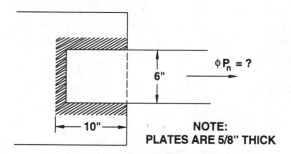

NOTE:
PLATES ARE 5/8" THICK

20. (LRFD). Calculate the adequacy of the welded connection shown below. The weld is a 9.5 mm fillet weld made from E70XX electrodes, using the SMAW process. The base steel is A36.

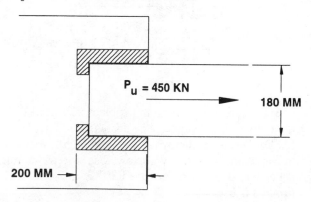

BOTH PLATES ARE 18 MM THICK

21. (LRFD). Calculate the adequacy of the welded connection shown below. The weld is a 1/4 in. fillet weld made from E60XX electrodes, using the SMAW process. The steel is A36.

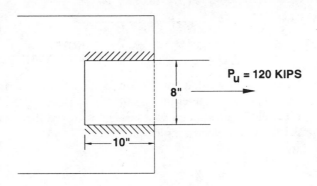

BOTH PLATE ARE 3/4" THICK

22. (LRFD). The connection used in exercise #20 changes to a 13mm fillet weld, how does this change the capacity?

23. (LRFD). Calculate the length of the full penetration groove weld shown below, if the connection is to hold a factored load of 80 kips. The electrode is an E60XX and the base steel is A36. The SMAW process is used.

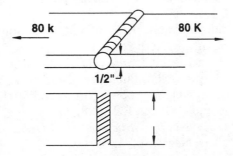

24. (LRFD). If the welded connection shown below is made from a 1/4 in. fillet weld around the total plate perimeter using the SMAW process; will the connection work? The load is factored, the steel is A36, and the electrodes are E70XX.

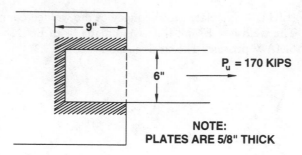

NOTE:
PLATES ARE 5/8" THICK

25. (LRFD). If a welded connection is made from a 6mm fillet weld around a total plate perimeter that is 300 mm long, using the SMAW process; will the weld be adequate in shear? The factored load is 200 KN, the steel is A36, and the electrodes are E70XX.

REFERENCES

1. AISC. *Manual of Steel Construction, Load and Resistance Factor Design*, First Edition, Chicago, American Institute of Steel Construction, 1986.

2. AISC. *Manual of Steel Construction, Allowable Stress Design*, Ninth Edition, Chicago, American Institute of Steel Construction, 1989.

3. AITC. *Timber Construction Manual*, Fourth Edition, Englewood, CO, American Institute of Timber Construction, 1994.

4. "Specifications for Structural Joints Using ASTM A325 or A490 Bolts", Research Council on Structural Connections of the Engineering Foundation, June 8, 1988.

5. Struik, John A., O. Oyeledun, Abayomi, and Fisher, John W. "Bolt Tension Control with a Direct Tension Indicator", Engineering Journal, AISC, 1973

6. J. O. Surtees and M. E. Ibrahim, "Load Indicating Washers", Civil Engineering, ASCE, April 1982.

7. McCormac, Jack C., *Structural Steel Design:* ASD Method, Fourth Edition, Harper Collins Publishers, New York, 1992, p. 308.

8. American Welding Society, Structural Welding Code-Steel (D1.1-88), 11th Edition, Miami, FL, 1988.

9. Burns, Thomas M., *Structured Steel Design-LRFD*, 1st Edition, Delmar Publishers, New York, 1995.

A

US AND SI STEEL SECTION TABLES

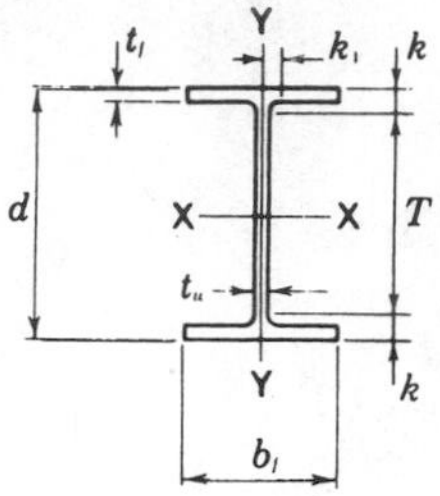

W SHAPES
Dimensions

Designation	Area A	Depth d		Web Thickness t_w		Web $\frac{t_w}{2}$	Flange Width b_f		Flange Thickness t_f		Distance T	Distance k	Distance k_1
	In.^{2}	In.		In.		In.	In.		In.		In.	In.	In.
W 24x117	34.4	24.26	24 1/4	0.550	9/16	5/16	12.800	12 3/4	0.850	7/8	21	1 5/8	1
x104	30.6	24.06	24	0.500	1/2	1/4	12.750	12 3/4	0.750	3/4	21	1 1/2	1
W 24x103[b]	30.3	24.53	24 1/2	0.550	9/16	5/16	9.000	9	0.980	1	21	1 3/4	13/16
x 94	27.7	24.31	24 1/4	0.515	1/2	1/4	9.065	9 1/8	0.875	7/8	21	1 5/8	1
x 84	24.7	24.10	24 1/8	0.470	1/2	1/4	9.020	9	0.770	3/4	21	1 9/16	15/16
x 76	22.4	23.92	23 7/8	0.440	7/16	1/4	8.990	9	0.680	11/16	21	1 7/16	15/16
x 68	20.1	23.73	23 3/4	0.415	7/16	1/4	8.965	9	0.585	9/16	21	1 3/8	15/16
W 24x 62	18.2	23.74	23 3/4	0.430	7/16	1/4	7.040	7	0.590	9/16	21	1 3/8	15/16
x 55	16.2	23.57	23 5/8	0.395	3/8	3/16	7.005	7	0.505	1/2	21	1 5/16	15/16
W 21x402[a]	118.0	26.02	26	1.730	1 3/4	7/8	13.405	13 3/8	3.130	3 1/8	18 1/4	3 7/8	1 7/16
x364[a]	107.0	25.47	25 1/2	1.590	1 9/16	13/16	13.265	13 1/4	2.850	2 7/8	18 1/4	3 5/8	1 3/8
x333[a]	97.9	25.00	25	1.460	1 7/16	3/4	13.130	13 1/8	2.620	2 5/8	18 1/4	3 3/8	1 5/16
x300[a]	88.2	24.53	24 1/2	1.320	1 5/16	11/16	12.990	13	2.380	2 3/8	18 1/4	3 1/8	1 1/4
x275[a]	80.8	24.13	24 1/8	1.220	1 1/4	5/8	12.890	12 7/8	2.190	2 3/16	18 1/4	3	1 3/16
x248[a]	72.8	23.74	23 3/4	1.100	1 1/8	9/16	12.775	12 3/4	1.990	2	18 1/4	2 3/4	1 1/8
x223	65.4	23.35	23 3/8	1.000	1	1/2	12.675	12 5/8	1.790	1 13/16	18 1/4	2 9/16	1 1/16
x201	59.2	23.03	23	0.910	15/16	1/2	12.575	12 5/8	1.630	1 5/8	18 1/4	2 3/8	1
x182	53.6	22.72	22 3/4	0.830	13/16	7/16	12.500	12 1/2	1.480	1 1/2	18 1/4	2 1/4	1
x166	48.8	22.48	22 1/2	0.750	3/4	3/8	12.420	12 3/8	1.360	1 3/8	18 1/4	2 1/8	15/16
x147	43.2	22.06	22	0.720	3/4	3/8	12.510	12 1/2	1.150	1 1/8	18 1/4	1 7/8	1 1/16
x132	38.8	21.83	21 7/8	0.650	5/8	5/16	12.440	12 1/2	1.035	1 1/16	18 1/4	1 13/16	1
x122	35.9	21.68	21 5/8	0.600	5/8	5/16	12.390	12 3/8	0.960	15/16	18 1/4	1 11/16	1
x111	32.7	21.51	21 1/2	0.550	9/16	5/16	12.340	12 3/8	0.875	7/8	18 1/4	1 5/8	15/16
x101	29.8	21.36	21 3/8	0.500	1/2	1/4	12.290	12 1/4	0.800	13/16	18 1/4	1 9/16	15/16
W 21x 93	27.3	21.62	21 5/8	0.580	9/16	5/16	8.420	8 3/8	0.930	15/16	18 1/4	1 11/16	1
x 83	24.3	21.43	21 3/8	0.515	1/2	1/4	8.355	8 3/8	0.835	13/16	18 1/4	1 9/16	15/16
x 73	21.5	21.24	21 1/4	0.455	7/16	1/4	8.295	8 1/4	0.740	3/4	18 1/4	1 1/2	15/16
x 68	20.0	21.13	21 1/8	0.430	7/16	1/4	8.270	8 1/4	0.685	11/16	18 1/4	1 7/16	7/8
x 62	18.3	20.99	21	0.400	3/8	3/16	8.240	8 1/4	0.615	5/8	18 1/4	1 3/8	7/8
W 21x 57	16.7	21.06	21	0.405	3/8	3/16	6.555	6 1/2	0.650	5/8	18 1/4	1 3/8	7/8
x 50	14.7	20.83	20 7/8	0.380	3/8	3/16	6.530	6 1/2	0.535	9/16	18 1/4	1 5/16	7/8
x 44	13.0	20.66	20 5/8	0.350	3/8	3/16	6.500	6 1/2	0.450	7/16	18 1/4	1 3/16	7/8

W SHAPES
Properties

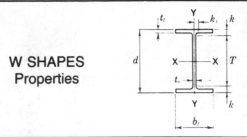

Nom-inal Wt. per Ft	Compact Section Criteria $\frac{b_f}{2t_f}$	$\frac{h_c}{t_w}$	F_y'''	X_1	$X_2 \times 10^6$	Elastic Properties Axis X-X I	S	r	Axis Y-Y I	S	r	Plastic Modulus Z_x	Z_y
Lb.			Ksi	Ksi	$(1/Ksi)^2$	In.^{4}	In.^{3}	In.	In.^{4}	In.^{3}	In.	In.^{3}	In.^{3}
117	7.5	39.2	42	2090	8190	3540	291	10.1	297	46.5	2.94	327	71.4
104	8.5	43.1	34	1860	12900	3100	258	10.1	259	40.7	2.91	289	62.4
103	4.6	39.2	42	2400	5280	3000	245	9.96	119	26.5	1.99	280	41.5
94	5.2	41.9	37	2180	7800	2700	222	9.87	109	24.0	1.98	254	37.5
84	5.9	45.9	30	1950	12200	2370	196	9.79	94.4	20.9	1.95	224	32.6
76	6.6	49.0	27	1760	18600	2100	176	9.69	82.5	18.4	1.92	200	28.6
68	7.7	52.0	24	1590	29000	1830	154	9.55	70.4	15.7	1.87	177	24.5
62	6.0	50.1	25	1700	25100	1550	131	9.23	34.5	9.80	1.38	153	15.7
55	6.9	54.6	21	1540	39600	1350	114	9.11	29.1	8.30	1.34	134	13.3
402	2.1	10.8	—	8000	41	12200	937	10.2	1270	189	3.27	1130	296
364	2.3	11.8	—	7340	57	10800	846	10.0	1120	168	3.23	1010	263
333	2.5	12.8	—	6790	78	9610	769	9.91	994	151	3.19	915	237
300	2.7	14.2	—	6200	111	8480	692	9.81	873	134	3.15	816	210
275	2.9	15.4	—	5720	150	7620	632	9.71	785	122	3.12	741	189
248	3.2	17.1	—	5210	215	6760	569	9.63	694	109	3.09	663	169
223	3.5	18.8	—	4700	319	5950	510	9.54	609	96.1	3.05	589	149
201	3.9	20.6	—	4290	453	5310	461	9.47	542	86.1	3.02	530	133
182	4.2	22.6	—	3910	649	4730	417	9.40	483	77.2	3.00	476	119
166	4.6	25.0	—	3590	904	4280	380	9.36	435	70.1	2.98	432	108
147	5.4	26.1	—	3140	1590	3630	329	9.17	376	60.1	2.95	373	92.6
132	6.0	28.9	—	2840	2350	3220	295	9.12	333	53.5	2.93	333	82.3
122	6.5	31.3	—	2630	3160	2960	273	9.09	305	49.2	2.92	307	75.6
111	7.1	34.1	55	2400	4510	2670	249	9.05	274	44.5	2.90	279	68.2
101	7.7	37.5	45	2200	6400	2420	227	9.02	248	40.3	2.89	253	61.7
93	4.5	32.3	61	2680	3460	2070	192	8.70	92.9	22.1	1.84	221	34.7
83	5.0	36.4	48	2400	5250	1830	171	8.67	81.4	19.5	1.83	196	30.5
73	5.6	41.2	38	2140	8380	1600	151	8.64	70.6	17.0	1.81	172	26.6
68	6.0	43.6	34	2000	10900	1480	140	8.60	64.7	15.7	1.80	160	24.4
62	6.7	46.9	29	1820	15900	1330	127	8.54	57.5	13.9	1.77	144	21.7
57	5.0	46.3	30	1960	13100	1170	111	8.36	30.6	9.35	1.35	129	14.8
50	6.1	49.4	26	1730	22600	984	94.5	8.18	24.9	7.64	1.30	110	12.2
44	7.2	53.6	22	1550	36600	843	81.6	8.06	20.7	6.36	1.26	95.4	10.2

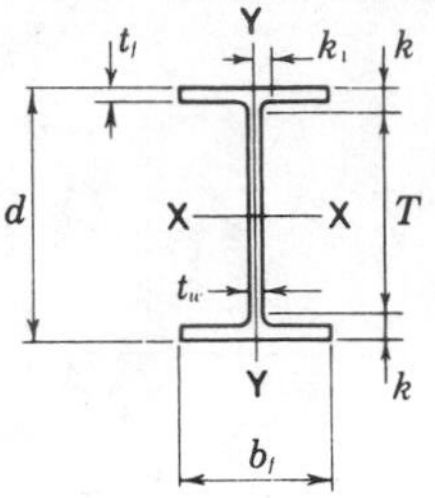

W SHAPES
Dimensions

Designation	Area A	Depth d		Web Thickness t_w		Web $\frac{t_w}{2}$	Flange Width b_f		Flange Thickness t_f		Distance T	Distance k	Distance k_1
	In.^{2}	In.		In.		In.	In.		In.		In.	In.	In.
W 18x311[a]	91.5	22.32	22 3/8	1.520	1 1/2	3/4	12.005	12	2.740	2 3/4	15 1/2	3 7/16	1 3/16
x283[a]	83.2	21.85	21 7/8	1.400	1 3/8	11/16	11.890	11 7/8	2.500	2 1/2	15 1/2	3 3/16	1 3/16
x258[a]	75.9	21.46	21 1/2	1.280	1 1/4	5/8	11.770	11 3/4	2.300	2 5/16	15 1/2	3	1 1/8
x234[a]	68.8	21.06	21	1.160	1 3/16	5/8	11.650	11 5/8	2.110	2 1/8	15 1/2	2 3/4	1
x211[a]	62.1	20.67	20 5/8	1.060	1 1/16	9/16	11.555	11 1/2	1.910	1 15/16	15 1/2	2 9/16	1
x192	56.4	20.35	20 3/8	0.960	1	1/2	11.455	11 1/2	1.750	1 3/4	15 1/2	2 7/16	15/16
x175	51.3	20.04	20	0.890	7/8	7/16	11.375	11 3/8	1.590	1 9/16	15 1/2	2 1/4	7/8
x158	46.3	19.72	19 3/4	0.810	13/16	7/16	11.300	11 1/4	1.440	1 7/16	15 1/2	2 1/8	7/8
x143	42.1	19.49	19 1/2	0.730	3/4	3/8	11.220	11 1/4	1.320	1 5/16	15 1/2	2	13/16
x130	38.2	19.25	19 1/4	0.670	11/16	3/8	11.160	11 1/8	1.200	1 3/16	15 1/2	1 7/8	13/16
x119	35.1	18.97	19	0.655	5/8	5/16	11.265	11 1/4	1.060	1 1/16	15 1/2	1 3/4	15/16
x106	31.1	18.73	18 3/4	0.590	9/16	5/16	11.200	11 1/4	0.940	15/16	15 1/2	1 5/8	15/16
x 97	28.5	18.59	18 5/8	0.535	9/16	5/16	11.145	11 1/8	0.870	7/8	15 1/2	1 9/16	7/8
x 86	25.3	18.39	18 3/8	0.480	1/2	1/4	11.090	11 1/8	0.770	3/4	15 1/2	1 7/16	7/8
x 76	22.3	18.21	18 1/4	0.425	7/16	1/4	11.035	11	0.680	11/16	15 1/2	1 3/8	13/16
W 18x 71	20.8	18.47	18 1/2	0.495	1/2	1/4	7.635	7 5/8	0.810	13/16	15 1/2	1 1/2	7/8
x 65	19.1	18.35	18 3/8	0.450	7/16	1/4	7.590	7 5/8	0.750	3/4	15 1/2	1 7/16	7/8
x 60	17.6	18.24	18 1/4	0.415	7/16	1/4	7.555	7 1/2	0.695	11/16	15 1/2	1 3/8	13/16
x 55	16.2	18.11	18 1/8	0.390	3/8	3/16	7.530	7 1/2	0.630	5/8	15 1/2	1 5/16	13/16
x 50	14.7	17.99	18	0.355	3/8	3/16	7.495	7 1/2	0.570	9/16	15 1/2	1 1/4	13/16
W 18x 46	13.5	18.06	18	0.360	3/8	13/16	6.060	6	0.605	5/8	15 1/2	1 1/4	13/16
x 40	11.8	17.90	17 7/8	0.315	5/16	3/16	6.015	6	0.525	1/2	15 1/2	1 3/16	13/16
x 35	10.3	17.70	17 3/4	0.300	5/16	3/16	6.000	6	0.425	7/16	15 1/2	1 1/8	3/4
W 16x100	29.4	16.97	17	0.585	9/16	5/16	10.425	10 3/8	0.985	1	13 5/8	1 11/16	15/16
x 89	26.2	16.75	16 3/4	0.525	1/2	1/4	10.365	10 3/8	0.875	7/8	13 5/8	1 9/16	7/8
x 77	22.6	16.52	16 1/2	0.455	7/16	1/4	10.295	10 1/4	0.760	3/4	13 5/8	1 7/16	7/8
x 67	19.7	16.33	16 3/8	0.395	3/8	3/16	10.235	10 1/4	0.665	11/16	13 5/8	1 3/8	13/16
W 16x 57	16.8	16.43	16 3/8	0.430	7/16	1/4	7.120	7 1/8	0.715	11/16	13 5/8	1 3/8	7/8
x 50	14.7	16.26	16 1/4	0.380	3/8	3/16	7.070	7 1/8	0.630	5/8	13 5/8	1 5/16	13/16
x 45	13.3	16.13	16 1/8	0.345	3/8	3/16	7.035	7	0.565	9/16	13 5/8	1 1/4	13/16
x 40	11.8	16.01	16	0.305	5/16	3/16	6.995	7	0.505	1/2	13 5/8	1 3/16	13/16
x 36	10.6	15.86	15 7/8	0.295	5/16	3/16	6.985	7	0.430	7/16	13 5/8	1 1/8	3/4
W 16x 31	9.12	15.88	15 7/8	0.275	1/4	1/8	5.525	5 1/2	0.440	7/16	13 5/8	1 1/8	3/4
x 26	7.68	15.69	15 3/4	0.250	1/4	1/8	5.500	5 1/2	0.345	3/8	13 5/8	1 1/16	3/4

W SHAPES
Properties

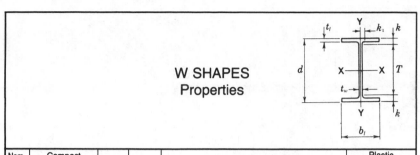

Nom-inal Wt. per Ft	Compact Section Criteria		F_y'''	X_1	$X_2 \times 10^6$	Elastic Properties						Plastic Modulus	
						Axis X-X			Axis Y-Y				
	$\frac{b_f}{2t_f}$	$\frac{h_c}{t_w}$		X_1	$X_2 \times 10^6$	I	S	r	I	S	r	Z_x	Z_y
Lb.			Ksi	Ksi	$(1/Ksi)^2$	In.^{4}	In.^{3}	In.	In.^{4}	In.^{3}	In.	In.^{3}	In.^{3}
311	2.2	10.6	—	8160	38	6960	624	8.72	795	132	2.95	753	207
283	2.4	11.5	—	7520	52	6160	564	8.61	704	118	2.91	676	185
258	2.6	12.5	—	6920	71	5510	514	8.53	628	107	2.88	611	166
234	2.8	13.8	—	6360	97	4900	466	8.44	558	95.8	2.85	549	149
211	3.0	15.1	—	5800	140	4330	419	8.35	493	85.3	2.82	490	132
192	3.3	16.7	—	5320	194	3870	380	8.28	440	76.8	2.79	442	119
175	3.6	18.0	—	4870	274	3450	344	8.20	391	68.8	2.76	398	106
158	3.9	19.8	—	4430	396	3060	310	8.12	347	61.4	2.74	356	94.8
143	4.2	21.9	—	4060	557	2750	282	8.09	311	55.9	2.72	322	85.4
130	4.6	23.9	—	3710	789	2460	256	8.03	278	49.9	2.70	291	76.7
119	5.3	24.5	—	3340	1210	2190	231	7.90	253	44.9	2.69	261	69.1
106	6.0	27.2	—	2990	1880	1910	204	7.84	220	39.4	2.66	230	60.5
97	6.4	30.0	—	2750	2580	1750	188	7.82	201	36.1	2.65	211	55.3
86	7.2	33.4	57	2460	4060	1530	166	7.77	175	31.6	2.63	186	48.4
76	8.1	37.8	45	2180	6520	1330	146	7.73	152	27.6	2.61	163	42.2
71	4.7	32.4	61	2680	3310	1170	127	7.50	60.3	15.8	1.70	145	24.7
65	5.1	35.7	50	2470	4540	1070	117	7.49	54.8	14.4	1.69	133	22.5
60	5.4	38.7	43	2290	6080	984	108	7.47	50.1	13.3	1.69	123	20.6
55	6.0	41.2	38	2110	8540	890	98.3	7.41	44.9	11.9	1.67	112	18.5
50	6.6	45.2	31	1920	12400	900	88.9	7.38	40.1	10.7	1.65	101	16.6
46	5.0	44.6	32	2060	10100	712	78.8	7.25	22.5	7.43	1.29	90.7	11.7
40	5.7	51.0	25	1810	17200	612	68.4	7.21	19.1	6.35	1.27	78.4	9.95
35	7.1	53.5	22	1590	30300	510	57.6	7.04	15.3	5.12	1.22	66.5	8.06
100	5.3	24.3	—	3450	1040	1490	175	7.10	186	35.7	2.51	198	54.9
89	5.9	27.0	—	3090	1630	1300	155	7.05	163	31.4	2.49	175	48.1
77	6.8	31.2	—	2680	2790	1110	134	7.00	138	26.9	2.47	150	41.1
67	7.7	35.9	50	2350	4690	954	117	6.96	119	23.2	2.46	130	35.5
57	5.0	33.0	59	2650	3400	758	92.2	6.72	43.1	12.1	1.60	105	18.9
50	5.6	37.4	46	2340	5530	659	81.0	6.68	37.2	10.5	1.59	92.0	16.3
45	6.2	41.2	38	2120	8280	586	72.7	6.65	32.8	9.34	1.57	82.3	14.5
40	6.9	46.6	30	1890	12900	518	64.7	6.63	28.9	8.25	1.57	72.9	12.7
36	8.1	48.1	28	1700	20800	448	56.5	6.51	24.5	7.00	1.52	64.0	10.8
31	6.3	51.6	24	1740	20000	375	47.2	6.41	12.4	4.49	1.17	54.0	7.03
26	8.0	56.8	20	1470	40900	301	38.4	6.26	9.59	3.49	1.12	44.2	5.48

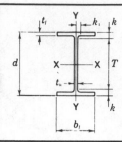

W SHAPES
Dimensions

Desig-nation	Area A	Depth d	Web Thickness t_w		$\frac{t_w}{2}$	Flange Width b_f		Flange Thickness t_f		Distance T	Distance k	Distance k_1
	In.²	In.	In.		In.	In.		In.		In.	In.	In.
W 14x730[a]	215.0	22.42	22⅜	3.070	3 1/16	1 9/16	17.890	17⅞	4.910	4 15/16	11¼	5 9/16	2 3/16
x665[a]	196.0	21.64	21⅝	2.830	2 13/16	1 7/16	17.650	17⅝	4.520	4½	11¼	5 3/16	2 1/16
x605[a]	178.0	20.92	20⅞	2.595	2⅝	1 5/16	17.415	17⅜	4.160	4 3/16	11¼	4 13/16	1 15/16
x550[a]	162.0	20.24	20¼	2.380	2⅜	1 3/16	17.200	17¼	3.820	3 13/16	11¼	4½	1 13/16
x500[a]	147.0	19.60	19⅝	2.190	2 3/16	1⅛	17.010	17	3.500	3½	11¼	4 3/16	1¾
x455[a]	134.0	19.02	19	2.015	2	1	16.835	16⅞	3.210	3 3/16	11¼	3⅞	1⅝
W 14x426[a]	125.0	18.67	18⅝	1.875	1⅞	15/16	16.695	16¾	3.035	3 1/16	11¼	3 11/16	1 9/16
x398[a]	117.0	18.29	18¼	1.770	1¾	⅞	16.590	16⅝	2.845	2⅞	11¼	3½	1½
x370[a]	109.0	17.92	17⅞	1.655	1⅝	13/16	16.475	16½	2.660	2 11/16	11¼	3 5/16	1 7/16
x342[a]	101.0	17.54	17½	1.540	1 9/16	13/16	16.360	16⅜	2.470	2½	11¼	3⅛	1⅜
x311[a]	91.4	17.12	17⅛	1.410	1 7/16	¾	16.230	16¼	2.260	2¼	11¼	2 15/16	1 5/16
x283[a]	83.3	16.74	16¾	1.290	1 5/16	11/16	16.110	16⅛	2.070	2 1/16	11¼	2¾	1¼
x257[a]	75.6	16.38	16⅜	1.175	1 3/16	⅝	15.995	16	1.890	1⅞	11¼	2 9/16	1 3/16
x233[a]	68.5	16.04	16	1.070	1 1/16	9/16	15.890	15⅞	1.720	1¾	11¼	2⅜	1 3/16
x211[a]	62.0	15.72	15¾	0.980	1	½	15.800	15¾	1.560	1 9/16	11¼	2¼	1⅛
x193	56.8	15.48	15½	0.890	⅞	7/16	15.710	15¾	1.440	1 7/16	11¼	2⅛	1 1/16
x176	51.8	15.22	15¼	0.830	13/16	7/16	15.650	15⅝	1.310	1 5/16	11¼	2	1 1/16
x159	46.7	14.98	15	0.745	¾	⅜	15.565	15⅝	1.190	1 3/16	11¼	1⅞	1
x145	42.7	14.78	14¾	0.680	11/16	⅜	15.500	15½	1.090	1 1/16	11¼	1¾	1

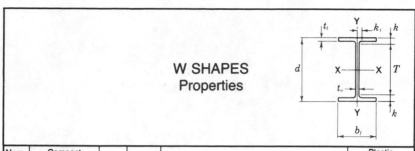

W SHAPES
Properties

Nominal Wt. per Ft	Compact Section Criteria			X_1	$X_2 \times 10^6$	Elastic Properties						Plastic Modulus	
						Axis X-X			Axis Y-Y				
	$\frac{b_f}{2t_f}$	$\frac{h_c}{t_w}$	F_y'''			I	S	r	I	S	r	Z_x	Z_y
Lb.			Ksi	Ksi	$(1/Ksi)^2$	In.⁴	In.³	In.	In.⁴	In.³	In.	In.³	In.³
730	1.8	3.7	—	17500	1.90	14300	1280	8.17	4720	527	4.69	1660	816
665	2.0	4.0	—	16300	2.46	12400	1150	7.98	4170	472	4.62	1480	730
605	2.1	4.4	—	15100	3.20	10800	1040	7.80	3680	423	4.55	1320	652
550	2.3	4.8	—	14200	4.15	9430	931	7.63	3250	378	4.49	1180	583
500	2.4	5.2	—	13100	5.49	8210	838	7.48	2880	339	4.43	1050	522
455	2.6	5.7	—	12200	7.30	7190	756	7.33	2560	304	4.38	936	468
426	2.8	6.1	—	11500	8.89	6600	707	7.26	2360	283	4.34	869	434
398	2.9	6.4	—	10900	11.0	6000	656	7.16	2170	262	4.31	801	402
370	3.1	6.9	—	10300	13.9	5440	607	7.07	1990	241	4.27	736	370
342	3.3	7.4	—	9600	17.9	4900	559	6.98	1810	221	4.24	672	338
311	3.6	8.1	—	8820	24.4	4330	506	6.88	1610	199	4.20	603	304
283	3.9	8.8	—	8120	33.4	3840	459	6.79	1440	179	4.17	542	274
257	4.2	9.7	—	7460	46.1	3400	415	6.71	1290	161	4.13	487	246
233	4.6	10.7	—	6820	64.9	3010	375	6.63	1150	145	4.10	436	221
211	5.1	11.6	—	6230	91.8	2660	338	6.55	1030	130	4.07	390	198
193	5.5	12.8	—	5740	125	2400	310	6.50	931	119	4.05	355	180
176	6.0	13.7	—	5280	173	2140	281	6.43	838	107	4.02	320	163
159	6.5	15.3	—	4790	249	1900	254	6.38	748	96.2	4.00	287	146
145	7.1	16.8	—	4400	348	1710	232	6.33	677	87.3	3.98	260	133

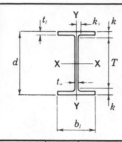

W SHAPES
Dimensions

Desig-nation	Area A	Depth d		Web Thickness t_w		$\frac{t_w}{2}$	Flange Width b_f		Thickness t_f		Distance T	k	k_1
	In.²	In.		In.		In.	In.		In.		In.	In.	In.
W 14x132	38.8	14.66	14⅝	0.645	⅝	⁵⁄₁₆	14.725	14¾	1.030	1	11¼	1¹¹⁄₁₆	¹⁵⁄₁₆
x120	35.3	14.48	14½	0.590	⁹⁄₁₆	⁵⁄₁₆	14.670	14⅝	0.940	¹⁵⁄₁₆	11¼	1⅝	¹⁵⁄₁₆
x109	32.0	14.32	14⅜	0.525	½	¼	14.605	14⅝	0.860	⅞	11¼	1⁹⁄₁₆	⅞
x 99	29.1	14.16	14⅛	0.485	½	¼	14.565	14⅝	0.780	¾	11¼	1⁷⁄₁₆	⅞
x 90	26.5	14.02	14	0.440	⁷⁄₁₆	¼	14.520	14½	0.710	¹¹⁄₁₆	11¼	1⅜	⅞
W 14x 82	24.1	14.31	14¼	0.510	½	¼	10.130	10⅛	0.855	⅞	11	1⅝	1
x 74	21.8	14.17	14⅛	0.450	⁷⁄₁₆	¼	10.070	10⅛	0.785	¹³⁄₁₆	11	1⁹⁄₁₆	¹⁵⁄₁₆
x 68	20.0	14.04	14	0.415	⁷⁄₁₆	¼	10.035	10	0.720	¾	11	1½	¹⁵⁄₁₆
x 61	17.9	13.89	13⅞	0.375	⅜	³⁄₁₆	9.995	10	0.645	⅝	11	1⁷⁄₁₆	¹⁵⁄₁₆
W 14x 53	15.6	13.92	13⅞	0.370	⅜	³⁄₁₆	8.060	8	0.660	¹¹⁄₁₆	11	1⁷⁄₁₆	¹⁵⁄₁₆
x 48	14.1	13.79	13¾	0.340	⁵⁄₁₆	³⁄₁₆	8.030	8	0.595	⅝	11	1⅜	⅞
x 43	12.6	13.66	13⅝	0.305	⁵⁄₁₆	³⁄₁₆	7.995	8	0.530	½	11	1⁵⁄₁₆	⅞
W 14x 38	11.2	14.10	14⅛	0.310	⁵⁄₁₆	³⁄₁₆	6.770	6¾	0.515	½	12	1¹⁄₁₆	⅝
x 34	10.0	13.98	14	0.285	⁵⁄₁₆	³⁄₁₆	6.745	6¾	0.455	⁷⁄₁₆	12	1	⅝
x 30	8.85	13.84	13⅞	0.270	¼	⅛	6.730	6¾	0.385	⅜	12	¹⁵⁄₁₆	⅝
W 14x 26	7.69	13.91	13⅞	0.255	¼	⅛	5.025	5	0.420	⁷⁄₁₆	12	¹⁵⁄₁₆	⁹⁄₁₆
x 22	6.49	13.74	13¾	0.230	¼	⅛	5.000	5	0.335	⁵⁄₁₆	12	⅞	⁹⁄₁₆

W SHAPES
Properties

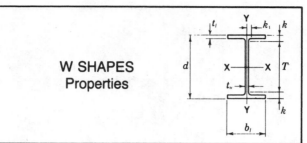

Nominal Wt. per Ft	Compact Section Criteria		F_y'''	X_1	$X_2 \times 10^6$	Elastic Properties						Plastic Modulus	
						Axis X-X			Axis Y-Y				
	$\frac{b_f}{2t_f}$	$\frac{h_c}{t_w}$				I	S	r	I	S	r	Z_x	Z_y
Lb.			Ksi	Ksi	$(1/Ksi)^2$	In.^{4}	In.^{3}	In.	In.^{4}	In.^{3}	In.	In.^{3}	In.^{3}
132	7.1	17.7	—	4180	428	1530	209	6.28	548	74.5	3.76	234	113
120	7.8	19.3	—	3830	601	1380	190	6.24	495	67.5	3.74	212	102
109	8.5	21.7	—	3490	853	1240	173	6.22	447	61.2	3.73	192	92.7
99	9.3	23.5	—	3190	1220	1110	157	6.17	402	55.2	3.71	173	83.6
90	10.2	25.9	—	2900	1750	999	143	6.14	362	49.9	3.70	157	75.6
82	5.9	22.4	—	3600	846	882	123	6.05	148	29.3	2.48	139	44.8
74	6.4	25.3	—	3290	1190	796	112	6.04	134	26.6	2.48	126	40.6
68	7.0	27.5	—	3020	1650	723	103	6.01	121	24.2	2.46	115	36.9
61	7.7	30.4	—	2720	2460	640	92.2	5.98	107	21.5	2.45	102	32.8
53	6.1	30.8	—	2830	2250	541	77.8	5.89	57.7	14.3	1.92	87.1	22.0
48	6.7	33.5	57	2580	3220	485	70.3	5.85	51.4	12.8	1.91	78.4	19.6
43	7.5	37.4	46	2320	4900	428	62.7	5.82	45.2	11.3	1.89	69.6	17.3
38	6.6	39.6	41	2190	6850	385	54.6	5.87	26.7	7.88	1.55	61.5	12.1
34	7.4	43.1	35	1970	10600	340	48.6	5.83	23.3	6.91	1.53	54.6	10.6
30	8.7	45.4	31	1750	17600	291	42.0	5.73	19.6	5.82	1.49	47.3	8.99
26	6.0	48.1	28	1890	13900	245	35.3	5.65	8.91	3.54	1.08	40.2	5.54
22	7.5	53.5	22	1610	27300	199	29.0	5.54	7.00	2.80	1.04	33.2	4.39

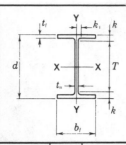

W SHAPES
Dimensions

Desig-nation	Area A	Depth d	Web Thickness t_w	$\frac{t_w}{2}$	Flange Width b_f	Flange Thickness t_f	Distance T	k	k_1
	In.²	In.	In.	In.	In.	In.	In.	In.	In.
W 12x336[a]	98.8	16.82	16⅞	1.775	1¾	⅞	13.385	13⅜	2.955	2¹⁵⁄₁₆	9½	3¹¹⁄₁₆	1½
x305[a]	89.6	16.32	16⅜	1.625	1⅝	¹³⁄₁₆	13.235	13¼	2.705	2¹¹⁄₁₆	9½	3⁷⁄₁₆	1⁷⁄₁₆
x279[a]	81.9	15.85	15⅞	1.530	1½	¾	13.140	13⅛	2.470	2½	9½	3³⁄₁₆	1⅜
x252[a]	74.1	15.41	15⅜	1.395	1⅜	¹¹⁄₁₆	13.005	13	2.250	2¼	9½	2¹⁵⁄₁₆	1⁵⁄₁₆
x230[a]	67.7	15.05	15	1.285	1⁵⁄₁₆	¹¹⁄₁₆	12.895	12⅞	2.070	2¹⁄₁₆	9½	2¾	1¼
x210[a]	61.8	14.71	14¾	1.180	1³⁄₁₆	⅝	12.790	12¾	1.900	1⅞	9½	2⅝	1¼
x190[a]	55.8	14.38	14⅜	1.060	1¹⁄₁₆	⁹⁄₁₆	12.670	12⅝	1.735	1¾	9½	2⁷⁄₁₆	1³⁄₁₆
x170[a]	50.0	14.03	14	0.960	¹⁵⁄₁₆	½	12.570	12⅝	1.560	1⁹⁄₁₆	9½	2¼	1⅛
x152	44.7	13.71	13¾	0.870	⅞	⁷⁄₁₆	12.480	12½	1.400	1⅜	9½	2⅛	1¹⁄₁₆
x136	39.9	13.41	13⅜	0.790	¹³⁄₁₆	⁷⁄₁₆	12.400	12⅜	1.250	1¼	9½	1¹⁵⁄₁₆	1
x120	35.3	13.12	13⅛	0.710	¹¹⁄₁₆	⅜	12.320	12⅜	1.105	1⅛	9½	1¹³⁄₁₆	1
x106	31.2	12.89	12⅞	0.610	⅝	⁵⁄₁₆	12.220	12¼	0.990	1	9½	1¹¹⁄₁₆	¹⁵⁄₁₆
x 96	28.2	12.71	12¾	0.550	⁹⁄₁₆	⁵⁄₁₆	12.160	12⅛	0.900	⅞	9½	1⅝	⅞
x 87	25.6	12.53	12½	0.515	½	¼	12.125	12⅛	0.810	¹³⁄₁₆	9½	1½	⅞
x 79	23.2	12.38	12⅜	0.470	½	¼	12.080	12⅛	0.735	¾	9½	1⁷⁄₁₆	⅞
x 72	21.1	12.25	12¼	0.430	⁷⁄₁₆	¼	12.040	12	0.670	¹¹⁄₁₆	9½	1⅜	⅞
x 65	19.1	12.12	12⅛	0.390	⅜	³⁄₁₆	12.000	12	0.605	⅝	9½	1⁵⁄₁₆	¹³⁄₁₆
W 12x 58	17.0	12.19	12¼	0.360	⅜	³⁄₁₆	10.010	10	0.640	⅝	9½	1⅜	¹³⁄₁₆
x 53	15.6	12.06	12	0.345	⅜	³⁄₁₆	9.995	10	0.575	⁹⁄₁₆	9½	1¼	¹³⁄₁₆
W 12x 50	14.7	12.19	12¼	0.370	⅜	³⁄₁₆	8.080	8⅛	0.640	⅝	9½	1⅜	¹³⁄₁₆
x 45	13.2	12.06	12	0.335	⁵⁄₁₆	³⁄₁₆	8.045	8	0.575	⁹⁄₁₆	9½	1¼	¹³⁄₁₆
x 40	11.8	11.94	12	0.295	⁵⁄₁₆	³⁄₁₆	8.005	8	0.515	½	9½	1¼	¾
W 12x 35	10.3	12.50	12½	0.300	⁵⁄₁₆	³⁄₁₆	6.560	6½	0.520	½	10½	1	⁹⁄₁₆
x 30	8.79	12.34	12⅜	0.260	¼	⅛	6.520	6½	0.440	⁷⁄₁₆	10½	¹⁵⁄₁₆	½
x 26	7.65	12.22	12¼	0.230	¼	⅛	6.490	6½	0.380	⅜	10½	⅞	½
W 12x 22	6.48	12.31	12¼	0.260	¼	⅛	4.030	4	0.425	⁷⁄₁₆	10½	⅞	½
x 19	5.57	12.16	12⅛	0.235	¼	⅛	4.005	4	0.350	⅜	10½	¹³⁄₁₆	½
x 16	4.71	11.99	12	0.220	¼	⅛	3.990	4	0.265	¼	10½	¾	½
x 14	4.16	11.91	11⅞	0.200	³⁄₁₆	⅛	3.970	4	0.225	¼	10½	¹¹⁄₁₆	½

W SHAPES
Properties

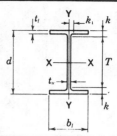

Nom-inal Wt. per Ft	Compact Section Criteria			X_1	$X_2 \times 10^6$	Elastic Properties						Plastic Modulus	
						Axis X-X			Axis Y-Y				
	$\frac{b_f}{2t_f}$	$\frac{h_c}{t_w}$	F_y'''	X_1	$X_2 \times 10^6$	I	S	r	I	S	r	Z_x	Z_y
Lb.			Ksi	Ksi	$(1/Ksi)^2$	In.^{4}	In.^{3}	In.	In.^{4}	In.^{3}	In.	In.^{3}	In.^{3}
336	2.3	5.5	—	12800	6.05	4060	483	6.41	1190	177	3.47	603	274
305	2.4	6.0	—	11800	8.17	3550	435	6.29	1050	159	3.42	537	244
279	2.7	6.3	—	11000	10.8	3110	393	6.16	937	143	3.38	481	220
252	2.9	7.0	—	10100	14.7	2720	353	6.06	828	127	3.34	428	196
230	3.1	7.6	—	9390	19.7	2420	321	5.97	742	115	3.31	386	177
210	3.4	8.2	—	8670	26.6	2140	292	5.89	664	104	3.28	348	159
190	3.7	9.2	—	7940	37.0	1890	263	5.82	589	93.0	3.25	311	143
170	4.0	10.1	—	7190	54.0	1650	235	5.74	517	82.3	3.22	275	126
152	4.5	11.2	—	6510	79.3	1430	209	5.66	454	72.8	3.19	243	111
136	5.0	12.3	—	5850	119	1240	186	5.58	398	64.2	3.16	214	98.0
120	5.6	13.7	—	5240	184	1070	163	5.51	345	56.0	3.13	186	85.4
106	6.2	15.9	—	4660	285	933	145	5.47	301	49.3	3.11	164	75.1
96	6.8	17.7	—	4250	405	833	131	5.44	270	44.4	3.09	147	67.5
87	7.5	18.9	—	3880	586	740	118	5.38	241	39.7	3.07	132	60.4
79	8.2	20.7	—	3530	839	662	107	5.34	216	35.8	3.05	119	54.3
72	9.0	22.6	—	3230	1180	597	97.4	5.31	195	32.4	3.04	108	49.2
65	9.9	24.9	—	2940	1720	533	87.9	5.28	174	29.1	3.02	96.8	44.1
58	7.8	27.0	—	3070	1470	475	78.0	5.28	107	21.4	2.51	86.4	32.5
53	8.7	28.1	—	2820	2100	425	70.6	5.23	95.8	19.2	2.48	77.9	29.1
50	6.3	26.2	—	3170	1410	394	64.7	5.18	56.3	13.9	1.96	72.4	21.4
45	7.0	29.0	—	2870	2070	350	58.1	5.15	50.0	12.4	1.94	64.7	19.0
40	7.8	32.9	59	2580	3110	310	51.9	5.13	44.1	11.0	1.93	57.5	16.8
35	6.3	36.2	49	2420	4340	285	45.6	5.25	24.5	7.47	1.54	51.2	11.5
30	7.4	41.8	37	2090	7950	238	38.6	5.21	20.3	6.24	1.52	43.1	9.56
26	8.5	47.2	29	1820	13900	204	33.4	5.17	17.3	5.34	1.51	37.2	8.17
22	4.7	41.8	37	2160	8640	156	25.4	4.91	4.66	2.31	0.847	29.3	3.66
19	5.7	46.2	30	1880	15600	130	21.3	4.82	3.76	1.88	0.822	24.7	2.98
16	7.5	49.4	26	1610	32000	103	17.1	4.67	2.82	1.41	0.773	20.1	2.26
14	8.8	54.3	22	1450	49300	88.6	14.9	4.62	2.36	1.19	0.753	17.4	1.90

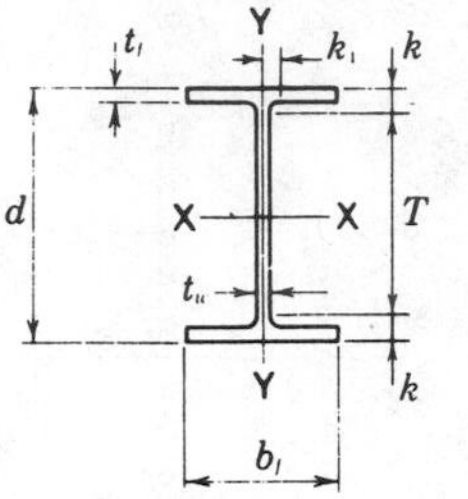

W SHAPES
Dimensions

Designation	Area A	Depth d		Web Thickness t_w		Web $\frac{t_w}{2}$	Flange Width b_f		Flange Thickness t_f		Distance T	Distance k	Distance k_1
	In.^{2}	In.		In.		In.	In.		In.		In.	In.	In.
W 10x112	32.9	11.36	11 3/8	0.755	3/4	3/8	10.415	10 3/8	1.250	1 1/4	7 5/8	1 7/8	15/16
x100	29.4	11.10	11 1/8	0.680	11/16	3/8	10.340	10 3/8	1.120	1 1/8	7 5/8	1 3/4	7/8
x 88	25.9	10.84	10 7/8	0.605	5/8	5/16	10.265	10 1/4	0.990	1	7 5/8	1 5/8	13/16
x 77	22.6	10.60	10 5/8	0.530	1/2	1/4	10.190	10 1/4	0.870	7/8	7 5/8	1 1/2	13/16
x 68	20.0	10.40	10 3/8	0.470	3/8	1/4	10.130	10 1/8	0.770	3/4	7 5/8	1 3/8	3/4
x 60	17.6	10.22	10 1/4	0.420	7/16	1/4	10.080	10 1/8	0.680	11/16	7 5/8	1 5/16	3/4
x 54	15.8	10.09	10 1/8	0.370	3/8	3/16	10.030	10	0.615	5/8	7 5/8	1 1/4	11/16
x 49	14.4	9.98	10	0.340	5/16	3/16	10.000	10	0.560	9/16	7 5/8	1 3/16	11/16
W 10x 45	13.3	10.10	10 1/8	0.350	3/8	3/16	8.020	8	0.620	5/8	7 5/8	1 1/4	11/16
x 39	11.5	9.92	9 7/8	0.315	5/16	3/16	7.985	8	0.530	1/2	7 5/8	1 1/8	11/16
x 33	9.71	9.73	9 3/4	0.290	5/16	3/16	7.960	8	0.435	7/16	7 5/8	1 1/16	11/16
W 10x 30	8.84	10.47	10 1/2	0.300	5/16	3/16	5.810	5 3/4	0.510	1/2	8 5/8	15/16	1/2
x 26	7.61	10.33	10 3/8	0.260	1/4	1/8	5.770	5 3/4	0.440	7/16	8 5/8	7/8	1/2
x 22	6.49	10.17	10 1/8	0.240	1/4	1/8	5.750	5 3/4	0.360	3/8	8 5/8	3/4	1/2
W 10x 19	5.62	10.24	10 1/4	0.250	1/4	1/8	4.020	4	0.395	3/8	8 5/8	13/16	1/2
x 17	4.99	10.11	10 1/8	0.240	1/4	1/8	4.010	4	0.330	5/16	8 5/8	3/4	1/2
x 15	4.41	9.99	10	0.230	1/4	1/8	4.000	4	0.270	1/4	8 5/8	11/16	7/16
x 12	3.54	9.87	9 7/8	0.190	3/16	1/8	3.960	4	0.210	3/16	8 5/8	5/8	7/16

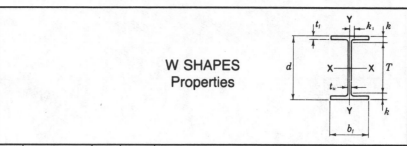

W SHAPES
Properties

Nom-inal Wt. per Ft	Compact Section Criteria					Elastic Properties						Plastic Modulus	
						Axis X-X			Axis Y-Y				
	$\frac{b_f}{2t_f}$	$\frac{h_c}{t_w}$	F_y'''	X_1	$X_2 \times 10^6$	I	S	r	I	S	r	Z_x	Z_y
Lb.			Ksi	Ksi	$(1/Ksi)^2$	In.^{4}	In.^{3}	In.	In.^{4}	In.^{3}	In.	In.^{3}	In.^{3}
112	4.2	10.4	—	7080	56.7	716	126	4.66	236	45.3	2.68	147	69.2
100	4.6	11.6	—	6400	83.8	623	112	4.60	207	40.0	2.65	130	61.0
88	5.2	13.0	—	5680	132	534	98.5	4.54	179	34.8	2.63	113	53.1
77	5.9	14.8	—	5010	213	455	85.9	4.49	154	30.1	2.60	97.6	45.9
68	6.6	16.7	—	4460	334	394	75.7	4.44	134	26.4	2.59	85.3	40.1
60	7.4	18.7	—	3970	525	341	66.7	4.39	116	23.0	2.57	74.6	35.0
54	8.2	21.2	—	3580	778	303	60.0	4.37	103	20.6	2.56	66.6	31.3
49	8.9	23.1	—	3280	1090	272	54.6	4.35	93.4	18.7	2.54	60.4	28.3
45	6.5	22.5	—	3650	758	248	49.1	4.32	53.4	13.3	2.01	54.9	20.3
39	7.5	25.0	—	3190	1300	209	42.1	4.27	45.0	11.3	1.98	46.8	17.2
33	9.1	27.1	—	2710	2510	170	35.0	4.19	36.6	9.20	1.94	38.8	14.0
30	5.7	29.5	—	2890	2160	170	32.4	4.38	16.7	5.75	1.37	36.6	8.84
26	6.6	34.0	55	2500	3790	144	27.9	4.35	14.1	4.89	1.36	31.3	7.50
22	8.0	36.9	47	2150	7170	118	23.2	4.27	11.4	3.97	1.33	26.0	6.10
19	5.1	35.4	51	2420	5160	96.3	18.8	4.14	4.29	2.14	0.874	21.6	3.35
17	6.1	36.9	47	2210	7820	81.9	16.2	4.05	3.56	1.78	0.844	18.7	2.80
15	7.4	38.5	43	1930	14300	68.9	13.8	3.95	2.89	1.45	0.810	16.0	2.30
12	9.4	46.6	30	1550	35400	53.8	10.9	3.90	2.18	1.10	0.785	12.6	1.74

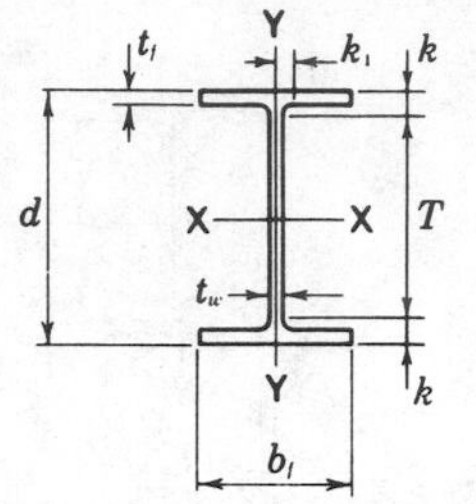

W SHAPES
Dimensions

Designation	Area A	Depth d		Web Thickness t_w		Web $\frac{t_w}{2}$	Flange Width b_f		Flange Thickness t_f		Distance T	Distance k	Distance k_1
	In.^{2}	In.		In.		In.	In.		In.		In.	In.	In.
W 8x 67	19.7	9.00	9	0.570	9/16	5/16	8.280	8 1/4	0.935	15/16	6 1/8	1 7/16	11/16
x 58	17.1	8.75	8 3/4	0.510	1/2	1/4	8.220	8 1/4	0.810	13/16	6 1/8	1 5/16	11/16
x 48	14.1	8.50	8 1/2	0.400	3/8	3/16	8.110	8 1/8	0.685	11/16	6 1/8	1 3/16	5/8
x 40	11.7	8.25	8 1/4	0.360	3/8	3/16	8.070	8 1/8	0.560	9/16	6 1/8	1 1/16	5/8
x 35	10.3	8.12	8 1/8	0.310	5/16	3/16	8.020	8	0.495	1/2	6 1/8	1	9/16
x 31	9.13	8.00	8	0.285	5/16	3/16	7.995	8	0.435	7/16	6 1/8	15/16	9/16
W 8x 28	8.25	8.06	8	0.285	5/16	3/16	6.535	6 1/2	0.465	7/16	6 1/8	15/16	9/16
x 24	7.08	7.93	7 7/8	0.245	1/4	1/8	6.495	6 1/2	0.400	3/8	6 1/8	7/8	9/16
W 8x 21	6.16	8.28	8 1/4	0.250	1/4	1/8	5.270	5 1/4	0.400	3/8	6 5/8	13/16	1/2
x 18	5.26	8.14	8 1/8	0.230	1/4	1/8	5.250	5 1/4	0.330	5/16	6 5/8	3/4	7/16
W 8 x15	4.44	8.11	8 1/8	0.245	1/4	1/8	4.015	4	0.315	5/16	6 5/8	3/4	1/2
x 13	3.84	7.99	8	0.230	1/4	1/8	4.000	4	0.255	1/4	6 5/8	11/16	7/16
x 10	2.96	7.89	7 7/8	0.170	3/16	1/8	3.940	4	0.205	3/16	6 5/8	5/8	7/16
W 6x 25	7.34	6.38	6 3/8	0.320	5/16	3/16	6.080	6 1/8	0.455	7/16	4 3/4	13/16	7/16
x 20	5.87	6.20	6 1/4	0.260	1/4	1/8	6.020	6	0.365	3/8	4 3/4	3/4	7/16
x 15	4.43	5.99	6	0.230	1/4	1/8	5.990	6	0.260	1/4	4 3/4	5/8	3/8
W 6x 16	4.74	6.28	6 1/4	0.260	1/4	1/8	4.030	4	0.405	3/8	4 3/4	3/4	7/16
x 12	3.55	6.03	6	0.230	1/4	1/8	4.000	4	0.280	1/4	4 3/4	5/8	3/8
x 9	2.68	5.90	5 7/8	0.170	3/16	1/8	3.940	4	0.215	3/16	4 3/4	9/16	3/8
W 5x 19	5.54	5.15	5 1/8	0.270	1/4	1/8	5.030	5	0.430	7/16	3 1/2	13/16	7/16
x 16	4.68	5.01	5	0.240	1/4	1/8	5.000	5	0.360	3/8	3 1/2	3/4	7/16
W 4x 13	3.83	4.16	4 1/8	0.280	1/4	1/8	4.060	4	0.345	3/8	2 3/4	11/16	7/16

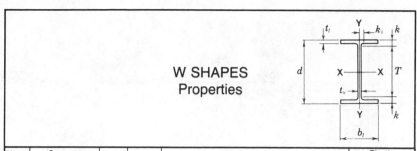

W SHAPES
Properties

Nom-inal Wt. per Ft	Compact Section Criteria			X_1	$X_2 \times 10^6$	Elastic Properties						Plastic Modulus	
	$\frac{b_f}{2t_f}$	$\frac{h_c}{t_w}$	F_y'''			Axis X-X			Axis Y-Y			Z_x	Z_y
						I	S	r	I	S	r		
Lb.			Ksi	Ksi	$(1/\text{Ksi})^2$	In.^{4}	In.^{3}	In.	In.^{4}	In.^{3}	In.	In.^{3}	In.^{3}
67	4.4	11.1	—	6620	73.9	272	60.4	3.72	88.6	21.4	2.12	70.2	32.7
58	5.1	12.4	—	5820	122	228	52.0	3.65	75.1	18.3	2.10	59.8	27.9
48	5.9	15.8	—	4860	238	184	43.3	3.61	60.9	15.0	2.08	49.0	22.9
40	7.2	17.6	—	4080	474	146	35.5	3.53	49.1	12.2	2.04	39.8	18.5
35	8.1	20.4	—	3610	761	127	31.2	3.51	42.6	10.6	2.03	34.7	16.1
31	9.2	22.2	—	3230	1180	110	27.5	3.47	37.1	9.27	2.02	30.4	14.1
28	7.0	22.2	—	3480	931	98.0	24.3	3.45	21.7	6.63	1.62	27.2	10.1
24	8.1	25.8	—	3020	1610	82.8	20.9	3.42	18.3	5.63	1.61	23.2	8.57
21	6.6	27.5	—	2890	2090	75.3	18.2	3.49	9.77	3.71	1.26	20.4	5.69
18	8.0	29.9	—	2490	3890	61.9	15.2	3.43	7.97	3.04	1.23	17.0	4.66
15	6.4	28.1	—	2670	3440	48.0	11.8	3.29	3.41	1.70	0.876	13.6	2.67
13	7.8	29.9	—	2370	5780	39.6	9.91	3.21	2.73	1.37	0.843	11.4	2.15
10	9.6	40.5	39	1760	17900	30.8	7.81	3.22	2.09	1.06	0.841	8.87	1.66
25	6.7	15.5	—	4410	369	53.4	16.7	2.70	17.1	5.61	1.52	18.9	8.56
20	8.2	19.1	—	3550	846	41.4	13.4	2.66	13.3	4.41	1.50	14.9	6.72
15	11.5	21.6	—	2740	2470	29.1	9.72	2.56	9.32	3.11	1.46	10.8	4.75
16	5.0	19.1	—	4010	591	32.1	10.2	2.60	4.43	2.20	0.966	11.7	3.39
12	7.1	21.6	—	3100	1740	22.1	7.31	2.49	2.99	1.50	0.918	8.30	2.32
9	9.2	29.2	—	2360	4980	16.4	5.56	2.47	2.19	1.11	0.905	6.23	1.72
19	5.8	14.0	—	5140	192	26.2	10.2	2.17	9.13	3.63	1.28	11.6	5.53
16	6.9	15.8	—	4440	346	21.3	8.51	2.13	7.51	3.00	1.27	9.59	4.57
13	5.9	10.6	—	5560	154	11.3	5.46	1.72	3.86	1.90	1.00	6.28	2.92

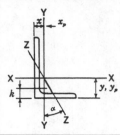

ANGLES
Equal legs and unequal legs
Properties for designing

Size and Thickness	k	Weight per Ft	Area	Axis X-X					
				I	S	r	y	Z	y_p
In.	In.	Lb.	In.²	In.⁴	In.³	In.	In.	In.³	In.
L 6x6 x1	1½	37.4	11.0	35.5	8.57	1.80	1.86	15.5	0.917
⅞	1⅜	33.1	9.73	31.9	7.63	1.81	1.82	13.8	0.811
¾	1¼	28.7	8.44	28.2	6.66	1.83	1.78	12.0	0.703
⅝	1⅛	24.2	7.11	24.2	5.66	1.84	1.73	10.2	0.592
⁹⁄₁₆	1¹⁄₁₆	21.9	6.43	22.1	5.14	1.85	1.71	9.26	0.536
½	1	19.6	5.75	19.9	4.61	1.86	1.68	8.31	0.479
⁷⁄₁₆	¹⁵⁄₁₆	17.2	5.06	17.7	4.08	1.87	1.66	7.34	0.422
⅜	⅞	14.9	4.36	15.4	3.53	1.88	1.64	6.35	0.363
⁵⁄₁₆	¹³⁄₁₆	12.4	3.65	13.0	2.97	1.89	1.62	5.35	0.304
L 6x4 x ⅞	1⅜	27.2	7.98	27.7	7.15	1.86	2.12	12.7	1.44
¾	1¼	23.6	6.94	24.5	6.25	1.88	2.08	11.2	1.38
⅝	1⅛	20.0	5.86	21.1	5.31	1.90	2.03	9.51	1.31
⁹⁄₁₆	1¹⁄₁₆	18.1	5.31	19.3	4,83	1.90	2.01	8.66	1.28
½	1	16.2	4.75	17.4	4.33	1.91	1.99	7.78	1.25
⁷⁄₁₆	¹⁵⁄₁₆	14.3	4.18	15.5	3.83	1.92	1.96	6.88	1.22
⅜	⅞	12.3	3.61	13.5	3.32	1.93	1.94	5.97	1.19
⁵⁄₁₆	¹³⁄₁₆	10.3	3.03	11.4	2.79	1.94	1.92	5.03	1.16
L 6x3½x ½	1	15.3	4.50	16.6	4.24	1.92	2.08	7.50	1.50
⅜	⅞	11.7	3.42	12.9	3.24	1.94	2.04	5.76	1.44
⁵⁄₁₆	¹³⁄₁₆	9.8	2.87	10.9	2.73	1.95	2.01	4.85	1.41
L 5x5 x ⅞	1⅜	27.2	7.98	17.8	5.17	1.49	1.57	9.33	0.798
¾	1¼	23.6	6.94	15.7	4.53	1.51	1.52	8.16	0.694
⅝	1⅛	20.0	5.86	13.6	3.86	1.52	1.48	6.95	0.586
½	1	16.2	4.75	11.3	3.16	1.54	1.43	5.68	0.475
⁷⁄₁₆	¹⁵⁄₁₆	14.3	4.18	10.0	2.79	1.55	1.41	5.03	0.418
⅜	⅞	12.3	3.61	8.74	2.42	1.56	1.39	4.36	0.361
⁵⁄₁₆	¹³⁄₁₆	10.3	3.03	7.42	2.04	1.57	1.37	3.68	0.303

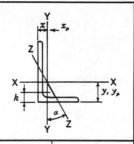

ANGLES
Equal legs and unequal legs
Properties for designing

Size and Thickness	Axis Y-Y						Axis Z-Z	
	I	S	r	x	Z	x_p	r	Tan α
In.	In.⁴	In.³	In.	In.	In.³	In.	In.	
L 6x6 x1	35.5	8.57	1.80	1.86	15.5	0.917	1.17	1.000
⅞	31.9	7.63	1.81	1.82	13.8	0.811	1.17	1.000
¾	28.2	6.66	1.83	1.78	12.0	0.703	1.17	1.000
⅝	24.2	5.66	1.84	1.73	10.2	0.592	1.18	1.000
⁹⁄₁₆	22.1	5.14	1.85	1.71	9.26	0.536	1.18	1.000
½	19.9	4.61	1.86	1.68	8.31	0.479	1.18	1.000
⁷⁄₁₆	17.7	4.08	1.87	1.66	7.34	0.422	1.19	1.000
⅜	15.4	3.53	1.88	1.64	6.35	0.363	1.19	1.000
⁵⁄₁₆	13.0	2.97	1.89	1.62	5.35	0.304	1.20	1.000
L 6x4 x ⅞	9.75	3.39	1.11	1.12	6.31	0.665	0.857	0.421
¾	8.68	2.97	1.12	1.08	5.47	0.578	0.860	0.428
⅝	7.52	2.54	1.13	1.03	4.62	0.488	0.864	0.435
⁹⁄₁₆	6.91	2.31	1.14	1.01	4.19	0.442	0.866	0.438
½	6.27	2.08	1.15	0.987	3.75	0.396	0.870	0.440
⁷⁄₁₆	5.60	1.85	1.16	0.964	3.30	0.349	0.873	0.443
⅜	4.90	1.60	1.17	0.941	2.85	0.301	0.877	0.446
⁵⁄₁₆	4.18	1.35	1.17	0.918	2.40	0.252	0.882	0.448
L 6x3½x ½	4.25	1.59	0.972	0.833	2.91	0.375	0.759	0.344
⅜	3.34	1.23	0.988	0.787	2.20	0.285	0.767	0.350
⁵⁄₁₆	2.85	1.04	0.996	0.763	1.85	0.239	0.772	0.352
L 5x5 x ⅞	17.8	5.17	1.49	1.57	9.33	0.798	0.973	1.000
¾	15.7	4.53	1.51	1.52	8.16	0.694	0.975	1.000
⅝	13.6	3.86	1.52	1.48	6.95	0.586	0.978	1.000
½	11.3	3.16	1.54	1.43	5.68	0.475	0.983	1.000
⁷⁄₁₆	10.0	2.79	1.55	1.41	5.03	0.418	0.986	1.000
⅜	8.74	2.42	1.56	1.39	4.36	0.361	0.990	1.000
⁵⁄₁₆	7.42	2.04	1.57	1.37	3.68	0.303	0.994	1.000

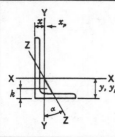

ANGLES
Equal legs and unequal legs
Properties for designing

Size and Thickness	k	Weight per Ft	Area	Axis X-X					
				I	S	r	y	Z	y_p
In.	In.	Lb.	In.^{2}	In.^{4}	In.^{3}	In.	In.	In.^{3}	In.
L 5x3½x¾	1¼	19.8	5.81	13.9	4.28	1.55	1.75	7.65	1.13
⅝	1⅛	16.8	4.92	12.0	3.65	1.56	1.70	6.55	1.06
½	1	13.6	4.00	9.99	2.99	1.58	1.66	5.38	1.00
⁷⁄₁₆	¹⁵⁄₁₆	12.0	3.53	8.90	2.64	1.59	1.63	4.77	0.969
⅜	⅞	10.4	3.05	7.78	2.29	1.60	1.61	4.14	0.938
⁵⁄₁₆	¹³⁄₁₆	8.7	2.56	6.60	1.94	1.61	1.59	3.49	0.906
¼	¾	7.0	2.06	5.39	1.57	1.62	1.56	2.83	0.875
L 5x3 x⅝	1	15.7	4.61	11.4	3.55	1.57	1.80	6.27	1.31
½	1	12.8	3.75	9.45	2.91	1.59	1.75	5.16	1.25
⁷⁄₁₆	¹⁵⁄₁₆	11.3	3.31	8.43	2.58	1.60	1.73	4.57	1.22
⅜	⅞	9.8	2.86	7.37	2.24	1.61	1.70	3.97	1.19
⁵⁄₁₆	¹³⁄₁₆	8.2	2.40	6.26	1.89	1.61	1.68	3.36	1.16
¼	¾	6.6	1.94	5.11	1.53	1.62	1.66	2.72	1.13
L 4x4 x¾	1⅛	18.5	5.44	7.67	2.81	1.19	1.27	5.07	0.680
⅝	1	15.7	4.61	6.66	2.40	1.20	1.23	4.33	0.576
½	⅞	12.8	3.75	5.56	1.97	1.22	1.18	3.56	0.469
⁷⁄₁₆	¹³⁄₁₆	11.3	3.31	4.97	1.75	1.23	1.16	3.16	0.414
⅜	¾	9.8	2.86	4.36	1.52	1.23	1.14	2.74	0.357
⁵⁄₁₆	¹¹⁄₁₆	8.2	2.40	3.71	1.29	1.24	1.12	2.32	0.300
¼	⅝	6.6	1.94	3.04	1.05	1.25	1.09	1.88	0.242
L 4x3½x⅝	1¹⁄₁₆	14.7	4.30	6.37	2.35	1.22	1.29	4.24	0.614
½	¹⁵⁄₁₆	11.9	3.50	5.32	1.94	1.23	1.25	3.50	0.500
⁷⁄₁₆	⅞	10.6	3.09	4.76	1.72	1.24	1.23	3.11	0.469
⅜	¹³⁄₁₆	9.1	2.67	4.18	1.49	1.25	1.21	2.71	0.438
⁵⁄₁₆	¾	7.7	2.25	3.56	1.26	1.26	1.18	2.29	0.406
¼	¹¹⁄₁₆	6.2	1.81	2.91	1.03	1.27	1.16	1.86	0.375

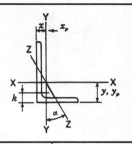

ANGLES
Equal legs and unequal legs
Properties for designing

Size and Thickness	Axis Y-Y						Axis Z-Z	
	I	S	r	x	Z	x_p	r	Tan
In.	In.^{4}	In.^{3}	In.	In.	In.^{3}	In.	In.	α
L 5x3½x¾	5.55	2.22	0.977	0.996	4.10	0.581	0.748	0.464
⅝	4.83	1.90	0.991	0.951	3.47	0.492	0.751	0.472
½	4.05	1.56	1.01	0.906	2.83	0.400	0.755	0.479
⁷⁄₁₆	3.63	1.39	1.01	0.883	2.49	0.353	0.758	0.482
⅜	3.18	1.21	1.02	0.861	2.16	0.305	0.762	0.486
⁵⁄₁₆	2.72	1.02	1.03	0.838	1.82	0.256	0.766	0.489
¼	2.23	0.830	1.04	0.814	1.47	0.206	0.770	0.492
L 5x3 x⅝	3.06	1.39	0.815	0.796	2.61	0.461	0.644	0.349
½	2.58	1.15	0.829	0.750	2.11	0.375	0.648	0.357
⁷⁄₁₆	2.32	1.02	0.837	0.727	1.86	0.331	0.651	0.361
⅜	2.04	0.888	0.845	0.704	1.60	0.286	0.654	0.364
⁵⁄₁₆	1.75	0.753	0.853	0.681	1.35	0.240	0.658	0.368
¼	1.44	0.614	0.861	0.657	1.09	0.194	0.663	0.371
L 4x4 x¾	7.67	2.81	1.19	1.27	5.07	0.680	0.778	1.000
⅝	6.66	2.40	1.20	1.23	4.33	0.576	0.779	1.000
½	5.56	1.97	1.22	1.18	3.56	0.469	0.782	1.000
⁷⁄₁₆	4.97	1.75	1.23	1.16	3.16	0.414	0.785	1.000
⅜	4.36	1.52	1.23	1.14	2.74	0.357	0.788	1.000
⁵⁄₁₆	3.71	1.29	1.24	1.12	2.32	0.300	0.791	1.000
¼	3.04	1.05	1.25	1.09	1.88	0.242	0.795	1.000
L 4x3½x⅝	4.52	1.84	1.03	1.04	3.33	0.537	0.719	0.745
½	3.79	1.52	1.04	1.00	2.73	0.438	0.722	0.750
⁷⁄₁₆	3.40	1.35	1.05	0.978	2.42	0.386	0.724	0.753
⅜	2.95	1.17	1.06	0.955	2.11	0.334	0.727	0.755
⁵⁄₁₆	2.55	0.994	1.07	0.932	1.78	0.281	0.730	0.757
¼	2.09	0.808	1.07	0.909	1.44	0.227	0.734	0.759

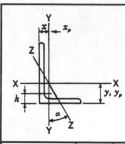

ANGLES
Equal legs and unequal legs
Properties for designing

Size and Thickness	k	Weight per Ft	Area	Axis X-X					
				I	S	r	y	Z	y_p
In.	In.	Lb.	In.²	In.⁴	In.³	In.	In.	In.³	In.
L 4 x3 x⅝	1¹⁄₁₆	13.6	3.98	6.03	2.30	1.23	1.37	4.12	0.813
½	¹⁵⁄₁₆	11.1	3.25	5.05	1.89	1.25	1.33	3.41	0.750
⁷⁄₁₆	⅞	9.8	2.87	4.52	1.68	1.25	1.30	3.03	0.719
⅜	¹³⁄₁₆	8.5	2.48	3.96	1.46	1.26	1.28	2.64	0.688
⁵⁄₁₆	¾	7.2	2.09	3.38	1.23	1.27	1.26	2.23	0.656
¼	¹¹⁄₁₆	5.8	1.69	2.77	1.00	1.28	1.24	1.82	0.625
L 3½x3½x½	⅞	11.1	3.25	3.64	1.49	1.06	1.06	2.68	0.464
⁷⁄₁₆	¹³⁄₁₆	9.8	2.87	3.26	1.32	1.07	1.04	2.38	0.410
⅜	¾	8.5	2.48	2.87	1.15	1.07	1.01	2.08	0.355
⁵⁄₁₆	¹¹⁄₁₆	7.2	2.09	2.45	0.976	1.08	0.990	1.76	0.299
¼	⅝	5.8	1.69	2.01	0.794	1.09	0.968	1.43	0.241
L 3½x3 x½	¹⁵⁄₁₆	10.2	3.00	3.45	1.45	1.07	1.13	2.63	0.500
⁷⁄₁₆	⅞	9.1	2.65	3.10	1.29	1.08	1.10	2.34	0.469
⅜	¹³⁄₁₆	7.9	2.30	2.72	1.13	1.09	1.08	2.04	0.438
⁵⁄₁₆	¾	6.6	1.93	2.33	0.954	1.10	1.06	1.73	0.406
¼	¹¹⁄₁₆	5.4	1.56	1.91	0.776	1.11	1.04	1.41	0.375
L 3½x2½x½	¹⁵⁄₁₆	9.4	2.75	3.24	1.41	1.09	1.20	2.53	0.750
⁷⁄₁₆	⅞	8.3	2.43	2.91	1.26	1.09	1.18	2.26	0.719
⅜	¹³⁄₁₆	7.2	2.11	2.56	1.09	1.10	1.16	1.97	0.688
⁵⁄₁₆	¾	6.1	1.78	2.19	0.927	1.11	1.14	1.67	0.656
¼	¹¹⁄₁₆	4.9	1.44	1.80	0.755	1.12	1.11	1.36	0.625
L 3 x3 x½	¹³⁄₁₆	9.4	2.75	2.22	1.07	0.898	0.932	1.93	0.458
⁷⁄₁₆	¾	8.3	2.43	1.99	0.954	0.905	0.910	1.72	0.406
⅜	¹¹⁄₁₆	7.2	2.11	1.76	0.833	0.913	0.888	1.50	0.352
⁵⁄₁₆	⅝	6.1	1.78	1.51	0.707	0.922	0.865	1.27	0.296
¼	⁹⁄₁₆	4.9	1.44	1.24	0.577	0.930	0.842	1.04	0.240
³⁄₁₆	½	3.71	1.09	0.962	0.441	0.939	0.820	0.794	0.182

ANGLES
Equal legs and unequal legs
Properties for designing

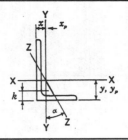

Size and Thickness	Axis Y-Y						Axis Z-Z	
	I	S	r	x	Z	x_p	r	Tan α
In.	In.⁴	In.³	In.	In.	In.³	In.	In.	
L 4 x3 x⅝	2.87	1.35	0.849	0.871	2.48	0.498	0.637	0.534
½	2.42	1.12	0.864	0.827	2.03	0.406	0.639	0.543
⁷⁄₁₆	2.18	0.992	0.871	0.804	1.79	0.359	0.641	0.547
⅜	1.92	0.866	0.879	0.782	1.56	0.311	0.644	0.551
⁵⁄₁₆	1.65	0.734	0.887	0.759	1.31	0.261	0.647	0.554
¼	1.36	0.599	0.896	0.736	1.06	0.211	0.651	0.558
L 3½x3½x½	3.64	1.49	1.06	1.06	2.68	0.464	0.683	1.000
⁷⁄₁₆	3.26	1.32	1.07	1.04	2.38	0.410	0.684	1.000
⅜	2.87	1.15	1.07	1.01	2.08	0.355	0.687	1.000
⁵⁄₁₆	2.45	0.976	1.08	0.990	1.76	0.299	0.690	1.000
¼	2.01	0.794	1.09	0.968	1.43	0.241	0.694	1.000
L 3½x3 x½	2.33	1.10	0.881	0.875	1.98	0.429	0.621	0.714
⁷⁄₁₆	2.09	0.975	0.889	0.853	1.76	0.379	0.622	0.718
⅜	1.85	0.851	0.897	0.830	1.53	0.328	0.625	0.721
⁵⁄₁₆	1.58	0.722	0.905	0.808	1.30	0.276	0.627	0.724
¼	1.30	0.589	0.914	0.785	1.05	0.223	0.631	0.727
L 3½x2½x½	1.36	0.760	0.704	0.705	1.40	0.393	0.534	0.486
⁷⁄₁₆	1.23	0.677	0.711	0.682	1.24	0.348	0.535	0.491
⅜	1.09	0.592	0.719	0.660	1.07	0.301	0.537	0.496
⁵⁄₁₆	0.939	0.504	0.727	0.637	0.907	0.254	0.540	0.501
¼	0.777	0.412	0.735	0.614	0.735	0.205	0.544	0.506
L 3 x3 x½	2.22	1.07	0.898	0.932	1.93	0.458	0.584	1.000
⁷⁄₁₆	1.99	0.954	0.905	0.910	1.72	0.406	0.585	1.000
⅜	1.76	0.833	0.913	0.888	1.50	0.352	0.587	1.000
⁵⁄₁₆	1.51	0.707	0.922	0.865	1.27	0.296	0.589	1.000
¼	1.24	0.577	0.930	0.842	1.04	0.240	0.592	1.000
³⁄₁₆	0.962	0.441	0.939	0.820	0.794	0.182	0.596	1.000

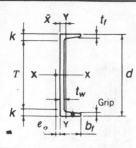

CHANNELS
AMERICAN STANDARD
Dimensions

Designation	Area A	Depth d	Web Thickness t_w		$\frac{t_w}{2}$	Flange Width b_f		Average thickness t_f		Distance T	k	Grip	Max. Flge. Fastener
	In.²	In.	In.		In.	In.		In.		In.	In.	In.	In.
C 15×50	14.7	15.00	0.716	11/16	3/8	3.716	3¾	0.650	⅝	12⅛	1 7/16	⅝	1
×40	11.8	15.00	0.520	½	¼	3.520	3½	0.650	⅝	12⅛	1 7/16	⅝	1
×33.9	9.96	15.00	0.400	⅜	3/16	3.400	3⅜	0.650	⅝	12⅛	1 7/16	⅝	1
C 12×30	8.82	12.00	0.510	½	¼	3.170	3⅛	0.501	½	9¾	1⅛	½	⅞
×25	7.35	12.00	0.387	⅜	3/16	3.047	3	0.501	½	9¾	1⅛	½	⅞
×20.7	6.09	12.00	0.282	5/16	⅛	2.942	3	0.501	½	9¾	1⅛	½	⅞
C 10×30	8.82	10.00	0.673	11/16	5/16	3.033	3	0.436	7/16	8	1	7/16	¾
×25	7.35	10.00	0.526	½	¼	2.886	2⅞	0.436	7/16	8	1	7/16	¾
×20	5.88	10.00	0.379	⅜	3/16	2.739	2¾	0.436	7/16	8	1	7/16	¾
×15.3	4.49	10.00	0.240	¼	⅛	2.600	2⅝	0.436	7/16	8	1	7/16	¾
C 9×20	5.88	9.00	0.448	7/16	¼	2.648	2⅝	0.413	7/16	7⅛	15/16	7/16	¾
×15	4.41	9.00	0.285	5/16	⅛	2.485	2½	0.413	7/16	7⅛	15/16	7/16	¾
×13.4	3.94	9.00	0.233	¼	⅛	2.433	2⅜	0.413	7/16	7⅛	15/16	7/16	¾
C 8×18.75	5.51	8.00	0.487	½	¼	2.527	2½	0.390	⅜	6⅛	15/16	⅜	¾
×13.75	4.04	8.00	0.303	5/16	⅛	2.343	2⅜	0.390	⅜	6⅛	15/16	⅜	¾
×11.5	3.38	8.00	0.220	¼	⅛	2.260	2¼	0.390	⅜	6⅛	15/16	⅜	¾
C 7×14.75	4.33	7.00	0.419	7/16	3/16	2.299	2¼	0.366	⅜	5¼	⅞	⅜	⅝
×12.25	3.60	7.00	0.314	5/16	3/16	2.194	2¼	0.366	⅜	5¼	⅞	⅜	⅝
× 9.8	2.87	7.00	0.210	3/16	⅛	2.090	2⅛	0.366	⅜	5¼	⅞	⅜	⅝
C 6×13	3.83	6.00	0.437	7/16	3/16	2.157	2⅛	0.343	5/16	4⅜	13/16	5/16	⅝
×10.5	3.09	6.00	0.314	5/16	3/16	2.034	2	0.343	5/16	4⅜	13/16	⅜	⅝
× 8.2	2.40	6.00	0.200	3/16	⅛	1.920	1⅞	0.343	5/16	4⅜	13/16	5/16	⅝
C 5× 9	2.64	5.00	0.325	5/16	3/16	1.885	1⅞	0.320	5/16	3½	¾	5/16	⅝
× 6.7	1.97	5.00	0.190	3/16	⅛	1.750	1¾	0.320	5/16	3½	¾	—	—
C 4× 7.25	2.13	4.00	0.321	5/16	3/16	1.721	1¾	0.296	5/16	2⅝	11/16	5/16	⅝
× 5.4	1.59	4.00	0.184	3/16	1/16	1.584	1⅝	0.296	5/16	2⅝	11/16	—	—
C 3× 6	1.76	3.00	0.356	⅜	3/16	1.596	1⅝	0.273	¼	1⅝	11/16	—	—
× 5	1.47	3.00	0.258	¼	⅛	1.498	1½	0.273	¼	1⅝	11/16	—	—
× 4.1	1.21	3.00	0.170	3/16	1/16	1.410	1⅜	0.273	¼	1⅝	11/16	—	—

CHANNELS
AMERICAN STANDARD
Properties

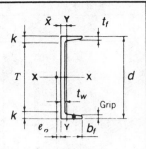

Nom- inal Wt. per Ft	$\bar{x}$	Shear Center Loca- tion e_o	$\dfrac{d}{A_f}$	Axis X-X			Axis Y-Y		
				I	S	r	I	S	r
Lb.	In.	In.		In.^{4}	In.^{3}	In.	In.^{4}	In.^{3}	In.
50	0.798	0.583	6.21	404	53.8	5.24	11.0	3.78	0.867
40	0.777	0.767	6.56	349	46.5	5.44	9.23	3.37	0.886
33.9	0.787	0.896	6.79	315	42.0	5.62	8.13	3.11	0.904
30	0.674	0.618	7.55	162	27.0	4.29	5.14	2.06	0.763
25	0.674	0.746	7.85	144	24.1	4.43	4.47	1.88	0.780
20.7	0.698	0.870	8.13	129	21.5	4.61	3.88	1.73	0.799
30	0.649	0.369	7.55	103	20.7	3.42	3.94	1.65	0.669
25	0.617	0.494	7.94	91.2	18.2	3.52	3.36	1.48	0.676
20	0.606	0.637	8.36	78.9	15.8	3.66	2.81	1.32	0.692
15.3	0.634	0.796	8.81	67.4	13.5	3.87	2.28	1.16	0.713
20	0.583	0.515	8.22	60.9	13.5	3.22	2.42	1.17	0.642
15	0.586	0.682	8.76	51.0	11.3	3.40	1.93	1.01	0.661
13.4	0.601	0.743	8.95	47.9	10.6	3.48	1.76	0.962	0.669
18.75	0.565	0.431	8.12	44.0	11.0	2.82	1.98	1.01	0.599
13.75	0.553	0.604	8.75	36.1	9.03	2.99	1.53	0.854	0.615
11.5	0.571	0.697	9.08	32.6	8.14	3.11	1.32	0.781	0.625
14.75	0.532	0.441	8.31	27.2	7.78	2.51	1.38	0.779	0.564
12.25	0.525	0.538	8.71	24.2	6.93	2.60	1.17	0.703	0.571
9.8	0.540	0.647	9.14	21.3	6.08	2.72	0.968	0.625	0.581
13	0.514	0.380	8.10	17.4	5.80	2.13	1.05	0.642	0.525
10.5	0.499	0.486	8.59	15.2	5.06	2.22	0.866	0.564	0.529
8.2	0.511	0.599	9.10	13.1	4.38	2.34	0.693	0.492	0.537
9	0.478	0.427	8.29	8.90	3.56	1.83	0.632	0.450	0.489
6.7	0.484	0.552	8.93	7.49	3.00	1.95	0.479	0.378	0.493
7.25	0.459	0.386	7.84	4.59	2.29	1.47	0.433	0.343	0.450
5.4	0.457	0.502	8.52	3.85	1.93	1.56	0.319	0.283	0.449
6	0.455	0.322	6.87	2.07	1.38	1.08	0.305	0.268	0.416
5	0.438	0.392	7.32	1.85	1.24	1.12	0.247	0.233	0.410
4.1	0.436	0.461	7.78	1.66	1.10	1.17	0.197	0.202	0.404

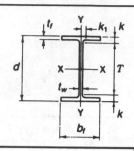

W SHAPES
Dimensions

Designation						Area A	Depth d	Web		Flange		Distance		
								Thickness t_w	$\dfrac{t_w}{2}$	Width b_f	Thickness t_f	T	k	k_1
mm	x	kg/m	in.	x	lb/ft	mm²	mm	mm	mm	mm	mm	mm	mm	mm
W	410 x	85	W	16 x	57	10800	417	10.90	5.45	181.0	18.2	347	35	16
W	410 x	74	W	16 x	50	9510	413	9.65	4.83	180.0	16.0	347	33	15
W	410 x	67	W	16 x	45	8560	410	8.76	4.38	179.0	14.4	346	32	15
W	410 x	60	W	16 x	40	7600	407	7.75	3.88	178.0	12.8	347	30	14
W	410 x	53	W	16 x	36	6820	403	7.49	3.75	177.0	10.9	345	29	14
W	410 x	46	W	16 x	31	5890	403	6.99	3.50	140.0	11.2	345	29	14
W	410 x	39	W	16 x	26	4960	399	6.35	3.18	140.0	8.8	345	27	13
W	360 x	1086ª	W	14 x	730	138000	569	78.00	39.00	454.0	125.0	287	141	54
W	360 x	990ª	W	14 x	665	126000	550	71.90	35.95	448.0	115.0	286	132	51
W	360 x	900ª	W	14 x	605	115000	531	65.90	32.95	442.0	106.0	287	122	48
W	360 x	818ª	W	14 x	550	104000	514	60.50	30.25	437.0	97.0	286	114	45
W	360 x	744ª	W	14 x	500	94800	498	55.60	27.80	432.0	88.9	286	106	43
W	360 x	677ª	W	14 x	455	86300	483	51.20	25.60	428.0	81.5	287	98	41
W	360 x	634ª	W	14 x	426	80800	474	47.60	23.80	424.0	77.1	286	94	39
W	360 x	592ª	W	14 x	398	75500	465	45.00	22.50	421.0	72.3	287	89	38
W	360 x	551ª	W	14 x	370	70200	455	42.00	21.00	418.0	67.6	287	84	36
W	360 x	509ª	W	14 x	342	64900	446	39.10	19.55	416.0	62.7	288	79	35
W	360 x	463ª	W	14 x	311	59000	435	35.80	17.90	412.0	57.4	285	75	33
W	360 x	421ª	W	14 x	283	53700	425	32.80	16.40	409.0	52.6	285	70	32
W	360 x	382ª	W	14 x	257	48800	416	29.80	14.90	406.0	48.0	286	65	30
W	360 x	347ª	W	14 x	233	44200	407	27.20	13.60	404.0	43.7	287	60	29
W	360 x	314	W	14 x	211	40000	399	24.90	12.45	401.0	39.6	285	57	28
W	360 x	287	W	14 x	193	36600	393	22.60	11.30	399.0	36.6	285	54	27
W	360 x	262	W	14 x	176	33400	387	21.10	10.55	398.0	33.3	285	51	26
W	360 x	237	W	14 x	159	30200	380	18.90	9.45	395.0	30.2	284	48	25
W	360 x	216	W	14 x	145	27500	375	17.30	8.65	394.0	27.7	287	44	24

W SHAPES
Properties

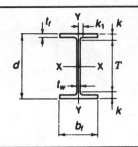

Nominal Mass per Meter kg/m	$\dfrac{b_f}{2t_f}$	$\dfrac{h_c}{t_w}$	F_y''' $\dfrac{N}{mm^2}$	X_1 $\dfrac{N}{mm^2}$	X_2 $\dfrac{10^{-8}mm^4}{N^2}$	I 10^6mm^4	S 10^3mm^3	r mm	I 10^6mm^4	S 10^3mm^3	r mm	Z_x 10^3mm^3	Z_y 10^3mm^3
85	5.0	33.0	407	18 300	7 150	315	1 510	171	18.0	199	40.8	1 730	310
74	5.6	37.4	317	16 100	11 600	275	1 330	170	15.6	173	40.5	1 510	269
67	6.2	41.2	262	14 600	17 400	245	1 200	169	13.8	154	40.2	1 350	239
60	6.9	46.6	207	13 000	27 100	216	1 060	169	12.0	135	39.7	1 200	209
53	8.1	48.1	193	11 700	43 800	186	923	165	10.1	114	38.5	1 050	177
46	6.3	51.6	165	12 000	42 100	156	774	163	5.14	73.4	29.5	884	115
39	8.0	56.8	138	10 100	86 000	126	632	159	4.02	57.4	28.5	727	90.2
1086	1.8	3.7	—	121 000	4	5 960	20 900	208	1 960	8 630	119	27 200	13 400
990	2.0	4.0	—	112 000	5	5 190	18 900	203	1 730	7 720	117	24 300	12 000
900	2.1	4.4	—	104 000	7	4 500	16 900	198	1 530	6 920	115	21 600	10 700
818	2.3	4.8	—	97 900	9	3 920	15 300	194	1 360	6 220	114	19 300	9 560
744	2.4	5.2	—	90 300	12	3 420	13 700	190	1 200	5 560	113	17 200	8 550
677	2.6	5.7	—	84 100	15	2 990	12 400	186	1 070	5 000	111	15 300	7 680
634	2.8	6.1	—	79 300	19	2 740	11 600	184	983	4 640	110	14 200	7 120
592	2.9	6.4	—	75 200	23	2 500	10 800	182	902	4 290	109	13 100	6 570
551	3.1	6.9	—	71 000	29	2 260	9 930	179	825	3 950	108	12 100	6 050
509	3.3	7.4	—	66 200	38	2 050	9 190	178	754	3 630	108	11 000	5 550
463	3.6	8.1	—	60 800	51	1 800	8 280	175	670	3 250	107	9 880	4 980
421	3.9	8.8	—	56 000	70	1 600	7 530	173	601	2 940	106	8 880	4 490
382	4.2	9.7	—	51 400	97	1 410	6 780	170	536	2 640	105	7 970	4 030
347	4.6	10.7	—	47 000	137	1 250	6 140	168	481	2 380	104	7 140	3 630
314	5.1	11.6	—	43 000	193	1 100	5 510	166	426	2 120	103	6 370	3 240
287	5.5	12.8	—	39 600	263	997	5 070	165	388	1 940	103	5 810	2 960
262	6.0	13.7	—	36 400	364	894	4 620	164	350	1 760	102	5 260	2 680
237	6.5	15.3	—	33 000	524	788	4 150	162	310	1 570	101	4 690	2 390
216	7.1	16.8	—	30 300	732	712	3 800	161	283	1 440	101	4 260	2 180

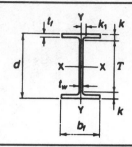

W SHAPES
Dimensions

Designation						Area A	Depth d	Web Thick- ness t_w	$\frac{t_w}{2}$	Flange Width b_f	Flange Thick- ness t_f	Distance T	Distance k	Distance k_1
mm x kg/m			in. x lb/ft			mm²	mm	mm	mm	mm	mm	mm	mm	mm
W	360	x	196	W	14 x 132	25 000	372	16.40	8.20	374.0	26.2	286	43	23
W	360	x	179	W	14 x 120	22 800	368	15.00	7.50	373.0	23.9	286	41	23
W	360	x	162	W	14 x 109	20 700	364	13.30	6.65	371.0	21.8	284	40	22
W	360	x	147	W	14 x 99	18 800	360	12.30	6.15	370.0	19.8	286	37	21
W	360	x	134	W	14 x 90	17 100	356	11.20	5.60	369.0	18.0	286	35	21
W	360	x	122	W	14 x 82	15 500	363	13.00	6.50	257.0	21.7	281	41	22
W	360	x	110	W	14 x 74	14 100	360	11.40	5.70	256.0	19.9	280	40	21
W	360	x	101	W	14 x 68	12 900	357	10.50	5.25	255.0	18.3	281	38	20
W	360	x	91	W	14 x 61	11 600	353	9.52	4.76	254.0	16.4	279	37	20
W	360	x	79	W	14 x 53	10 100	354	9.40	4.70	205.0	16.8	280	37	20
W	360	x	72	W	14 x 48	9 130	350	8.64	4.32	204.0	15.1	280	35	20
W	360	x	64	W	14 x 43	8 150	347	7.75	3.88	203.0	13.5	281	33	19
W	360	x	57	W	14 x 38	7 200	358	7.87	3.94	172.0	13.1	304	27	14
W	360	x	51	W	14 x 34	6 450	355	7.24	3.62	171.0	11.6	305	25	14
W	360	x	45	W	14 x 30	5 710	352	6.86	3.43	171.0	9.8	304	24	14
W	360	x	39	W	14 x 26	4 960	353	6.48	3.24	128.0	10.7	305	24	13
W	360	x	33	W	14 x 22	4 190	349	5.84	2.92	127.0	8.5	305	22	13

W SHAPES
Properties

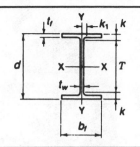

Nom-	Compact					Elastic	Properties					Plastic	
inal	Section			X_1	X_2	Axis X-X			Axis Y-Y			Modulus	
Mass	Criteria					I	S	r	I	S	r	Z_x	Z_y
per	$\dfrac{b_f}{2t_f}$	$\dfrac{h_c}{t_w}$	F_y'''										
Meter			$\dfrac{N}{mm^2}$	$\dfrac{N}{mm^2}$	$\dfrac{10^{-8}mm^4}{N^2}$								
kg/m						10^6mm^4	10^3mm^3	mm	10^6mm^4	10^3mm^3	mm	10^3mm^3	10^3mm^3
196	7.1	17.7	—	28800	900	636	3420	159	229	1220	95.7	3840	1860
179	7.8	19.3	—	26400	1260	575	3130	159	207	1110	95.3	3480	1680
162	8.5	21.7	—	24100	1790	516	2840	158	186	1000	94.8	3140	1520
147	9.3	23.5	—	22000	2570	463	2570	157	167	903	94.2	2840	1370
134	10.2	25.9	—	20000	3680	415	2330	156	151	818	94.0	2560	1240
122	5.9	22.4	—	24800	1780	365	2010	153	61.5	479	63.0	2270	732
110	6.4	25.3	—	22700	2500	331	1840	153	55.7	435	62.9	2060	664
101	7.0	27.5	—	20800	3470	302	1690	153	50.6	397	62.6	1880	606
91	7.7	30.4	—	18800	5170	267	1510	152	44.8	353	62.1	1680	538
79	6.1	30.8	—	19500	4730	227	1280	150	24.2	236	48.9	1430	362
72	6.7	33.5	393	17800	6770	201	1150	148	21.4	210	48.4	1280	322
64	7.5	37.4	317	16000	10300	179	1030	148	18.8	185	48.0	1140	284
57	6.6	39.6	283	15100	14400	160	894	149	11.1	129	39.3	1010	199
51	7.4	43.1	241	13600	22300	141	794	148	9.68	113	38.7	895	174
45	8.7	45.4	214	12100	37000	121	688	146	8.16	95.4	37.8	776	147
39	6.0	48.1	193	13000	29200	102	578	143	3.75	58.6	27.5	661	91.6
33	7.5	53.3	152	11100	57400	82.9	475	141	2.91	45.8	26.4	544	71.9

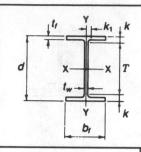

W SHAPES
Dimensions

Designation								Area A	Depth d	Web Thickness t_w	$\frac{t_w}{2}$	Flange Width b_f	Flange Thickness t_f	Distance T	k	k_1
mm	x	kg/m		in.	x	lb/ft		mm²	mm	mm	mm	mm	mm	mm	mm	mm
W	310	x	500ᵃ	W	12	x	336	63700	427	45.10	22.55	340.0	75.1	239	94	38
W	310	x	454ᵃ	W	12	x	305	57800	415	41.30	20.65	336.0	68.7	241	87	36
W	310	x	415ᵃ	W	12	x	279	52800	403	38.90	19.45	334.0	62.7	241	81	35
W	310	x	375ᵃ	W	12	x	252	47800	391	35.40	17.70	330.0	57.2	241	75	33
W	310	x	342ᵃ	W	12	x	230	43700	382	32.60	16.30	328.0	52.6	242	70	32
W	310	x	313ᵃ	W	12	x	210	39900	374	30.00	15.00	325.0	48.3	240	67	30
W	310	x	283	W	12	x	190	36000	365	26.90	13.45	322.0	44.1	241	62	29
W	310	x	253	W	12	x	170	32300	356	24.40	12.20	319.0	39.6	242	57	27
W	310	x	226	W	12	x	152	28900	348	22.10	11.05	317.0	35.6	240	54	26
W	310	x	202	W	12	x	136	25800	341	20.10	10.05	315.0	31.8	243	49	25
W	310	x	179	W	12	x	120	22800	333	18.00	9.00	313.0	28.1	241	46	24
W	310	x	158	W	12	x	106	20100	327	15.50	7.75	310.0	25.1	241	43	23
W	310	x	143	W	12	x	96	18200	323	14.00	7.00	309.0	22.9	241	41	22
W	310	x	129	W	12	x	87	16500	318	13.10	6.55	308.0	20.6	242	38	22
W	310	x	117	W	12	x	79	15000	314	11.90	5.95	307.0	18.7	240	37	21
W	310	x	107	W	12	x	72	13600	311	10.90	5.45	306.0	17.0	241	35	21
W	310	x	97	W	12	x	65	12300	308	9.91	4.96	305.0	15.4	242	33	20
W	310	x	86	W	12	x	58	11000	310	9.14	4.57	254.0	16.3	240	35	20
W	310	x	79	W	12	x	53	10000	306	8.76	4.38	254.0	14.6	242	32	20
W	310	x	74	W	12	x	50	9480	310	9.40	4.70	205.0	16.3	240	35	20
W	310	x	67	W	12	x	45	8530	306	8.51	4.26	204.0	14.6	242	32	19
W	310	x	60	W	12	x	40	7600	303	7.49	3.75	203.0	13.1	239	32	19
W	310	x	52	W	12	x	35	6670	318	7.62	3.81	167.0	13.2	268	25	11
W	310	x	45	W	12	x	30	5670	313	6.60	3.30	166.0	11.2	265	24	11
W	310	x	39	W	12	x	26	4930	310	5.84	2.92	165.0	9.7	266	22	11
W	310	x	33	W	12	x	22	4180	313	6.60	3.30	102.0	10.8	269	22	11
W	310	x	28	W	12	x	19	3600	309	5.97	2.99	102.0	8.9	267	21	11
W	310	x	24	W	12	x	16	3040	305	5.59	2.80	101.0	6.7	267	19	10
W	310	x	21	W	12	x	14	2680	303	5.08	2.54	101.0	5.7	269	17	10

W SHAPES
Properties

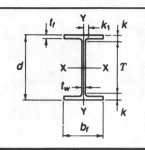

Nom-inal Mass per Meter kg/m	Compact Section Criteria			X_1	X_2	Elastic Properties						Plastic Modulus	
						Axis X-X			Axis Y-Y				
	$\dfrac{b_f}{2t_f}$	$\dfrac{h_c}{t_w}$	F_y''' $\dfrac{N}{mm^2}$	$\dfrac{N}{mm^2}$	$\dfrac{10^{-8}mm^4}{N^2}$	I 10^6mm^4	S 10^3mm^3	r mm	I 10^6mm^4	S 10^3mm^3	r mm	Z_x 10^3mm^3	Z_y 10^3mm^3
500	2.3	5.5	—	88 300	13	1 690	7 920	163	494	2 910	88.1	9 880	4 490
454	2.4	6.0	—	81 400	17	1 480	7 130	160	436	2 600	86.9	8 820	4 000
415	2.7	6.3	—	75 800	23	1 300	6 450	157	391	2 340	86.1	7 900	3 610
375	2.9	7.0	—	69 600	31	1 130	5 780	154	344	2 080	84.8	7 000	3 210
342	3.1	7.6	—	64 700	41	1 010	5 290	152	310	1 890	84.2	6 330	2 910
313	3.4	8.2	—	59 800	56	896	4 790	150	277	1 700	83.3	5 720	2 620
283	3.7	9.2	—	54 700	78	787	4 310	148	246	1 530	82.7	5 100	2 340
253	4.0	10.1	—	49 600	114	682	3 830	145	215	1 350	81.6	4 490	2 060
226	4.5	11.2	—	44 900	167	596	3 430	144	189	1 190	80.9	3 980	1 830
202	5.0	12.3	—	40 300	250	520	3 050	142	166	1 050	80.2	3 510	1 610
179	5.6	13.7	—	36 100	387	445	2 670	140	144	920	79.5	3 050	1 400
158	6.2	15.9	—	32 100	600	386	2 360	139	125	806	78.9	2 670	1 220
143	6.8	17.7	—	29 300	852	348	2 150	138	113	731	78.8	2 420	1 110
129	7.5	18.9	—	26 800	1 230	308	1 940	137	100	649	77.8	2 160	991
117	8.2	20.7	—	24 300	1 760	275	1 750	135	90.2	588	77.5	1 950	893
107	9.0	22.6	—	22 300	2 480	248	1 590	135	81.2	531	77.3	1 770	806
97	9.9	24.9	—	20 300	3 620	222	1 440	134	72.9	478	77.0	1 590	725
86	7.8	27.0	—	21 200	3 090	199	1 280	135	44.6	351	63.7	1 420	533
79	8.7	28.1	—	19 400	4 420	177	1 160	133	39.9	314	63.2	1 280	478
74	6.3	26.2	—	21 900	2 970	165	1 060	132	23.4	228	49.7	1 190	350
67	7.0	29.0	—	19 800	4 350	145	948	130	20.7	203	49.3	1 060	310
60	7.8	32.9	407	17 800	6 540	129	851	130	18.3	180	49.1	941	275
52	6.3	36.2	338	16 700	9 130	119	748	134	10.3	123	39.3	841	189
45	7.4	41.8	255	14 400	16 700	99.2	634	132	8.55	103	38.8	708	158
39	8.5	47.2	200	12 500	29 200	84.8	547	131	7.23	87.6	38.3	609	134
33	4.7	41.8	255	14 900	18 200	65.0	415	125	1.92	37.6	21.4	480	59.6
28	5.7	46.2	207	13 000	32 800	54.2	351	123	1.58	31.0	20.9	406	49.1
24	7.5	49.4	179	11 100	67 300	42.8	281	119	1.16	23.0	19.5	329	36.8
21	8.8	54.3	152	10 002	104 000	37.0	244	117	0.986	19.5	19.2	287	31.2

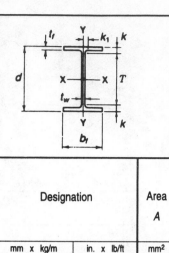

W SHAPES
Dimensions

Designation						Area	Depth	Web			Flange		Distance		
								Thick-ness t_w	$\dfrac{t_w}{2}$		Width b_f	Thick-ness t_f	T	k	k_1
						A	d								
mm	x	kg/m	in.	x	lb/ft	mm²	mm	mm	mm		mm	mm	mm	mm	mm
W	250	x	167	W	10	x	112	21 300	289	19.20	9.60	265.0	31.8	193	48	22
W	250	x	149	W	10	x	100	19 000	282	17.30	8.65	263.0	28.4	194	44	21
W	250	x	131	W	10	x	88	16 700	275	15.40	7.70	261.0	25.1	193	41	20
W	250	x	115	W	10	x	77	14 600	269	13.50	6.75	259.0	22.1	193	38	19
W	250	x	101	W	10	x	68	12 900	264	11.90	5.95	257.0	19.6	194	35	19
W	250	x	89	W	10	x	60	11 400	260	10.70	5.35	256.0	17.3	194	33	18
W	250	x	80	W	10	x	54	10 200	256	9.40	4.70	255.0	15.6	192	32	17
W	250	x	73	W	10	x	49	9 310	253	8.64	4.32	254.0	14.2	193	30	17
W	250	x	67	W	10	x	45	8 560	257	8.89	4.45	204.0	15.7	193	32	17
W	250	x	58	W	10	x	39	7 400	252	8.00	4.00	203.0	13.5	194	29	17
W	250	x	49	W	10	x	33	6 260	247	7.37	3.69	202.0	11.0	193	27	16
W	250	x	45	W	10	x	30	5 700	266	7.62	3.81	148.0	13.0	218	24	11
W	250	x	39	W	10	x	26	4 910	262	6.60	3.30	147.0	11.2	218	22	11
W	250	x	33	W	10	x	22	4 180	258	6.10	3.05	146.0	9.1	220	19	11
W	250	x	28	W	10	x	19	3 620	260	6.35	3.18	102.0	10.0	218	21	11
W	250	x	25	W	10	x	17	3 220	257	6.10	3.05	102.0	8.4	219	19	11
W	250	x	22	W	10	x	15	2 850	254	5.84	2.92	102.0	6.9	220	17	11
W	250	x	18	W	10	x	12	2 280	251	4.83	2.42	101.0	5.3	219	16	10

W SHAPES
Properties

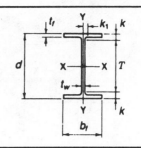

Nominal Mass per Meter kg/m	Compact Section Criteria			X_1	X_2	Elastic Properties						Plastic Modulus	
						Axis X-X			Axis Y-Y				
	$\dfrac{b_f}{2t_f}$	$\dfrac{h_c}{t_w}$	F_y''' $\dfrac{N}{mm^2}$	$\dfrac{N}{mm^2}$	$\dfrac{10^{-8}mm^4}{N^2}$	I 10^6mm^4	S 10^3mm^3	r mm	I 10^6mm^4	S 10^3mm^3	r mm	Z_x 10^3mm^3	Z_y 10^3mm^3
167	4.2	10.4	—	48 800	119	300	2 080	119	98.8	746	68.1	2 430	1 140
149	4.6	11.6	—	44 100	176	259	1 840	117	86.2	656	67.4	2 130	1 000
131	5.2	13.0	—	39 200	278	221	1 610	115	74.5	571	66.8	1 850	870
115	5.9	14.8	—	34 500	448	189	1 410	114	64.1	495	66.3	1 600	753
101	6.6	16.7	—	30 800	703	164	1 240	113	55.5	432	65.6	1 400	656
89	7.4	18.7	—	27 400	1 100	143	1 100	112	48.4	378	65.2	1 230	574
80	8.2	21.2	—	24 700	1 640	126	984	111	43.1	338	65.0	1 090	513
73	8.9	23.1	—	22 600	2 290	113	893	110	38.8	306	64.6	985	463
67	6.5	22.5	—	25 200	1 590	104	809	110	22.2	218	50.9	901	332
58	7.5	25.0	—	22 000	2 730	87.3	693	109	18.8	185	50.4	770	283
49	9.1	27.1	—	18 700	5 280	70.6	572	106	15.1	150	49.1	633	228
45	5.7	29.5	—	19 900	4 540	71.1	535	112	7.03	95	35.1	602	146
39	6.6	34.0	379	17 200	7 970	60.1	459	111	5.94	80.8	34.8	514	124
33	8.0	36.9	324	14 800	15 100	49.1	381	108	4.75	65.1	33.7	426	99.9
28	5.1	35.4	352	16 700	10 900	39.9	307	105	1.78	34.9	22.2	352	54.7
25	6.1	36.9	324	15 200	16 500	34.2	266	103	1.49	29.2	21.5	306	46.1
22	7.4	38.5	296	13 300	30 100	28.8	227	101	1.22	23.9	20.7	263	38.0
18	9.4	46.6	207	10 700	74 500	22.5	179	99.3	0.919	18.2	20.1	208	28.8

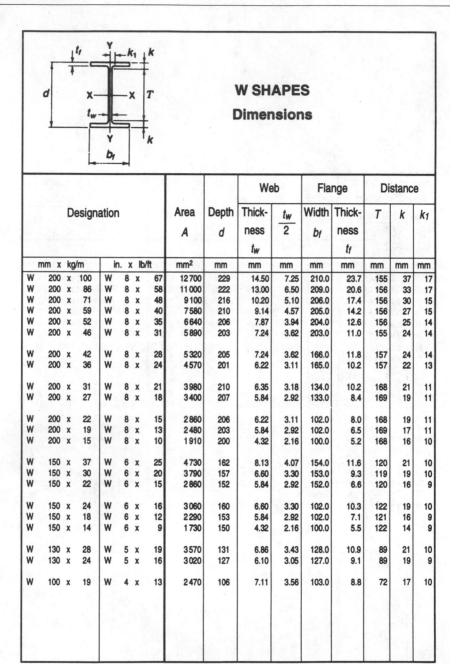

W SHAPES
Dimensions

Designation						Area A	Depth d	Web		Flange		Distance		
								Thickness t_w	$\dfrac{t_w}{2}$	Width b_f	Thickness t_f	T	k	k_1
mm x kg/m			in. x lb/ft			mm²	mm	mm	mm	mm	mm	mm	mm	mm
W	200 x	100	W	8 x	67	12 700	229	14.50	7.25	210.0	23.7	155	37	17
W	200 x	86	W	8 x	58	11 000	222	13.00	6.50	209.0	20.6	156	33	17
W	200 x	71	W	8 x	48	9 100	216	10.20	5.10	206.0	17.4	156	30	15
W	200 x	59	W	8 x	40	7 580	210	9.14	4.57	205.0	14.2	156	27	15
W	200 x	52	W	8 x	35	6 640	206	7.87	3.94	204.0	12.6	156	25	14
W	200 x	46	W	8 x	31	5 890	203	7.24	3.62	203.0	11.0	155	24	14
W	200 x	42	W	8 x	28	5 320	205	7.24	3.62	166.0	11.8	157	24	14
W	200 x	36	W	8 x	24	4 570	201	6.22	3.11	165.0	10.2	157	22	13
W	200 x	31	W	8 x	21	3 980	210	6.35	3.18	134.0	10.2	168	21	11
W	200 x	27	W	8 x	18	3 400	207	5.84	2.92	133.0	8.4	169	19	11
W	200 x	22	W	8 x	15	2 860	206	6.22	3.11	102.0	8.0	168	19	11
W	200 x	19	W	8 x	13	2 480	203	5.84	2.92	102.0	6.5	169	17	11
W	200 x	15	W	8 x	10	1 910	200	4.32	2.16	100.0	5.2	168	16	10
W	150 x	37	W	6 x	25	4 730	162	8.13	4.07	154.0	11.6	120	21	10
W	150 x	30	W	6 x	20	3 790	157	6.60	3.30	153.0	9.3	119	19	10
W	150 x	22	W	6 x	15	2 860	152	5.84	2.92	152.0	6.6	120	16	9
W	150 x	24	W	6 x	16	3 060	160	6.60	3.30	102.0	10.3	122	19	10
W	150 x	18	W	6 x	12	2 290	153	5.84	2.92	102.0	7.1	121	16	9
W	150 x	14	W	6 x	9	1 730	150	4.32	2.16	100.0	5.5	122	14	9
W	130 x	28	W	5 x	19	3 570	131	6.86	3.43	128.0	10.9	89	21	10
W	130 x	24	W	5 x	16	3 020	127	6.10	3.05	127.0	9.1	89	19	9
W	100 x	19	W	4 x	13	2 470	106	7.11	3.56	103.0	8.8	72	17	10

W SHAPES
Properties

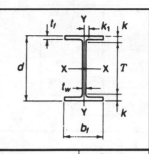

Nom-inal Mass per Meter kg/m	Compact Section Criteria $\frac{b_f}{2t_f}$	$\frac{h_c}{t_w}$	F_y''' $\frac{N}{mm^2}$	X_1 $\frac{N}{mm^2}$	X_2 $\frac{10^{-8}mm^4}{N^2}$	Elastic Properties Axis X-X I 10^6mm^4	S 10^3mm^3	r mm	Axis Y-Y I 10^6mm^4	S 10^3mm^3	r mm	Plastic Modulus Z_x 10^3mm^3	Z_y 10^3mm^3
100	4.4	11.1	—	45 600	155	113	987	94.3	36.6	349	53.7	1 150	533
86	5.1	12.4	—	40 100	257	94.7	853	92.8	31.4	300	53.4	981	458
71	5.9	15.8	—	33 500	501	76.6	709	91.7	25.4	247	52.8	803	375
59	7.2	17.6	—	28 100	997	61.2	583	89.9	20.4	199	51.9	653	303
52	8.1	20.4	—	24 900	1 600	52.7	512	89.1	17.8	175	51.8	569	266
46	9.2	22.2	—	22 300	2 480	45.5	448	87.9	15.3	151	51.0	496	230
42	7.0	22.2	—	24 000	1 960	40.9	399	87.7	9.01	109	41.2	446	165
36	8.1	25.8	—	20 800	3 390	34.4	342	86.8	7.64	92.6	40.9	380	141
31	6.6	27.5	—	19 900	4 400	31.3	298	88.7	4.10	61.2	32.1	335	93.7
27	8.0	29.9	—	17 200	8 180	25.8	249	87.1	3.29	49.5	31.1	279	76.0
22	6.4	28.1	—	18 400	7 240	20.0	194	83.6	1.42	27.8	22.3	222	43.7
19	7.8	29.9	—	16 300	12 200	16.5	163	81.6	1.15	22.5	21.5	187	35.5
15	9.6	40.5	269	12 100	37 700	12.8	128	81.9	.870	17.4	21.3	145	27.1
37	6.7	15.5	—	30 400	776	22.2	274	68.5	7.07	91.8	38.7	310	140
30	8.2	19.1	—	24 500	1 780	17.1	218	67.2	5.54	72.4	38.2	244	110
22	11.5	21.6	—	18 900	5 200	12.1	159	65.0	3.87	50.9	36.8	176	77.6
24	5.0	19.1	—	27 600	1 240	13.4	168	66.2	1.83	35.9	24.5	192	55.3
18	7.1	21.6	—	21 400	3 660	9.19	120	63.3	1.26	24.7	23.5	136	38.3
14	9.2	29.2	—	16 300	10 500	6.84	91.2	62.9	0.912	18.2	23.0	102	28.1
28	5.8	14.0	—	35 400	404	10.9	166	55.3	3.81	59.5	32.7	190	90.7
24	6.9	15.8	—	30 600	728	8.83	139	54.1	3.12	49.1	32.1	157	74.9
19	5.9	10.6	—	38 300	324	4.75	89.6	43.9	1.60	31.1	25.5	103	47.8

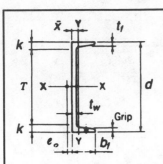

CHANNELS
AMERICAN STANDARD
Dimensions

Designation				Area	Depth	Web			Flange		Distance		Grip	Max. Flg. Fastener
				A	d	Thickness t_w	$\dfrac{t_w}{2}$		Width b_f	Thickness t_f	T	k		
mm	x	kg/m	in. x lb/ft	mm²	mm	mm	mm		mm	mm	mm	mm	mm	mm
C 380 x 74			C 15 x 50	9480	381.0	18.20	9.10		94.4	16.50	307.0	37	16	M24
C 380 x 60			C 15 x 40	7610	381.0	13.20	6.60		89.4	16.50	307.0	37	16	M24
C 380 x 50			C 15 x 33.9	6430	381.0	10.20	5.10		86.4	16.50	307.0	37	16	M24
C 310 x 45			C 12 x 30	5690	305.0	13.00	6.50		80.5	12.70	247.0	29	13	M22
C 310 x 37			C 12 x 25	4740	305.0	9.83	4.92		77.4	12.70	247.0	29	13	M22
C 310 x 31			C 12 x 20.7	3930	305.0	7.16	3.58		74.7	12.70	247.0	29	13	M22
C 250 x 45			C 10 x 30	5690	254.0	17.10	8.55		77.0	11.10	204.0	25	11	M20
C 250 x 37			C 10 x 25	4740	254.0	13.40	6.70		73.3	11.10	204.0	25	11	M20
C 250 x 30			C 10 x 20	3790	254.0	9.63	4.82		69.6	11.10	204.0	25	11	M20
C 250 x 23			C 10 x 15.3	2900	254.0	6.10	3.05		66.0	11.10	204.0	25	11	M20
C 230 x 30			C 9 x 20	3790	229.0	11.40	5.70		67.3	10.50	181.0	24	11	M20
C 230 x 22			C 9 x 15	2850	229.0	7.24	3.62		63.1	10.50	181.0	24	11	M20
C 230 x 20			C 9 x 13.4	2540	229.0	5.92	2.96		61.8	10.50	181.0	24	11	M20
C 200 x 28			C 8 x 18.75	3550	203.0	12.40	6.20		64.2	9.90	155.0	24	10	M20
C 200 x 20			C 8 x 13.75	2610	203.0	7.70	3.85		59.5	9.90	155.0	24	10	M20
C 200 x 17			C 8 x 11.5	2180	203.0	5.59	2.80		57.4	9.90	155.0	24	10	M20
C 180 x 22			C 7 x 14.75	2790	178.0	10.60	5.30		58.4	9.30	134.0	22	10	M16
C 180 x 18			C 7 x 12.25	2320	178.0	7.98	3.99		55.7	9.30	134.0	22	10	M16
C 180 x 15			C 7 x 9.8	1850	178.0	5.33	2.67		53.1	9.30	134.0	22	10	M16
C 150 x 19			C 6 x 13	2470	152.0	11.10	5.55		54.8	8.70	110.0	21	8	M16
C 150 x 16			C 6 x 10.5	1990	152.0	7.98	3.99		51.7	8.70	110.0	21	10	M16
C 150 x 12			C 6 x 8.2	1550	152.0	5.08	2.54		48.8	8.70	110.0	21	8	M16
C 130 x 13			C 5 x 9	1700	127.0	8.25	4.13		47.9	8.10	89.0	19	8	M16
C 130 x 10			C 5 x 6.7	1270	127.0	4.83	2.42		44.5	8.10	89.0	19	—	—
C 100 x 11			C 4 x 7.25	1370	102.0	8.15	4.08		43.7	7.50	68.0	17	8	M16
C 100 x 8			C 4 x 5.4	1030	102.0	4.67	2.34		40.2	7.50	68.0	17	—	—
C 75 x 9			C 3 x 6	1140	76.2	9.04	4.52		40.5	6.90	42.2	17	—	—
C 75 x 7			C 3 x 5	948	76.2	6.55	3.28		38.0	6.90	42.2	17	—	—
C 75 x 6			C 3 x 4.1	781	76.2	4.32	2.16		35.8	6.90	42.2	17	—	—

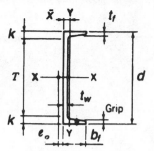

CHANNELS
AMERICAN STANDARD
Properties

Nominal Mass. per Meter	$\bar{x}$	Shear Center Location e_o	PNA Location x_p	Axis X-X I	Z	S	r	Axis Y-Y I	Z	S	r
kg/m	mm	mm	mm	10^6mm^4	10^3mm^3	10^3mm^3	mm	10^6mm^4	10^3mm^3	10^3mm^3	mm
74	1110	132	12.4	168	1120	882	133	4.58	134	61.8	22.0
60	609	110	9.91	145	937	761	138	3.84	113	55.1	22.5
50	425	96.1	8.38	131	826	688	143	3.38	102	50.9	22.9
45	364	40.5	9.30	67.4	551	442	109	2.14	71.0	33.8	19.4
37	226	34.9	7.75	59.9	479	393	112	1.86	62.9	30.9	19.8
31	154	30.1	6.40	53.7	416	352	117	1.61	57.2	28.3	20.2
45	513	21.3	11.2	42.9	436	338	86.8	1.64	61.9	27.1	17.0
37	291	18.4	9.02	38.0	377	299	89.5	1.40	52.3	24.3	17.2
30	155	15.3	7.42	32.8	316	258	93.0	1.17	44.4	21.6	17.6
23	88.2	12.2	5.66	28.1	259	221	98.4	0.949	38.5	19.0	18.1
30	181	10.6	8.25	25.3	275	221	81.7	1.01	40.5	19.2	16.3
22	87.5	8.32	6.17	21.2	221	185	86.2	0.803	33.6	16.7	16.8
20	70.8	7.57	5.51	19.9	205	174	88.5	0.733	32.0	15.8	17.0
28	183	6.74	8.71	18.3	226	180	71.8	0.824	35.6	16.5	15.2
20	78.4	5.16	6.38	15.0	179	148	75.8	0.637	28.3	14.0	15.6
17	54.6	4.43	5.31	13.6	156	134	79.0	0.549	25.9	12.8	15.9
22	112	3.52	7.82	11.3	159	127	63.6	0.574	26.9	12.8	14.3
18	67.6	3.01	6.48	10.1	138	113	66.0	0.487	23.4	11.5	14.5
15	41.9	2.47	5.16	8.87	117	99.7	69.2	0.403	20.6	10.2	14.8
19	101	1.94	8.05	7.24	119	95.3	54.1	0.437	22.3	10.5	13.3
16	54.7	1.60	6.48	6.33	101	83.3	56.4	0.360	18.8	9.22	13.5
12	31.6	1.27	5.03	5.45	84.1	71.7	59.3	0.288	16.3	8.04	13.6
13	46	0.787	6.65	3.70	71.4	58.3	46.7	0.263	15.0	7.35	12.4
10	23.3	0.596	5.51	3.12	57.5	49.1	49.6	0.199	12.5	6.18	12.5
11	34.7	0.333	6.71	1.91	46.0	37.5	37.3	0.180	11.4	5.62	11.5
8	16.9	0.247	6.12	1.60	37.0	31.4	39.4	0.133	9.32	4.65	11.4
9	30.7	0.124	7.39	0.862	28.2	22.6	27.5	0.127	8.91	4.39	10.6
7	18.1	0.102	6.15	0.770	24.6	20.2	28.5	0.103	7.64	3.83	10.4
6	11.4	0.083	12.0	0.691	21.3	18.1	29.8	0.082	6.57	3.32	10.2

B

TABLE OF TYPICAL PROPERTIES FOR SELECTED MATERIALS AND RADII OF GYRATION

US Customary Units

Material	Density (lb/ft³)	Modulus of Elasticity 10³(ksi)	Shear Modulus (ksi)	Tensile Yield Strength (ksi)	Tensile Ultimate Strength (ksi)	Shear Yield Strength (ksi)	Thermal Coefficient, α (in/in/°F)
Steel 1020	500	29 to 30	11.5	48	65		6.5×10^{-6}
Steel A36	490	29	11.5	36	58	21	6.5×10^{-6}
Steel A572	490	29	11.5	42–65	60–80	—	6.5×10^{-6}
Gray Cast Iron	450	15	6	—	20	32	6.4×10^{-6}
6061-T6 Aluminum	170	10	3.7	35–40	42–45	23–27	12.8×10^{-6}
C26800 Copper Alloy	550	16	6	20	44	11.4	9.0×10^{-6}
Brass (Rolled)	540	14	6	50	60	50	10.4×10^{-6}
Bronze (Heat Treated)	535	12	5	55	75	37	10.1×10^{-6}
Concrete	150	3–3.6	—	—	—	—	5.5×10^{-6}

SI Units

Material	Density (KN/m³)	Modulus of Elasticity 10³(MPa)	Shear Modulus (MPa)	Tensile Yield Strength (MPa)	Tensile Ultimate Strength (MPa)	Shear Yield Strength (MPa)	Thermal Coefficient, α (m/m/°C)
Steel 1020	77	200–210	80	330	448		11.7×10^{-6}
Steel A36	77	207	80	248	400	145	11.7×10^{-6}
Steel A572	77	207	80	290–448	414–552	—	11.7×10^{-6}
Gray Cast Iron	71	100	41	—	138	220	11.5×10^{-6}
6061-T6 Aluminum	26	70	25.5	241–276	290–310	159–186	23×10^{-6}
C26800 Copper Alloy	84	105	41	138	303	79	16.2×10^{-6}
Brass (Rolled)	84	97	41	345	414	345	18.7×10^{-6}
Bronze (Heat Treated)	84	83	35	380	518	255	18.2×10^{-6}
Concrete	23.5	21–25	—	—	—	—	10.9×10^{-6}

Radius of Gyration Formulas

Shape	Radii of Gyration
h — rectangle (b × h) — $\bar{x}$	.289h
triangle (b, h) — $\bar{x}$	.2357h
circle — $\bar{x}$	d/4
half circle (r) — $\bar{x}$	.2643r

C

BEAM LOADING TABLES

BEAM DIAGRAMS AND FORMULAS
For various static loading conditions

SIMPLE BEAM—UNIFORMLY DISTRIBUTED LOAD

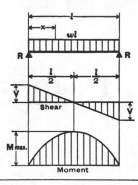

Total Equiv. Uniform Load . . . . . $= wl$

$R = V$ . . . . $= \dfrac{wl}{2}$

V_x . . . . $= w\left(\dfrac{l}{2} - x\right)$

M max. (at center) . . . . $= \dfrac{wl^2}{8}$

M_x . . . . $= \dfrac{wx}{2}(l - x)$

Δmax. (at center) . . . . $= \dfrac{5\,wl^4}{384\,EI}$

Δ_x . . . . $= \dfrac{wx}{24EI}(l^3 - 2lx^2 + x^3)$

SIMPLE BEAM—LOAD INCREASING UNIFORMLY TO ONE END

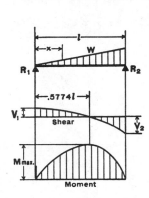

Total Equiv. Uniform Load . . . $= \dfrac{16W}{9\sqrt{3}} = 1.0264W$

$R_1 = V_1$ . . . . $= \dfrac{W}{3}$

$R_2 = V_2$ max. . . . . $= \dfrac{2W}{3}$

V_x . . . . $= \dfrac{W}{3} - \dfrac{Wx^2}{l^2}$

M max. $\left(\text{at } x = \dfrac{l}{\sqrt{3}} = .5774l\right)$ . . $= \dfrac{2Wl}{9\sqrt{3}} = .1283\,Wl$

M_x . . . . $= \dfrac{Wx}{3l^2}(l^2 - x^2)$

Δmax. $\left(\text{at } x = l\sqrt{1 - \sqrt{\dfrac{8}{15}}} = .5193l\right) = .01304\,\dfrac{Wl^3}{EI}$

Δ_x . . . . $= \dfrac{Wx}{180EI\,l^2}(3x^4 - 10l^2x^2 + 7l^4)$

SIMPLE BEAM—LOAD INCREASING UNIFORMLY TO CENTER

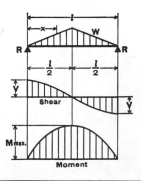

Total Equiv. Uniform Load . . . $= \dfrac{4W}{3}$

$R = V$ . . . . $= \dfrac{W}{2}$

V_x $\left(\text{when } x < \dfrac{l}{2}\right)$ . . . . $= \dfrac{W}{2l^2}(l^2 - 4x^2)$

M max. (at center) . . . . $= \dfrac{Wl}{6}$

M_x $\left(\text{when } x < \dfrac{l}{2}\right)$ . . . . $= Wx\left(\dfrac{1}{2} - \dfrac{2x^2}{3l^2}\right)$

Δmax. (at center) . . . . $= \dfrac{Wl^3}{60EI}$

Δ_x $\left(\text{when } x < \dfrac{l}{2}\right)$ . . . . $= \dfrac{Wx}{480\,EI\,l^2}(5l^2 - 4x^2)^2$

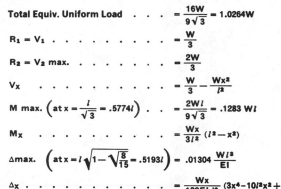

BEAM DIAGRAMS AND FORMULAS
For various static loading conditions

SIMPLE BEAM—CONCENTRATED LOAD AT CENTER

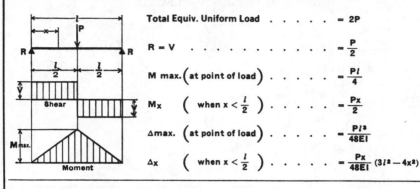

Total Equiv. Uniform Load . . . . . $= 2P$

$R = V$ . . . . $= \dfrac{P}{2}$

M max. $\left(\text{at point of load}\right)$ . . . . . $= \dfrac{Pl}{4}$

M_x $\left(\text{when } x < \dfrac{l}{2}\right)$ . . . . . $= \dfrac{Px}{2}$

Δmax. $\left(\text{at point of load}\right)$ . . . . . $= \dfrac{Pl^3}{48EI}$

Δ_x $\left(\text{when } x < \dfrac{l}{2}\right)$ . . . . . $= \dfrac{Px}{48EI}(3l^2 - 4x^2)$

SIMPLE BEAM—CONCENTRATED LOAD AT ANY POINT

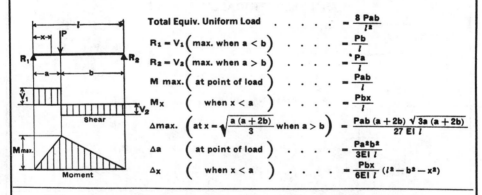

Total Equiv. Uniform Load . . . . . $= \dfrac{8\,Pab}{l^2}$

$R_1 = V_1\left(\text{max. when } a < b\right)$ . . . . $= \dfrac{Pb}{l}$

$R_2 = V_2\left(\text{max. when } a > b\right)$ . . . . $= \dfrac{Pa}{l}$

M max. $\left(\text{at point of load}\right)$ . . . . $= \dfrac{Pab}{l}$

M_x $\left(\text{when } x < a\right)$ . . . . . $= \dfrac{Pbx}{l}$

Δmax. $\left(\text{at } x = \sqrt{\dfrac{a(a+2b)}{3}}\ \text{when } a > b\right) = \dfrac{Pab(a+2b)\sqrt{3a(a+2b)}}{27\,EI\,l}$

Δa $\left(\text{at point of load}\right)$ . . . . . $= \dfrac{Pa^2b^2}{3EI\,l}$

Δ_x $\left(\text{when } x < a\right)$ . . . . . $= \dfrac{Pbx}{6EI\,l}(l^2 - b^2 - x^2)$

SIMPLE BEAM—TWO EQUAL CONCENTRATED LOADS SYMMETRICALLY PLACED

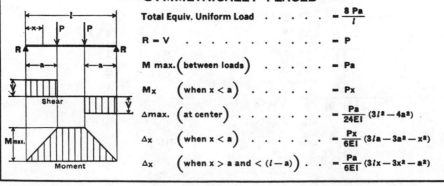

Total Equiv. Uniform Load . . . . . $= \dfrac{8\,Pa}{l}$

$R = V$ . . . . $= P$

M max. $\left(\text{between loads}\right)$ . . . . . $= Pa$

M_x $\left(\text{when } x < a\right)$ . . . . . $= Px$

Δmax. $\left(\text{at center}\right)$ . . . . . . $= \dfrac{Pa}{24EI}(3l^2 - 4a^2)$

Δ_x $\left(\text{when } x < a\right)$ . . . . . $= \dfrac{Px}{6EI}(3la - 3a^2 - x^2)$

Δ_x $\left(\text{when } x > a \text{ and } < (l-a)\right)$ . . $= \dfrac{Pa}{6EI}(3lx - 3x^2 - a^2)$

BEAM DIAGRAMS AND FORMULAS
For various static loading conditions

SIMPLE BEAM—TWO EQUAL CONCENTRATED LOADS UNSYMMETRICALLY PLACED

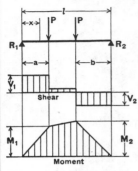

$R_1 = V_1 \left(\text{max. when } a < b \right)$ . . . . . $= \dfrac{P}{l}(l - a + b)$

$R_2 = V_2 \left(\text{max. when } a > b \right)$ . . . . . $= \dfrac{P}{l}(l - b + a)$

$V_x \quad \left(\text{when } x > a \text{ and } < (l - b) \right)$ . . . $= \dfrac{P}{l}(b - a)$

$M_1 \quad \left(\text{max. when } a > b \right)$ . . . . . $= R_1 a$

$M_2 \quad \left(\text{max. when } a < b \right)$ . . . . . $= R_2 b$

$M_x \quad \left(\text{when } x < a \right)$ . . . . . . $= R_1 x$

$M_x \quad \left(\text{when } x > a \text{ and } < (l - b) \right)$ . . $= R_1 x - P(x - a)$

SIMPLE BEAM—TWO UNEQUAL CONCENTRATED LOADS UNSYMMETRICALLY PLACED

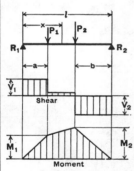

$R_1 = V_1$ . . . . $= \dfrac{P_1(l - a) + P_2 b}{l}$

$R_2 = V_2$ . . . . $= \dfrac{P_1 a + P_2(l - b)}{l}$

$V_x \quad \left(\text{when } x > a \text{ and } < (l - b) \right)$ . . $= R_1 - P_1$

$M_1 \quad \left(\text{max. when } R_1 < P_1 \right)$ . . . $= R_1 a$

$M_2 \quad \left(\text{max. when } R_2 < P_2 \right)$ . . . $= R_2 b$

$M_x \quad \left(\text{when } x < a \right)$ . . . . . . $= R_1 x$

$M_x \quad \left(\text{when } x > a \text{ and } < (l - b) \right)$ . . $= R_1 x - P_1(x - a)$

BEAM FIXED AT ONE END, SUPPORTED AT OTHER—UNIFORMLY DISTRIBUTED LOAD

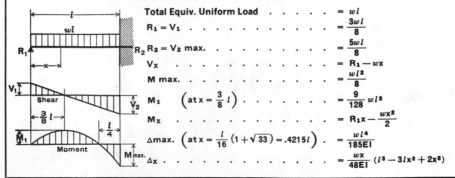

Total Equiv. Uniform Load . . . . . $= wl$

$R_1 = V_1$ . . . . $= \dfrac{3wl}{8}$

$R_2 = V_2 \text{ max.}$ . . . . $= \dfrac{5wl}{8}$

V_x . . . . $= R_1 - wx$

$M \text{ max.}$ . . . . $= \dfrac{wl^2}{8}$

$M_1 \quad \left(\text{at } x = \dfrac{3}{8}l \right)$ . . . . $= \dfrac{9}{128}wl^2$

M_x . . . . $= R_1 x - \dfrac{wx^2}{2}$

$\Delta \text{max.} \left(\text{at } x = \dfrac{l}{16}\left(1 + \sqrt{33}\right) = .4215l \right)$. $= \dfrac{wl^4}{185EI}$

Δ_x . . . . $= \dfrac{wx}{48EI}(l^3 - 3lx^2 + 2x^3)$

BEAM DIAGRAMS AND FORMULAS
For various static loading conditions

CANTILEVER BEAM—LOAD INCREASING UNIFORMLY TO FIXED END

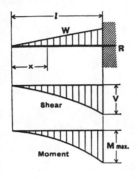

Total Equiv. Uniform Load . . . . . $= \dfrac{8}{3} W$

$R = V$ . . . . $= W$

V_x . . . . $= W \dfrac{x^2}{l^2}$

M max. $\left(\text{at fixed end}\right)$ . . . . . $= \dfrac{Wl}{3}$

M_x . . . . $= \dfrac{Wx^3}{3l^2}$

Δmax. $\left(\text{at free end}\right)$ . . . . . $= \dfrac{Wl^3}{15EI}$

Δ_x . . . . $= \dfrac{W}{60EIl^2}(x^5 - 5l^4x + 4l^5)$

CANTILEVER BEAM—UNIFORMLY DISTRIBUTED LOAD

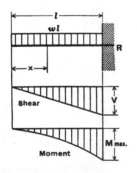

Total Equiv. Uniform Load . . . . . $= 4wl$

$R = V$ . . . . $= wl$

V_x . . . . $= wx$

M max. $\left(\text{at fixed end}\right)$ . . . . . $= \dfrac{wl^2}{2}$

M_x . . . . $= \dfrac{wx^2}{2}$

Δmax. $\left(\text{at free end}\right)$ . . . . . $= \dfrac{wl^4}{8EI}$

Δ_x . . . . $= \dfrac{w}{24EI}(x^4 - 4l^3x + 3l^4)$

BEAM FIXED AT ONE END, FREE TO DEFLECT VERTICALLY BUT NOT ROTATE AT OTHER—UNIFORMLY DISTRIBUTED LOAD

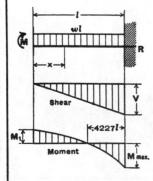

Total Equiv. Uniform Load . . . . . $= \dfrac{8}{3} wl$

$R = V$ . . . . $= wl$

V_x . . . . $= wx$

M max. $\left(\text{at fixed end}\right)$ . . . . . $= \dfrac{wl^2}{3}$

M_1 $\left(\text{at deflected end}\right)$ . . . . $= \dfrac{wl^2}{6}$

M_x . . . . $= \dfrac{w}{6}(l^2 - 3x^2)$

Δmax. $\left(\text{at deflected end}\right)$ . . . . $= \dfrac{wl^4}{24EI}$

Δ_x . . . . $= \dfrac{w(l^2 - x^2)^2}{24EI}$

BEAM DIAGRAMS AND FORMULAS
For various static loading conditions

CANTILEVER BEAM—CONCENTRATED LOAD AT ANY POINT

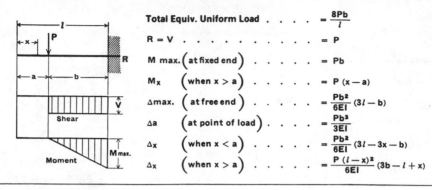

Total Equiv. Uniform Load . . . . . $= \dfrac{8Pb}{l}$

$R = V$ . . . . $= P$

M max. $\left(\text{at fixed end}\right)$ . . . . . $= Pb$

M_x $\left(\text{when } x > a\right)$ . . . . . $= P(x-a)$

Δmax. $\left(\text{at free end}\right)$ . . . . . $= \dfrac{Pb^2}{6EI}(3l-b)$

Δa $\left(\text{at point of load}\right)$ . . . $= \dfrac{Pb^3}{3EI}$

Δ_x $\left(\text{when } x < a\right)$ . . . . . $= \dfrac{Pb^2}{6EI}(3l-3x-b)$

Δ_x $\left(\text{when } x > a\right)$ . . . . . $= \dfrac{P(l-x)^2}{6EI}(3b-l+x)$

CANTILEVER BEAM—CONCENTRATED LOAD AT FREE END

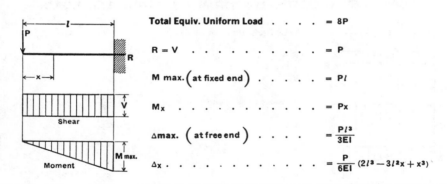

Total Equiv. Uniform Load . . . . . $= 8P$

$R = V$ . . . . $= P$

M max. $\left(\text{at fixed end}\right)$ . . . . . $= Pl$

M_x . . . . $= Px$

Δmax. $\left(\text{at free end}\right)$ . . . . $= \dfrac{Pl^3}{3EI}$

Δ_x . . . . $= \dfrac{P}{6EI}(2l^3 - 3l^2x + x^3)$

BEAM FIXED AT ONE END, FREE TO DEFLECT VERTICALLY BUT NOT ROTATE AT OTHER—CONCENTRATED LOAD AT DEFLECTED END

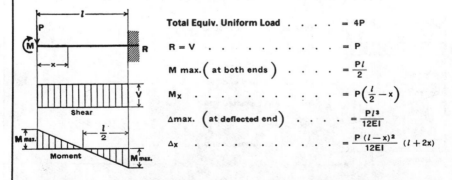

Total Equiv. Uniform Load . . . . . $= 4P$

$R = V$ . . . . $= P$

M max. $\left(\text{at both ends}\right)$ . . . . . $= \dfrac{Pl}{2}$

M_x . . . . $= P\left(\dfrac{l}{2} - x\right)$

Δmax. $\left(\text{at deflected end}\right)$ . . . . $= \dfrac{Pl^3}{12EI}$

Δ_x . . . . $= \dfrac{P(l-x)^2}{12EI}(l+2x)$

BEAM DIAGRAMS AND FORMULAS
For various static loading conditions

CONTINUOUS BEAM—TWO EQUAL SPANS—UNIFORM LOAD ON ONE SPAN

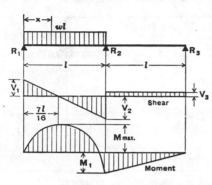

Total Equiv. Uniform Load $= \dfrac{49}{64} wl$

$R_1 = V_1 \quad \ldots \ldots \ldots = \dfrac{7}{16} wl$

$R_2 = V_2 + V_3 \quad \ldots \ldots = \dfrac{5}{8} wl$

$R_3 = V_3 \quad \ldots \ldots \ldots = -\dfrac{1}{16} wl$

$V_2 \quad \ldots \ldots \ldots = \dfrac{9}{16} wl$

$M \text{ max.} \left(\text{at } x = \dfrac{7}{16} l\right) \ldots = \dfrac{49}{512} wl^2$

$M_1 \quad \left(\text{at support } R_2\right) \ldots = \dfrac{1}{16} wl^2$

$M_x \quad \left(\text{when } x < l\right) \ldots = \dfrac{wx}{16}(7l - 8x)$

$\Delta \text{ Max. } (0.472\, l \text{ from } R_1) = 0.0092\, wl^4/EI$

CONTINUOUS BEAM—TWO EQUAL SPANS—CONCENTRATED LOAD AT CENTER OF ONE SPAN

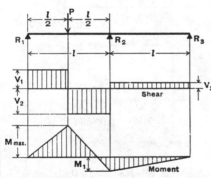

Total Equiv. Uniform Load $. = \dfrac{13}{8} P$

$R_1 = V_1 \quad \ldots \ldots \ldots = \dfrac{13}{32} P$

$R_2 = V_2 + V_3 \quad \ldots \ldots = \dfrac{11}{16} P$

$R_3 = V_3 \quad \ldots \ldots \ldots = -\dfrac{3}{32} P$

$V_2 \quad \ldots \ldots \ldots = \dfrac{19}{32} P$

$M \text{ max.} \left(\text{at point of load}\right) \ldots = \dfrac{13}{64} Pl$

$M_1 \quad \left(\text{at support } R_2\right) \ldots = \dfrac{3}{32} Pl$

$\Delta \text{ Max. } (0.480\, l \text{ from } R_1) = 0.015\, Pl^3/EI$

CONTINUOUS BEAM—TWO EQUAL SPANS—CONCENTRATED LOAD AT ANY POINT

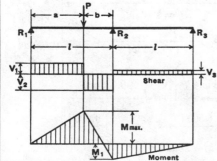

$R_1 = V_1 \quad \ldots \ldots \ldots = \dfrac{Pb}{4l^3}\left(4l^2 - a(l+a)\right)$

$R_2 = V_2 + V_3 \quad \ldots \ldots = \dfrac{Pa}{2l^3}\left(2l^2 + b(l+a)\right)$

$R_3 = V_3 \quad \ldots \ldots \ldots = -\dfrac{Pab}{4l^3}(l+a)$

$V_2 \quad \ldots \ldots \ldots = \dfrac{Pa}{4l^3}\left(4l^2 + b(l+a)\right)$

$M \text{ max.}\left(\text{at point of load}\right) . = \dfrac{Pab}{4l^3}\left(4l^2 - a(l+a)\right)$

$M_1 \quad \left(\text{at support } R_2\right) \ldots = \dfrac{Pab}{4l^2}(l+a)$

BEAM DIAGRAMS AND DEFLECTIONS
For various static loading conditions

CONTINUOUS BEAM—THREE EQUAL SPANS—ONE END SPAN UNLOADED

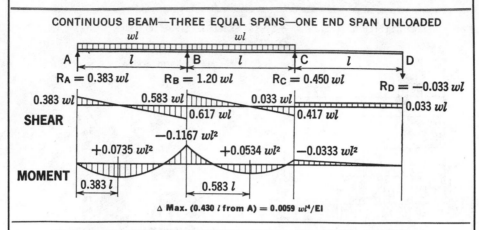

$R_A = 0.383\ wl$ $R_B = 1.20\ wl$ $R_C = 0.450\ wl$ $R_D = -0.033\ wl$

SHEAR

$0.383\ wl$ $0.583\ wl$ $0.617\ wl$ $0.033\ wl$ $0.417\ wl$ $0.033\ wl$

MOMENT

$-0.1167\ wl^2$ $+0.0735\ wl^2$ $+0.0534\ wl^2$ $-0.0333\ wl^2$

$0.383\ l$ $0.583\ l$

Δ Max. (0.430 l from A) = 0.0059 wl^4/EI

CONTINUOUS BEAM—THREE EQUAL SPANS—END SPANS LOADED

$R_A = 0.450\ wl$ $R_B = 0.550\ wl$ $R_C = 0.550\ wl$ $R_D = 0.450\ wl$

SHEAR

$0.450\ wl$ $0.550\ wl$ $0.550\ wl$ $0.450\ wl$

MOMENT

$-0.050\ wl^2$ $+0.1013\ wl^2$ $+0.1013\ wl^2$

$0.450\ l$ $0.450\ l$

Δ Max. (0.479 l from A or D) = 0.0099 wl^4/EI

CONTINUOUS BEAM—THREE EQUAL SPANS—ALL SPANS LOADED

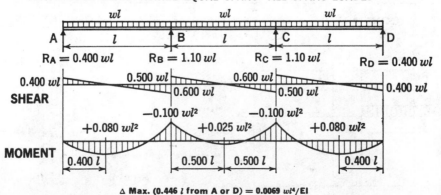

$R_A = 0.400\ wl$ $R_B = 1.10\ wl$ $R_C = 1.10\ wl$ $R_D = 0.400\ wl$

SHEAR

$0.400\ wl$ $0.500\ wl$ $0.600\ wl$ $0.600\ wl$ $0.500\ wl$ $0.400\ wl$

MOMENT

$-0.100\ wl^2$ $-0.100\ wl^2$ $+0.080\ wl^2$ $+0.025\ wl^2$ $+0.080\ wl^2$

$0.400\ l$ $0.500\ l$ $0.500\ l$ $0.400\ l$

Δ Max. (0.446 l from A or D) = 0.0069 wl^4/EI

D

STANDARD TIMBER SIZES

Table D.1   Properties of Sawn Lumber and Timber

Nominal Size (in.) b d	Standard Dressed Size (in.) (S4S) b d	Area of Section A (in.²)	Moment of Inertia I (in.⁴)	Section Modulus S (in.³)	Weight[a] in Pounds per Linear Foot of Piece When Weight of Wood per Cubic Foot Equals:					
					25 lb	30 lb	35 lb	40 lb	45 lb	50 lb
1 × 3	¾ × 2½	1.875	0.9766	0.7812	0.3	0.4	0.5	0.5	0.6	0.7
1 × 4	¾ × 3½	2.625	2.680	1.531	0.5	0.5	0.6	0.7	0.8	0.9
1 × 6	¾ × 5½	4.125	10.40	3.781	0.7	0.9	1.0	1.1	1.3	1.4
1 × 8	¾ × 7¼	5.438	23.82	6.570	0.9	1.1	1.3	1.5	1.7	1.9
1 × 10	¾ × 9¼	6.938	49.47	10.70	1.2	1.4	1.7	1.9	2.2	2.4
1 × 12	¾ × 11¼	8.438	88.99	15.82	1.5	1.8	2.1	2.3	2.6	2.9
2 × 3	1½ × 2½	3.750	1.953	1.563	0.7	0.8	0.9	1.0	1.2	1.3
2 × 4	1½ × 3½	5.250	5.359	3.063	0.9	1.1	1.3	1.5	1.6	1.8
2 × 5	1½ × 4½	6.750	11.39	5.063	1.2	1.4	1.6	1.9	2.1	2.3
2 × 6	1½ × 5½	8.250	20.80	7.563	1.4	1.7	2.0	2.3	2.6	2.9
2 × 8	1½ × 7¼	10.88	47.64	13.14	1.9	2.3	2.6	3.0	3.4	3.8
2 × 10	1½ × 9¼	13.88	98.93	21.39	2.4	2.9	3.4	3.9	4.3	4.8
2 × 12	1½ × 11¼	16.88	178.0	31.64	2.9	3.5	4.1	4.7	5.3	5.9
2 × 14	1½ × 12¼	19.88	290.8	43.89	3.5	4.1	4.8	5.5	6.2	6.9
3 × 1	2½ × ¾	1.875	0.08789	0.2344	0.3	0.4	0.5	0.5	0.6	0.7
3 × 2	2½ × 1½	3.750	0.7031	0.9375	0.7	0.8	0.9	1.0	1.2	1.3
3 × 4	2½ × 3½	8.750	8.932	5.104	1.5	1.8	2.1	2.4	2.7	3.0
3 × 5	2½ × 4½	11.25	18.98	8.438	2.0	2.3	2.7	3.1	3.5	3.9
3 × 6	2½ × 5½	13.75	34.66	12.60	2.4	2.9	3.3	3.8	4.3	4.8
3 × 8	2½ × 7¼	18.12	79.39	21.90	3.1	3.8	4.4	5.0	5.7	6.3
3 × 10	2½ × 9¼	23.12	164.9	35.65	4.0	4.8	5.6	6.4	7.2	8.0
3 × 12	2½ × 11¼	28.12	296.6	52.73	4.9	5.9	6.8	7.8	8.7	9.8
3 × 14	2½ × 13¼	33.12	484.6	73.15	5.8	6.9	8.1	9.2	10.4	11.5
3 × 16	2½ × 15¼	38.12	738.9	96.90	6.6	7.9	9.3	10.6	11.9	13.2
4 × 1	3½ × ¾	2.625	0.1230	0.3281	0.5	0.5	0.6	0.7	0.8	0.9
4 × 2	3½ × 1½	5.250	0.9844	1.313	0.9	1.1	1.3	1.5	1.6	1.8
4 × 3	3½ × 2½	8.750	4.557	3.646	1.5	1.8	2.1	2.4	2.7	3.0
4 × 4	3½ × 3½	12.25	12.50	7.146	2.1	2.6	3.0	3.4	3.8	4.3
4 × 5	3½ × 4½	15.75	26.58	11.81	2.7	3.3	3.8	4.4	4.9	5.5
4 × 6	3½ × 5½	19.25	48.53	17.65	3.3	4.0	4.7	5.3	6.0	6.7
4 × 8	3½ × 7¼	25.38	111.1	30.66	4.4	5.3	6.2	7.0	7.9	8.8
4 × 10	3½ × 9¼	32.38	230.8	49.91	5.6	6.7	7.9	8.9	10.1	12.2
4 × 12	3½ × 11¼	39.38	415.3	73.83	6.8	8.2	9.6	10.9	12.3	13.7
4 × 14	3½ × 13¼	46.38	678.5	102.4	8.0	9.7	11.3	12.9	14.5	16.1
4 × 16	3½ × 15¼	53.38	1034	135.7	9.3	11.1	13.0	14.8	16.7	18.6
5 × 2	4½ × 1½	6.750	1.266	1.688	1.2	1.4	1.6	1.9	2.1	2.3
5 × 3	4¼ × 2½	11.25	5.859	4.688	2.0	2.3	2.7	3.1	3.5	3.9
5 × 4	4½ × 3½	15.75	16.08	9.188	2.7	3.3	3.8	4.4	4.9	5.5
5 × 5	4½ × 4½	20.25	34.17	15.19	3.5	4.2	4.9	5.7	6.3	7.0
6 × 1	5½ × ¾	4.125	0.1933	0.5156	0.7	0.9	1.0	1.1	1.3	1.4
6 × 2	5½ × 1½	8.250	1.547	2.063	1.4	1.7	2.0	2.3	2.6	2.9

Table D.2 Typical Design Values for Visually Graded Lumber (psi)

Species	Size Classification	Bending	Shear Parallel to Grain	Compression Perpendicular to Grain	Compression Parallel to Grain
eastern softwood					
select struct.		1250	70	335	1200
no. 1	2"–4" thick	900	70	335	1000
no.2	2"–4" thick	575	70	335	825
construction	2"–4" thick	675	70	335	1050
eastern white pine					
select struct.		1250	70	350	1200
no. 1	2"–4" thick	775	70	350	1000
no.2	2"–4" thick	575	70	350	825
construction	2"–4" thick	675	70	350	1050
northern white cedar					
select struct.		775	60	370	750
no. 1	2"–4" thick	575	60	370	600
no.2	2"–4" thick	550	60	370	475
construction	2"–4" thick	625	60	370	625
red oak					
select struct.		1150	85	820	1000
no. 1	2"–4" thick	825	85	820	825
no.2	2"–4" thick	800	85	820	625
construction	2"–4" thick	925	85	820	850
spruce-pine-fir					
select struct.	2"–4" thick	1250	70	425	1400
no. 1	2"–4" thick	875	70	425	1100
no.2	2"–4" thick	875	70	425	1100
construction	2"–4" thick	975	70	425	1350
western cedar					
select struct.		1000	75	425	1000
no. 1	2"–4" thick	725	75	425	825
no.2	2"–4" thick	700	75	425	650
construction	2"–4" thick	800	75	425	850

E

USE OF INTEGRATION TO FIND CENTROIDS, MOMENTS OF INERTIA, AND BENDING MOMENTS

E.1   Introduction

Integral calculus has many useful applications throughout the fields of statics and strength of materials. One of the most useful applications of integral calculus is finding the area under complx curves or other functions. In statics, the area within a complex shape is used as a basis for finding the centroid and moment of inertia of that shape. This appendix will attempt to briefly highlight some of the more common applications of integral calculus.

E.2   Finding Centroids by Integration

The method used to find the centroid of a composite area, which was discussed in Chapter 6, focused on breaking the shape into several standard geometric shapes. Since the centroid of each standard shape was already known, the centroid was found by summing the first moment of area for each standard shape and dividing that sum by the total area. This was defined by the following formula:

$$\bar{x} = \Sigma\, ax\, /\, A \qquad\qquad\qquad \text{(Eq. 6-3)}$$

$$\bar{y} = \Sigma\, ay\, /\, A \qquad\qquad\qquad \text{(Eq. 6-4)}$$

Integration is a mathematical process of summing up all the minute increments of area under a line or a curve. Therefore, if a shape can be described mathematically by a line having an equation, the aforementioned equations (6-3 and 6-4) can be rewitten as follows:

$$\bar{x} = \int x\ dA\, /\, dA \qquad\qquad\qquad \text{(Eq. E-1)}$$

$$\bar{y} = \int y\ dA\, /\, dA \qquad\qquad\qquad \text{(Eq. E-2)}$$

where

$$dA = \text{increment of area}$$

When using integration to find a centroid, it is still necessary to know the area of the incremental area, dA, as well as its centroid, and to define the limits of integration properly. The limits of integration for both the $\bar{x}$ and $\bar{y}$ distances must be taken relative to the same point of reference. The following example will illustrate the use of integral calculus to find the centroid of a composite shape.

EXAMPLE E.1

Find the centroid of the shape shown below. The point of origin is given as the origin of the axes shown.

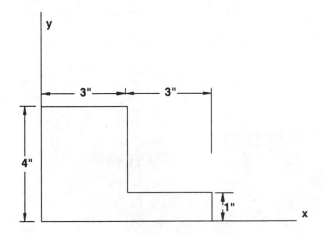

Start the process by breaking up the shape into simple areas. This may be done as shown.

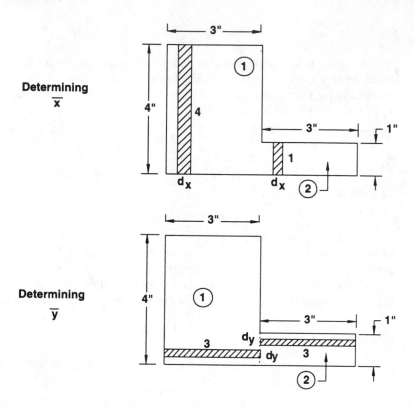

Use integration to find the x distance from the point of origin to the centroid of the composite shape by applying Equation F-1. The limits of integration will be the x distance from the beginning of the shape to the end of the shape.

Area 1 - Large rectangle

$$\bar{x} = \int x \; dA/dA \qquad \text{(Eq. E-1)}$$

$$\bar{x} = \int_{0}^{3} x \; 4dx/4dx$$

Area 2 - Small rectangle

$$\bar{x} = \int x \; dA/dA \qquad \text{(Eq. E-1)}$$

$$\bar{x} = \int_{3}^{6} x \; 1dx/dx$$

Since the centroid is found by taking the sum of all areas and dividing by the total area, the individual shapes should be combined as follows:

$$\overline{x} = \int_0^3 \frac{x\,(4dx)}{4dx} + \int_3^6 \frac{x\,(1dx)}{dx}$$

Integrating over the limits for each:

$$\overline{x} = [2x^2]_0^3 + [x^2/2]_3^6 / [4x]_0^3 + [x]_3^6$$

$$\overline{x} = [18 - 0] + [18 - 4.5]/[24 - 12] + [6 - 3] = 31.5/15 = 2.1 \text{ in.}$$

Use intregration to find the y distance from the point of origin to the centroid of the composite shape by applying Equation F-2. The limits of integration will be the y distance from the beginning of the shape to the end of the shape.

Area 1 - Large Rectangle

$$\overline{y} = \int y \; dA/dA \qquad\qquad \text{(Eq. F-2)}$$

$$\overline{y} = \int_0^4 y \; 3dy/3dy$$

Area 2 - Small Rectangle

$$\overline{y} = \int y \; dA/dA \qquad\qquad \text{(Eq. F-2)}$$

$$\overline{y} = \int_0^1 y \; 3dy/3dy$$

Since the centroid is found by taking the sum of all areas and dividing by the total area, the individual shapes should be combined as follows:

$$\overline{y} = \int_0^4 \frac{y\,(3dy)}{3dy} + \int_0^1 \frac{y\,(3dy)}{3dy}$$

Integrating over the limits for each:

$$y = [1.5y^2]_0^4 + [1.5y]_0^1 / [3y]_0^4 + [3y]_0^1$$

$$y = [24 - 0] + [1.5 - 0] / [12 - 0] + [3 - 0] = 25.5/15 = 1.7 \text{ in.}$$

Therefore, the centroid of this composite shape is located at $\overline{x} = 2.1$ in. and $\overline{y} = 1.7$ in. from the origin. Notce that these are exactly the same results found in Example 6.1.

E.3 Finding Moment of Inertia by Integration

The moment of inertia about an x axis of a plane area was defined in Chapter 7 by the following formula:

$$I_x = \Sigma y^2 \, \Delta A \qquad \text{(Eq. 7-1)}$$

where

I_x = moment of inertia about the x axis

y = distance from the neutral axis or the axis under consideration

ΔA = incremental area being considered

The incremental area, ΔA, can be thought of as various small areas that, when added together, give the total area, A. If the incremental area under consideration can be defined mathematically by an equation, then integration can be used by summing the product of the differential areas, dA, and the square of the distance from that differential area to the axis under consideration. Equation 7-1 can be expressed as follows:

$$I_x = \int y^2 \, dA \qquad \text{(Eq. E-3)}$$

Similarly, the moment of inertia about a y axis under consideration can be expressed as an integral, as follows:

$$I_y = \int x^2 \, dA \qquad \text{(Eq. E-4)}$$

The following example will illustrate this method.

EXAMPLE E.2

For the shape shown below, calculate the moment of inertia about the x-axis located 8 in. from the centroid of the rectangle.

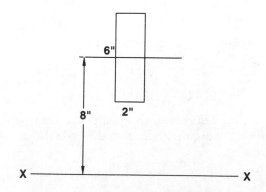

Moment of inertia for the rectangle shown was described in Chapter 7 by the following equation:

$$I_x = \Sigma y^2 \ \Delta A \qquad \text{(Eq. 7-1)}$$

The moment of inertia for the shape can be calculated from integration, using the following formula:

$$I_x = \int y^2 \ dA \qquad \text{(Eq. E-3)}$$

For this problem, the incremental slice of area within the rectangle can be expressed as follows:

$$dA = 2 \ dy$$

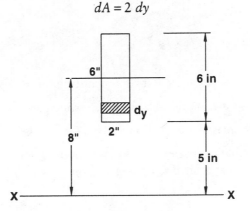

and therefore Equation E-3 can be rewritten as:

$$I_x = \int_5^{11} y^2 \ (2dy) \qquad \text{(Eq. E-3)}$$

Notice the limits of integration are 5 and 11, since the rectangle begins 5 in. way from the axis in question and ends at a distance of 11 inches. Integrating this becomes:

$$I_x = \int_5^{11} y^2 \ (2dy)$$

$$= [(2/3)y^3]_5^{11} \ = [887.33 - 83.33]$$

$$= 804 \ \text{in.}^4$$

Therefore, the moment of inertia for the rectangle, abount the x-axis, is 804 in.[4] Notice that this is exactly the same answer found by using the parallel axis theorem in Example 7.1.

E.4   Finding Bending Moments by Integration

In Chapter 11, the relationship between shear and bending moments within a beam was discussed. The change of bending moment between any two points is equal to the moment at the beginning of that section plus the moment caused by the shear force over that section. This change can be expressed as follows:

$$\Delta M = V_x \Delta x$$

Since $V_1 \Delta x$ is the area under the shear diagram, this statement may be expressed in a similar manner by using integral calculus and integrating between the appropriate limits. This can be expressed as follows:

$$dM = \int_1^2 V_x \, dx \qquad \text{(Eq. E-5)}$$

where

dM = change in bending moment from point 1 to point 2

V_x = function describing shear through area to be integrated

dx = incremental x distance of the area being considered

The following example will illustrate the use of integral calculus to find the moment at a particlar point along a beam.

EXAMPLE E.3

Find the bending moment at a point located 6 ft. from the left support for the beam shown below.

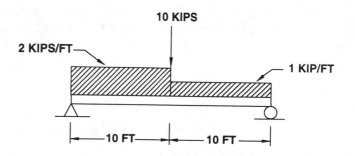

This problem uses the same beam that was used in Example 11.5, for which we have already solved and constructed the shear diagram. The shear diagram for this beam is as follows:

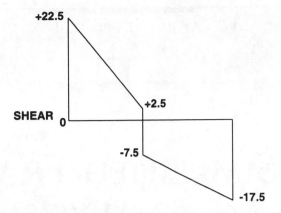

Since the moment at 6 ft. from the left support is needed, the limits of integration are from the beginning to the end of the area, which is 0 to 6. The function describing the shear diagram in this region is that for a straight line and is as follows:

$$22.5 - 2x$$

Therefore the the integral is as follows:

$$dM = \int_0^6 V_x \, dx$$

$$dM = \int_0^6 (22.5 - 2x) \, dx$$

Integrating, we find:

$$M = [22.5x]_0^6 - [x^2]_0^6$$

$$= [\,135 - 0\,] - [\,36 - 0\,] = 99 \text{ kip-ft.}$$

Therefore, the bending moment at the 6 ft. mark is 99 kip-ft.

APPENDIX

F

SIMPLIFIED FRAME ANALYSIS

F.1   Introduction

Rigid framed structures are widely used in construction today because of their increased load carrying capability. The frame's increased strength is due to the moment resistance of the joints, which causes moments at the member's midspan to be reduced. Although this increased capacity is a definite advantage, a rigid frame is more complicated to analyze because it is indeterminate. This indeterminancy requires the designer to be familiar with techniques of calculating the moments and forces on the members of a rigid frame. These forces and moments must first be determined, so that the member design can be performed. Therefore, rigid frame analysis is an extremely important skill that the structural designer will utilize frequently.

Although this book only strives to cover the fundamentals of force analysis, a brief summarization of a very common frame analysis technique will be illustrated in the following paragraphs.

Frames can generally be broken down into two broad categories—braced and unbraced. Braced frames are those having some element or elements that restrict lateral movement, or sidesway. Unbraced frames are those that have insufficient structural elements to restrict lateral movement, and therefore are free to exhibit sidesway. The handling of frames subjected to sidesway involves somewhat more complicated analysis procedure, and therefore is left to other texts. One of the most common, manual frame analysis techniques is the moment distribution method. This technique will be the subject of the following section.

F.2   Moment Distribution

The *moment distribution method* was pioneered by Hardy Cross in the 1920's and is the predominant method of analyzing the forces in indeterminate structures. This method is based on redistributing moments to the ends of a member, in a series of successive iterations. The method is based on calculating the sum of moments in an indeterminate structure by first considering the moments at the member ends, assuming that all the joints were fixed. The moments are then redistributed as one joint (at a time) becomes free to rotate. This procedure may go through a number of iterations, depending on the precision required.

The procedure begins by assuming all joints in the structure are fixed. The loads on the structure would then impart *fixed end moments (FEM's)* to the ends of the member. Typical fixed-end moments are shown in Table F-1.

The procedure continues by unlocking one joint at a time, thus allowing the unbalanced moment at that joint to cause a moment in the opposite, fixed end of

Table F-1 Typical Fixed End Moment Cases

the member. This moment that is transferred to the opposite end is referred to as the *carry-over moment*. The carry-over moment to the opposite end of a member will be 1/2 of the moment at the unlocked end, and of the same sign convention (clockwise or counterclockwise). This 1/2 carry-over moment holds true if the member is prismatic (of constant cross-section), as illustrated in Figure F-1. The examples shown in this text will always consider the members to be of constant cross-section, since this is typically the case.

Throughout this structure, this sequence of locking and unlocking will continue, one joint at a time. There must be an accounting for any variation of rotational stiffnesses in the members that frame into a particular joint. An applied moment will split itself among the members framing into that joint, based upon the relative rotational stiffness of each member. For instance, if two members frame into a joint, each having the same rotational stiffness, a moment applied to that joint will split equally among the two members. A *distribution factor, (DF)* can be calculated for each member at a joint, realizing that the sum of all distribution factors should equal 1. The distribution factor can be calculated as follows:

$$DF_{AB} = (I/L)_{AB} / \Sigma(I/L)$$

where

DF_{AB} = distribution factor for member AB at joint A

$(I/L)_{AB}$ = stiffness of member AB

$\Sigma(I/L)$ = sum of all stiffnesses framing into a particular joint, joint A

It should be noted that for simple (pinned) supports, the final moment must always be balanced to zero. To consolidate the moment distribution procedure, it is convenient to modify the distribution factor for a pinned member by 3/4. This is because the stiffness of a pinned end has only 3/4 the stiffness of a fixed end. In a similar manner, it is sometimes convenient to modify the distribution factor of a symmetrical member by 1/2, again to consolidate the procedure. It should be remembered that neither of the modifications has to be performed in a moment

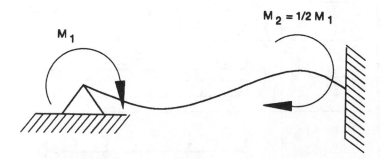

Figure F-1 Carryover Effect of Moment from a Pinned End to the Fixed End

distribution, and that such modifications are only done in order to shorten the procedure. In the following examples, neither modification will be used, in order to fully concentrate on the procedure.

The procedure for moment distribution can be outlined in a step-by-step procedure as follows:

1. Calculate distribution factors at the joints.
2. Lock all joints and calculate fixed-end moments.
3. Unlock a joint and calculate the carry-over moment.
4. Calculate new moment at each joint.
5. Determine the distribution of new moment at each joint, using distribution factors.
6. Repeat steps 3 through 6.

The procedure as outlined will typically converage on a solution after three or four iterations, although good approximate solutions may be calculated after one or two iterations. The following example will illustrate this technique.

EXAMPLE F.1

Analyze the braced frame shown below using moment distribution. The length of all members is 10 feet. The moment of inertia of the beam is 1/2 that of the columns (or $I_{col} = 2I$, while $I_{beam} = 1I$).

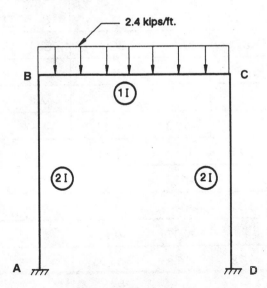

To begin, the distribution factors should be calculated for joints B and C, since those joints have two members framing into them. The distribution factors for joints A and D can be assumed as 1.0, since there are no other members to receive a distributed moment at these locations.

At joint B:

$$DF_{BA} = (2I/10 \text{ ft.}) / (2I/10 \text{ ft.} + 1I/10 \text{ ft.}) = .67$$

$$DF_{BC} = (1I/10 \text{ ft.}) / (2I/10 \text{ ft.} + 1I/10 \text{ ft.}) = .33$$

Similarly, at joint C:

$$DF_{CB} = (1I/10 \text{ ft.}) / (2I/10 \text{ ft.} + 1I/10 \text{ ft.}) = .33$$

$$DF_{CD} = (2I/10 \text{ ft.}) / (2I/10 \text{ ft.} + 1I/10 \text{ ft.}) = .67$$

Next, calculate the fixed end moments. Only member BC has a load on it, therefore the fixed end moments will be applied at the B end of member BC (termed BC) and the C end of member BC (termed CB). From table A-1, the fixed end moments are as follows:

$$FEM = wl^2/12 = (2.4 \text{ kips/ft.})(10 \text{ ft.})^2/12$$

$$FEM = 20 \text{ kip-ft. (CCW at } BC, CW \text{ at } CB)$$

The moment distributions proceeds by balancing the FEM's at each joint at a time, redistributing those carry-over moments, and balancing again. A tabular format as shown is typically the most convenient method for performing, and clearly demonstrating, this procedure and the iteration process.

	JT. A	JT. B		JT. C		JT. D
	AB	BA	BC	CB	CD	DC
DF	1	.67	.33	.33	.67	1
FEM			−20	+20		
BAL		+13.3	+6.67	−6.67	−13.33	
CO	+6.67		−3.33	+3.33		−6.67
BAL		+2.23	+1.1	−1.1	−2.23	
CO	+1.11		−0.55	+0.55		−1.11
BAL		+0.37	+0.18	−0.18	−0.37	
SUM	+7.78	+15.93	−15.93	+15.93	−15.93	−7.78

From the moments that were found in the technique, the designer can create shear and moment diagrams for each member in the frame. Thus, the maximum shears and moment could be found and the design process initiated.

ANSWERS TO SELECTED PROBLEMS

Chapter 1

1. a) 10.81   b) 132.5   c) 8.81   d) 459
3. a) 74.8   b)9.6   c) 10.26   d) 8.8
5. 433.1 ft.
7. 820.4 ft. from reference
9. 11.4 m

Chapter 2

1. 89.8 N @ 48.26°
2. 261.8 lb @ 81.830°
5. 25 N vector; $x = 19.15$ N $y = -16.07$ N
    35 N vector; $x = 29.35$ N $y = 19.06$ N
    70 N vector; $x = -69.96$ N $y = 3.44$ N
7. 69.93 @ 83.37°
9. 21.23 lb @ −3.24°
11. 113.57 lb @ 142.4°
13. $R = 215$ lb $x = 6.42$ ft.
15. $R = 448$ kips x 28.1 ft. from front axle
17. $R = 41$ kips $x = 10$ ft.
19. $x = 3.8$ m
21. $R = 15$ kips $x = 4.7$ ft. off front
23. $R = 17.83$ KN @ 3.28° $X_A = 68.6$ mm

Chapter 3

7. $R_{BV} = 35$ kips $R_{AV} = 10$ kips
9. $T_A = 205.8$ lb $T_B = 260$ lb
11. $R_{wall} = 20$ kips $M_{wall} = 125$ kip-ft. (CCW)

13. $R_{CV} = 10$ kips $R_{CH} = 7.54$ kips $T = 30.94$ kips
15. $R_{AH} = 35.5$ KN $R_{AV} = 50.48$ KN $R_{DV} =$
    40.23 KN
17. $CO = -5.72$ kips $AO = 2$ kips $BO = 1.57$ kips

Chapter 4

1. $AH = 31.25$ C $AB = 18.75$ T $BH = 25$ T
    $HI = 18.75$ C $BI = 31.25$ C $BC = 37.5$ T
    $IJ = 37.5$ C $IC = 25$ T $CD = 41.25$ T
    $CJ = 6.25$ C $DJ = 0$
3. $AG = 10$ T $GF = 37.95$ T $AF = 18.03$ C
    $AB = 21$ C $BF = 15$ T $BC = O$
    $BE = 22.13$ C $EF = 22.13$ T $EC = 14$ T
    $DC = ED = 0$
5. $AE = 8.97$ C $AB = 6.34$ T $BC = 6.34$ T
    $BE = 5$ T $EF = 5.45$ C $EC = 3.55$ C
    $FC = 11.51$ T $CD = 3.83$ T $GD = 7.66$ T
    $FG = 0$ $FD = 8.56$ T
7. $AC = 5$ C $AB = 5$ C $AD = 21.2$ T
    $CD = 0$ $DE = 21.2$ T $BD = 30$ C
    $BE = 7.1$ C
9. $BC = 1.5$ T $FG = 6.59$ C $FC = 5.37$ C
11. $BC = 7.5$ C $EC = 7.5$ C $ED = 10.6$ T
13. $BG = 14.14$ C $GF = 20$ T $BC = 40$ C
15. $D_y = 3.42$ $C_y = 3.58$ $E_y = 0$
    $E_x = 1.78$
17. $E_y = 25$ $D_y = 35$ $C_y = 10$
    $E_x = 60$ $C_x = 60$

Chapter 5

1. $P = 66.7$ lb
3. $P = 22$ KN
5. $P = 701.9$ lb
7. $18.78°$
9. 6 ft.
11. $\mu = 0.49$ or greater
13. P 37.5 KN
15. $P = 27.63$ KN
17. $P = 130$ lb
19. $T_o = 639$ lb
21. $T_{max} = 937$ lb
23. $T_{max} = 170.7$ lb

Chapter 6

1. $x = .91$ mm $y = 2.27$ mm
3. $x = 33$ in. $y = 36$ in.
5. $x = 2.62$ in. $y = 3.07$ in.
7. $x = 3.87$ in. y 2.37 in.
9. $x = 4$ in. $y = 8.33$ in.
11. $x = 7.5$ in. $y = 3.59$ in.
13. $x = -.09$ mm $y = -.73$ mm
15. $x = 0$ $y = 36$ in.
17. $x = 2.62$ in. $y = 1.57$ in.
19. $x = -5.13$ in. $y = -.625$ in.
21. $x = 4$ in. $y = 3.33$ in.
23. $x = 0$ $y = -.13$ in.

Chapter 7

1. $I_x = 558.67$ in.^{4} $I_y = 174.67$ in. 4
3. $I_x = I_y = 18.7$ in.^{4}
5. $I_x = I_y = 86.67$ in.^{4}
7. $I_x = I_y = 7.66$ in.^{4}
9. $I_x = 5.35 \times 10^6$ mm^4 $I_y = 2.947 \times 10^6$ mm^4
12. $I_x = 698.9$ in.^{4} $I_y = 11.61$ in.^{4}
13. $I_x = 34.22$ in.^{4} $I_y = 116.38$ in.^{4}
15. $I_x = 646.5$ in.^{4} $I_y = 188.4$ in.^{4}

Chapter 8

1. 5000 psi
3. 60.3 ksi
5. 79617 Kpa
7. 0.68 mm
9. 15,750 lb
11. 1.73 in.
13. $BC = 3.33$ ksi T $AC = 4.71$ ksi C
15. 22.76 KN
17. 3.08 ft.
19. 0.5P
22. +.00099 mm
24. 0.16

Chapter 9

1. .507 in. each direction
3. 16.6×10^{-5} mm/mm/°C
5. 4.86 Mpa
7. 141.5 °F, 146.6 °F
9. 38 °C. 609 °C
11. 42.7 °F
13. $F_s = 49,814$ lb $F_c = 15,187$ lb
15. $F_w = 23.$lKN $F_s = 76.9$KN
17. $F_a = 7.3$ KN $F_s = 32.7$ KN
19. 0
21. $F_s = 70$kip $F_a = 30$kip
23. .00051 in.
25. @ 0°, $\sigma_n = 2.76$ Kpa, $\sigma_s = 0$

Chapter 10

1. 2970 psi, 2546 psi
3. 1273 KPa, 1018 Kpa
5. 165,625 psi
7. 68.9 mm
9. 2.03 m
11. .736 Pa
13. $\varphi_{AB} = .000173$ rad $\varphi_{BC} = .00116$ rad
   $\varphi_{CD} = .000252$ rad
15. $\varphi_{AB} = .000535$ rad $\varphi_{BC} = .00256$ rad
   $\varphi_{CD} = .0008$ rad
17. 2.51 in.
19. 861 kW
21. 4.17 Hp

Chapter I I

1. $V_{max} = 13.1$ kips M_{max} 90.3 kip-ft.
3. $V_{max} = 16.1$ KN M_{max} 48.3 KN-m
5. $V_{max} = 23.4$ kips M_{max} 94.89 kip-ft.
7. $V_{max} = 22.5$ kips M_{max} 95.62 kip-ft.
9. $V_{max} = 9.2$ KN $M_{max} = 32.2$ Kn-m
11. $V_{max} = 13.93$ KN $M_{max} = 28.28$ KN-m
13. $V_{max} = 54.4$ kips $M_{max} = 246.6$ kip-ft.
15. $V_{max} = 17.58$ kips $M_{max} = 37.97$ kip-ft.
17. $V_{max} = 18$ kips $M_{max} = 60.75$ kip-ft.
19. $V_{max} = 31$ kips $M_{max} = 315.6$ kip-ft.

Chapter 12

1. 2.66 ksi, .89 ksi
3. 3755 kip-in.
5. 126 ksi, 63 ksi
7. .81 ksi, .54 ksi, .27 ksi
9. 18.3 ksi
11. 47,910 KPa
13. $\sigma_{top} = 45210$ psi, σ_{btm} 88,250 psi
15. $\sigma_{top} = 12.75$ ksi, σ_{btm} 28.64 ksi
17. .88 in.

19. 667 psi, 593 psi
21. 555 kips
23. 4.58 ksi
25. 4.55 ksi, .9 ksi
27. 25,590 KPa, 1347 Kpa, 14,810 Kpa
29. 6.85 ksi
31. 9420 psi
33. 13.7 in. × 13.7 in.
35. 2 − 2 × 12's

Chapter 13

1. 1.02 in.
3. 1.05 in.
5. 0.22 in.
7. 0.67 mm
9. 0.295 in.
11. 242.6 KN
13. 4.46 kip
15. 2.7 mm
17. 4.3 in.
19. 130×10^6 mm^4
21. 0.39 in.

Chapter 14

1. max = 1816.7 psi (C), min = 1483.3 psi (T)
3. 396,000 lb
5. max = 185,000 psi(C), min = 175,000 psi (T)
7. max = 166.7 Kpa, min = 0
9. A = 434 psi (T), B = 260.4 psi (C), C = 781.2 psi (C), D = 86.8 psi (C)
11. 21.6 in.
13. 6.08 in.
15. max = 1.995×10^6 KPa, min = 1.915×10^6 KPa
17. at 0 degrees; σ_n = 5.19 KPa σ_s = 1.79 KPa
    at 10 degrees; σ_n = 5.06 KPa σ_s = 2.37 KPa
    at 20 degrees; σ_n = 4.62 KPa σ_s = 2.66 KPa
21. Max σ_n = 2600 psi (T) @ 14.03° CW
    Max σ_s = 850 psi with 1750 psi (T) @ 30.97° CCW
23. Max σ_n = 3052 psi (C) @ 7.47° CCW

Chapter 15

1. Yes
3. Increases design capacity but still inadequate
5. No good, M_u = 390.6 kip-ft. and ϕM_n = 315 kip-ft.

7. W 12 × 50 works
9. W 14 × 48 works
11. 5.86 kip/ft.
13. It works M_{allow} = 352.8 kip-ft.
15. W 21 × 68 works
17. W 18 × 55, W 18 × 35, W 18 × 35
19. 142 lb/ft.
21. No
23. No
25. 7 kip/ft.
27. 15 in.

Chapter 16

1. 92527 lb
3. 81.1
5. 319 kip
7. W 12 × 58
9. W 10 × 33
11. 12 × 6 × 1/2
13. 3667 KN
15. Yes
17. W 12 × 79
19. W 12 × 72
21. 2882 KN
23. W 360 × 162

Chapter 17

1. 25.2 kip
3. fails in. tension
5. still fails in. tension
7. 281.5 KN
9. 889 KN
11. 14 bolts
13. Increase plate per tension requirements
15. 8.9 in.
17. Yes, still fails
19. 121.5 kip (tension)
21. 95.3 kip
23. 5.5 in.
    Max σ_s = 1552 psi with 1500 psi (C) @ 37.53° CW

INDEX